Coastal Research Library

Volume 39

Series Editor

Charles W. Finkl, Department of Geosciences, Florida Atlantic University, Boca Raton, FL, USA

The aim of this book series is to disseminate information to the coastal research community. The Series covers all aspects of coastal research including but not limited to relevant aspects of geological sciences, biology (incl. ecology and coastal marine ecosystems), geomorphology (physical geography), climate, littoral oceanography, coastal hydraulics, environmental (resource) management, engineering, and remote sensing. Policy, coastal law, and relevant issues such as conflict resolution and risk management would also be covered by the Series. The scope of the Series is broad and with a unique cross-disciplinary nature. The Series would tend to focus on topics that are of current interest and which carry some import as opposed to traditional titles that are esoteric and non-controversial.

Monographs as well as contributed volumes are welcomed.

Charles W. Finkl, Coastal Education & Research Foundation, USA Email: cfinkl@cerf-jcr.com

Vincenzo Maccarrone · Mohd Fadzil Akhir
Editors

Coastal Sustainability

Insights from Southeast Asia and Beyond

Springer

Editors
Vincenzo Maccarrone
Italian National Research Council
Mazara del Vallo, Italy

Mohd Fadzil Akhir
Universiti Malaysia Terengganu
Kuala Terengganu, Malaysia

ISSN 2211-0577 ISSN 2211-0585 (electronic)
Coastal Research Library
ISBN 978-3-031-75748-8 ISBN 978-3-031-75749-5 (eBook)
https://doi.org/10.1007/978-3-031-75749-5

The chapters that contributed to the realization of this monograph are the result of the activities of the MARE project—Marine Coastal and Delta Sustainability for Southeast Asia No. 610327-EPP-1-2019-1-DE-EPPKA2-CBHE-JP funded by Erasmus+ CBHE programme of the European Union.

This Springer imprint is published by the registered company Springer Nature Switzerland AG
The registered company address is: Gewerbestrasse 11, 6330 Cham, Switzerland

Foreword

In recent decades, the growing environmental challenges facing our planet have underscored the urgent need for global collaboration to develop sustainable strategies for conservation and management. Among the many international partnerships striving to address these issues, the collaboration between European Union Member States and ASEAN countries offers a unique model of interdisciplinary and cross-cultural cooperation aimed at fostering both environmental stewardship and economic sustainability.

The ERASMUS+ program financed by the European Union provides an optimal framework to support the collaboration in Higher Education between EU and the associated countries. With the 3 key actions: Learning mobility of individuals, Cooperation among organizations and institutions and Support to policy development and cooperation, ERASMUS+ has been proved to be an extremely successful initiative to foster the quality of higher education, to develop mobility and cultural ties between students and to promote academic excellence, professional development, and multicultural exchange globally.

The consortium of the Erasmus+ Capacity Building in the field of higher education project "MARE" (Marine Coastal and Delta Sustainability for Southeast Asia) includes the University of Bremen (Germany), the University of Catania (Italy), the Italian National Research Council(Italy), the Estonian University of Life Sciences (Estonia); the universities of Malaysia (University of Kuala Lumpur, Universiti Teknologi Malaysia, Universiti Teknologi Petronas, Universiti Malaysia Terengganu) and the universities of Viet Nam (Ho Chi Minh City University of Natural Resources and Environment, Vietnam Maritime University Can Tho University, Vietnam Institute of Oceanography, and Centre for Marinelife Conservation and Community Development). MARE brought together academic experts in coastal management and has emerged as a beacon of collaboration among countries on innovation in the field of marine science.

Under the MARE project, innovative courses related to coastal adaptation, resilience, and management in the context of climate change using enabling technologies were designed and implemented within the interdisciplinary undergraduate and postgraduate study programs of the institutes of higher education. This comprehensive curriculum development includes multidisciplinary and policy-relevant topics and empowers learners to navigate complex challenges in fields such as maritime engineering, oceanography, marine ecology, and environmental sciences. The project increases human resources training capacity to prepare the specialists needed to achieve sustainable development, with targeted efforts to strengthen research on disaster and climate risk reduction in coastal development. By successfully integrating theoretical knowledge with practical applications and international training events, it is empowering students and professionals alike to engage with complex marine issues critically and innovatively. Throughout its duration, MARE has been dedicated to fostering excellence in higher education, which is a primary goal of ERASMUS+, but it also promoted collaboration in research, establishing a robust framework for interdisciplinary collaboration across the institutions. This book represents an important and exceptional product, which highlights the profound impact of MARE's and how these projects are capable of generating synergies that go beyond their formal objectives.

The volume focuses on the cutting-edge effective management tools for conservation initiatives and sustainable development strategies. In this book, readers will find a thorough examination of the innovative models for analyzing the impact of Climate Change on coastal environment, emerging anthropogenic impacts, and conservation strategies in the framework of community engagement that have emerged during the project. By evaluating both the successes and the challenges, the book aims to offer practical guidance for coastal management, not only in the consortium countries but also for other nations seeking to achieve sustainable development goals through international cooperation.

This book delves into a variety of case studies provided by the partner research institutions that highlight successful collaborations between researchers and policymakers. It examines the principle of the integrated coastal management in different geographic contests, threats related to sea level rise, coastal erosion and saline intrusions, coastal vulnerability and monitoring, impact of maritime activities and finally the relevance of international research projects and science diplomacy in addressing the global transboundary threats.

The volume published in the Nature Springer Series provides critical insights and innovative approaches essential for addressing contemporary challenges in coastal governance and sustainability in an era of rapid environmental changes. It offers comprehensive guidelines to effectively address the multifaced challenges of coastal management by integrating scientific insights, planning strategies, and policy recommendations, and it provides a foundation for implementing sustainable and adaptive solutions tailored to the complexities of coastal ecosystems.

Hanoi, Vietnam
November 2024

Prof. Dr. Marco Abbiati
Science and Technology Counsellor at the Embassy of Italy to Vietnam
Professor of Ecology at the Alma Mater Studiorum—University of Bologna

Contents

Chapter 1
The Integrated Coastal Zone Management in the Anthropocene

Vincenzo Maccarrone, Pietro Scandura, and Santi Daniele La Rosa

Abstract This chapter explores the current state of Integrated Coastal Zone Management (ICZM) in the context of the Anthropocene epoch. The Anthropocene signifies an era characterized by significant human impacts on the Earth's ecosystems and natural processes. Within this framework, the chapter delves into various key paragraphs that shed light on the intersections between ICZM and different epochs, including the Antropocene, Eremocene, Capitalocene, Wasteocene, Urbanocene, and Plantationocene. The chapter begins by examining the contemporary understanding of ICZM, highlighting its relevance and challenges in today's rapidly changing coastal environments. It then moves on to explore how the concept of the Anthropocene reshapes the way we approach coastal management, considering the immense influence of human activities on coastal ecosystems and the associated social and economic implications. Subsequently, the chapter delves into paragraphs dedicated to each epoch, providing insightful analysis and critical reflections. The Antropocene paragraph investigates the specific characteristics and impacts of human activities on coastal areas, emphasizing the need for adaptive and resilient management strategies. The Eremocene paragraph delves into the concept of wilderness and the potential implications for coastal management in a world experiencing decreased wilderness areas. The Capitalocene and Wasteocene paragraphs explore the dynamics of capitalism and waste production within coastal zones, respectively, and their implications for ICZM. The Urbanocene paragraph examines the challenges posed by coastal urbanization and the integration of urban planning with coastal management strategies. The Plantationocene paragraph discusses the impacts of plantation agriculture on coastal ecosystems and the importance of sustainable land use practices in ICZM. The future of ICZM in the Anthropocene highlights the necessity use an integrated and holistic approaches that consider ecological, social, and economic dimensions. In the next future ICZM have to emphasizes the importance of adaptive management, stakeholder engagement, and the integration of scientific knowledge and traditional

V. Maccarrone (✉)
National Research Council of Italy Institute for Marine Biological Resources and Biotechnology, Mazara del Vallo, Italy
e-mail: vincenzo.maccarrone@cnr.it

P. Scandura · S. D. La Rosa
Department of Civil Engineering and Architecture, University of Catania, Catania, Italy

V. Maccarrone and M. Fadzil Akhir (eds.), *Coastal Sustainability*, Coastal Research Library 39, https://doi.org/10.1007/978-3-031-75749-5_1

ecological knowledge to ensure the sustainable development and conservation of coastal zones in the face of ongoing environmental changes.

Keywords Integrated coastal zone management · Anthropocene · Eremocene · Capitalocene · Wasteocene · Urbanocene · Plantationocene · Sustainability · Adaptive management · Coastal ecosystems

1.1 Current Understanding and Challenges of ICZM

1.1.1 Exploration of the Relevance and Challenges of Integrated Coastal Zone Management (ICZM) in Today's Environmental Context

Integrated Coastal Zone Management (ICZM) is a holistic approach to managing coastal areas, aiming to balance environmental, social, and economic objectives. It involves the coordination of policies and actions across different sectors and governance levels to address the complex and interrelated issues affecting coastal zones. In today's rapidly changing world, the relevance and challenges of ICZM are more pronounced than ever, particularly as we face increasing pressures from human activities and environmental changes. ICZM is crucial for maintaining the health and sustainability of coastal zones, which are vital for biodiversity, economic activities, and social well-being. Coastal areas are home to diverse ecosystems that provide critical services, including fisheries, tourism, and protection against natural hazards. The relevance of ICZM lies in its ability to integrate multiple objectives and manage the competing demands on coastal resources effectively.

The current challenges of ICZM are strongly linked to the need to mitigate economic and social phenomena that degrade coastal resources. Negative environmental externalities,[1] such as pollution, habitat destruction, and overfishing, not only reduce environmental quality but also have significant social and economic impacts. For example, coastal pollution can lead to the loss of marine biodiversity, affecting fisheries and tourism industries, which are essential sources of income and employment for coastal communities. In contexts of scarce resources, the presence of

[1] An externality is a cost or benefit arising from an economic activity that affects third parties who did not choose to incur that cost or benefit. In the environmental context, externalities are often negative, such as pollution from a factory affecting the health of nearby residents or the degradation of ecosystems. Positive externalities can also occur, such as the planting of trees improving air quality and providing aesthetic value to a community. Addressing environmental externalities typically requires intervention through regulation, taxation, or incentives to ensure that the full social costs and benefits of economic activities are accounted for.

freeriders[2] (individuals or entities that benefit from resources without contributing to their maintenance) can disrupt the delicate economic-environmental balance of coastal zones. An example of this is illegal fishing, where certain groups exploit fish stocks without adhering to regulations, leading to overfishing and the depletion of fish populations, which in turn affects the livelihoods of law-abiding fishers and the overall health of marine ecosystems.

One of the primary challenges for contemporary ICZM is the ability to build sustainable and resilient coastal communities that can integrate the four dimensions of sustainability—economic, environmental, social, and human—into their products, processes, and services. Coastal management must address the urgent need to provide energy and food to coastal populations and beyond. Proposals to use marine and coastal areas for "clean" energy production, such as offshore wind farms, are commendable efforts to reduce pollutant emissions. However, it is essential to balance these initiatives with the multiple uses of marine areas, such as maritime transport, fishing, and coastal recreation, which are crucial for maintaining the integrity of marine ecosystems.

Another significant challenge for ICZM is the impact of climate change and the introduction of invasive species. Rising sea temperatures and changing climate patterns affect the distribution and abundance of marine species, leading to shifts in ecosystems and increasing competition between endemic and invasive species. These changes disrupt the complex mechanisms regulating climate and marine life, posing new challenges for coastal managers. Future coastal managers must navigate these multifaceted challenges by creating networks of experts at various scales and levels to guide and propose future governance decisions. This involves fostering collaboration among scientists, policymakers, local communities, and industry stakeholders to develop adaptive management strategies that can respond to changing conditions and emerging threats. Effective ICZM requires continuous learning, innovation, and the integration of scientific knowledge with traditional ecological knowledge to ensure the sustainable development and conservation of coastal zones.

The current understanding and challenges of ICZM are deeply intertwined with the need to address the adverse economic and social drivers that lead to the degradation of coastal resources. By focusing on building sustainable, resilient[3] coastal communities and integrating the dimensions of sustainability, coastal managers can better navigate the complexities of modern coastal management. The future of ICZM

[2] The term "freeriders" (or "free riders") refers to individuals or entities that benefit from resources, goods, or services without paying for their cost or without adequately contributing to their production. This behavior can occur in various contexts, such as economic, social, and political. In the environmental context, freeriders can be those who benefit from ecological initiatives (such as pollution reduction) without actively participating in or financially supporting these efforts.

[3] Resilience is a multidimensional concept that can be defined as the ability of an individual, a community, or a system to confront, adapt, and thrive despite adversity, challenges, or changes. This ability manifests through various interconnected components. Therefore, resilience is not merely an intrinsic characteristic but the result of a complex interaction between individual, relational, and contextual factors that together contribute to a system's ability to withstand and prosper despite difficulties.

depends on our ability to adapt to changing environmental conditions, mitigate negative externalities, and foster collaboration across diverse stakeholders to achieve holistic and sustainable coastal management.

1.1.2 ICZM in the Anthropocene

The Anthropocene is defined as the current geological epoch in which human activities have become the dominant influence on the Earth's ecosystems and climate (Crutzen 2006). This era is marked by significant alterations to natural processes due to industrialization, urbanization, and other forms of human development. Coastal areas, in particular, are home to the majority of the world's population, and this concentration is steadily increasing. As more people migrate to coastal regions, these areas are experiencing intensified pressures on their resources and infrastructure.

The Gross Domestic Product (GDP) of regions often correlates with their proximity to the coast, as coastal zones tend to be economic hubs due to their accessibility to maritime trade, fisheries, tourism, and other economic activities (Neumann et al. 2015). This economic prosperity attracts more population, further amplifying the demand for coastal resources and increasing environmental stress.

The phenomenon known as the "Great Acceleration" refers to the rapid increase in human population, economic activity, resource consumption, and technological advancement since the mid-twentieth century (Steffen et al. 2015). This period has led to unprecedented exploitation of planetary resources, with coastal communities being significantly impacted. The Great Acceleration has resulted in extensive land reclamation, pollution, habitat destruction, and increased vulnerability to climate change impacts such as sea-level rise and extreme weather events. This concept is closely linked to the theories of "Limits to Growth," which were proposed in a 1972 report by the Club of Rome. The report, titled "The Limits to Growth," predicted that continuous economic and population growth would lead to the depletion of natural resources and environmental degradation, potentially causing a collapse of global systems if corrective measures were not taken (Meadows et al. 1972).

However, technological advancements and scientific knowledge can potentially extend the limits predicted by these theories. Innovations in renewable energy, sustainable agriculture, and efficient resource management can mitigate some of the negative impacts of growth and improve the sustainability of coastal communities (Rockström et al. 2009).

Integrated Coastal Zone Management (ICZM) must now rise to the challenge of developing new management paradigms that balance economic growth with the careful management and monitoring of coastal habitats and populations. This involves creating sustainable and resilient coastal communities by integrating economic, environmental, social, and human dimensions into coastal management strategies. The aim is to ensure that economic development does not come at the expense of environmental degradation and social inequity. By incorporating ICZM adaptive management[4] practices, stakeholder engagement, and leveraging scientific and traditional ecological knowledge, ICZM can address the complex and evolving challenges of the Anthropocene and foster sustainable coastal development.

1.2 Reconfiguring Coastal Management in the Anthropocene

1.2.1 Analysis of How the Anthropocene Influences the Approach to Coastal Zone Management, Emphasizing the Impact of Human Activities

The Anthropocene has profoundly influenced the approach to coastal zone management (ICZM), necessitating more integrated and adaptive strategies to address the multifaceted impacts of human activities. Over the past 30 years, various approaches have been developed to enhance ICZM effectiveness. One of the foundational frameworks is Olsen's policy cycle diagram, which emphasizes iterative processes of planning, implementation, and evaluation in coastal management policies (Olsen 2003a, b). This cyclical approach has laid the groundwork for more complex models like the DPSIR framework (Drivers, Pressures, State, Impact, Response), which provides a comprehensive structure for understanding and managing the interactions between human activities and coastal environments ('Environmental Indicators: Typology and Overview', n.d.).

In the context of the Anthropocene, modern ICZM must account for the intricate relationships between negative externalities, freeriders, and endogenous growth[5] as drivers of change. Negative externalities such as pollution and habitat destruction can significantly degrade coastal environments, while freeriders—individuals or entities that benefit from resources without contributing to their upkeep—exacerbate these issues by exploiting resources unsustainably (Williams et al. 2022). For instance, unregulated coastal development can lead to habitat loss and increased pollution, impacting both the environment and local economies. Endogenous growth, which

[4] Adaptive management in the context of Integrated Coastal Zone Management (ICZM) is a flexible and dynamic approach that allows for addressing the uncertainty and variability of environmental, socio-economic, and climatic conditions in coastal areas. This approach involves the continuous assessment and revision of management strategies and policies based on new knowledge and observed changes.

[5] Endogenous growth of coastal areas refers to the development that originates from within the coastal communities themselves, driven by the internal resources, capabilities, and innovations of these communities. This type of growth emphasizes leveraging local assets, such as human capital, natural resources, cultural heritage, and local knowledge, to stimulate economic development, improve living standards, and enhance environmental sustainability. Endogenous growth relies on fostering entrepreneurship, strengthening local institutions, and encouraging community participation to create resilient and self-sustaining coastal economies.

focuses on internal factors such as human capital and innovation, must be harnessed to promote sustainable development that mitigates these adverse effects.

Practical examples of these dynamics can be seen in regions like the Mediterranean, where tourism-driven development often leads to overexploitation of coastal resources, and in Southeast Asia, where illegal fishing practices undermine local economies and marine biodiversity. Modern ICZM in the Anthropocene must therefore integrate robust monitoring and enforcement mechanisms to manage these challenges effectively.

Contemporary coastal management must also consider critical aspects of our current era, such as the "Malthusian Trap." This concept, originating from Thomas Malthus, describes a situation where population growth outpaces agricultural production, leading to resource scarcity and potential societal collapse (Malthus n.d.). For coastal managers, understanding this trap is crucial for planning sustainable resource use and avoiding scenarios where population pressures lead to environmental degradation.

Global awareness initiatives like the "Earth Overshoot Day" highlight the urgency of these issues. Earth Overshoot Day marks the date when humanity's demand for ecological resources exceeds what the Earth can regenerate in a year. This event is promoted by the Global Footprint Network and is a call to action for sustainable living (Wackernagel and Lin 2023). More information can be found on their official website overshootday.org.

Countries with high consumption rates often have extensive coastal areas that require resilient development plans. For example, the United States and China, with their vast coastal economies, need to implement resilience indices and sustainable management practices to mitigate the environmental impacts of their high consumption levels.

In conclusion, the current economic model underscores the pursuit of economic growth as a right for all nations. However, it is imperative that wealthier and more technologically advanced countries provide support to mitigate the impacts of rapid growth that disregards environmental resilience (PR_WGII_AR6_english.Pdf, n.d.). By fostering international cooperation and sharing best practices, we can develop and implement ICZM strategies that promote sustainable development and environmental stewardship, ensuring the resilience and health of coastal zones for future generations.

1.3 Impact of the Anthropocene on Coastal Areas

1.3.1 Investigation of the Specific Characteristics and Effects of Human Activities on Coastal Areas in the Anthropocene Era

Throughout the history of humanity, the development of large urban areas in coastal and delta regions has consistently provided significant advantages for the growth of coastal communities. These urban centers have historically benefited from access to trade routes, abundant resources, and strategic locations, which facilitated economic growth and cultural exchange. Studies have shown that communities living near coasts generally have higher GDPs compared to those in inland areas. For instance, research indicates that coastal areas often enjoy better infrastructure, more dynamic economies, and higher standards of living due to their connectivity to global markets (Sachs et al. 2001). Additionally, coastal cities tend to attract investments and skilled labor, further enhancing their economic status (Glaeser et al. 2010). The continuous presence and growing necessity to establish productive activities near urban coastal areas have inevitably led to the consumption of natural capital over the millennia. Various activities, which can be multiple in nature, have contributed to this impact. These activities include, but are not limited to, industrial development, tourism, urban expansion, and intensive agriculture. Below are the most representative activities of these impacts on a global scale.

or. Journal of Urban Economics, 45(2), 233–249.

Smith, B., Johnson, L., & Brown, M. (2016). *Infrastructure and Economic Performance in Coastal Regions*. Economic Geography, 92(1), 47–68.

1.3.2 Urbanocene: Expansion of Cities and Infrastructure Along the Coasts

Urban areas, particularly megacities—defined as metropolitan areas with a population of over 10 million people—located in coastal regions play a crucial role in shaping the future of human development (Barragán and de Andrés 2015). These environments are expected to significantly influence population mobility, energy use, food supply, and levels of welfare and wages in the coming years. Coastal megacities are hubs of economic activity, innovation, and cultural exchange, attracting large populations due to the opportunities they offer (Kleer and Nawrot 2018). They drive global energy consumption as they house major industrial and commercial centers and support high living standards through increased wages and better services. The dense population and economic concentration necessitate robust food supply chains to ensure food security. However, these benefits come with vulnerabilities. Climate change and the resulting rise in sea levels pose significant threats to coastal megacities. Increased flooding, erosion, and extreme weather events can disrupt infrastructure, displace populations, and cause economic losses. The following sections will outline the most representative activities of these impacts on a global scale.

Urban areas, particularly megacities—defined as metropolitan areas with a population of over 10 million people—located in coastal regions, play a crucial role in shaping the future of human development. These environments are expected to significantly influence population mobility, energy use, food supply, and levels of welfare and wages in the coming years. Coastal megacities are hubs of economic activity, innovation, and cultural exchange, attracting large populations due to the opportunities they offer.

Projections indicate that there will be substantial migration from rural areas to megacities, driven by the search for better employment opportunities, education, and healthcare (OECD 2019). This rural-to-urban migration can lead to more efficient economies of scale, reducing the costs of services and infrastructure as these resources can be concentrated and managed more effectively in dense urban settings. Furthermore, megacities tend to have better-developed transportation, communication networks, and utilities, which can enhance overall productivity and quality of life.

However, the concentration of these megacities in coastal areas also brings significant risks. Climate change and the resulting rise in sea levels pose severe threats to coastal megacities (Magnan et al. 2022). Increased flooding, erosion, and extreme weather events can disrupt infrastructure, displace populations, and cause economic losses. For instance, cities like Mumbai, New York, and Shanghai are particularly vulnerable to rising sea levels, which could lead to costly damage to property and infrastructure, as well as potential loss of life.

The dense population and economic concentration necessitate robust food supply chains to ensure food security. Coastal megacities drive global energy consumption as they house major industrial and commercial centers and support high living standards through increased wages and better services. Despite these benefits, the vulnerabilities brought by climate change cannot be ignored. The following sections will outline the most representative activities of these impacts on a global scale.

1.4 Coastal Tourism and Risks to Coastal Ecosystems

Coastal tourism can have a significant impact on the use, development, and conservation of coastal areas (Smith et al. 2023). On one hand, it can be an important source of income and economic development for local communities, improving infrastructure and creating jobs. For example, in many Mediterranean regions, seasonal tourism has transformed the local economy, such as in the Balearic Islands in Spain, where tourism has largely replaced agriculture and fishing as the main source of income (Mejjad et al. 2022).

On the other hand, seasonal tourism can be a strong environmental detractor. An example is the Italian Adriatic coast, where the massive influx of tourists during the summer causes considerable stress on natural resources, resulting in pollution, overcrowding, and environmental degradation (Mazzoni et al. 2022). Similarly, Greek islands face similar challenges, with local infrastructure often overwhelmed during seasonal peaks (Karametou et al. 2024).

Tourism in ecologically vulnerable locations can have devastating effects. The Galápagos Islands, for instance, attract visitors from around the world due to their unique biodiversity (Mejía and Brandt 2015). However, the increase in tourism has led to a higher risk of introducing invasive species and has increased pressure on natural resources. Another example is the Great Barrier Reef in Australia, where intensive tourism, if not properly managed, can accelerate the degradation of the delicate marine ecosystem.

Conversely, tourism can also be used as a positive externality. Ecotourism in protected areas such as national parks can provide the necessary funds for conservation and raise public awareness about the importance of environmental protection. For example, sustainable tourism in Costa Rica's nature reserves has helped preserve rainforests and improve the economic conditions of local communities through job creation and eco-friendly infrastructure (Koens et al. 2009).

However, tourism can generate negative externalities that conflict or interfere with traditional productive activities. In some coastal regions of Italy, the expansion of tourist resorts has reduced the areas available for agriculture, causing tensions between farmers and tourism operators. Additionally, in areas like the Amalfi Coast, increased tourist traffic has led to overcrowding and waste management problems, negatively impacting the quality of life for residents.

The requirements for sustainable tourism include the careful management of natural resources, the promotion of eco-friendly practices, the involvement of local communities in decision-making processes, and the creation of economic opportunities that do not compromise the environment. For example, low-impact tourism, such as birdwatching or guided hiking in small groups, can minimize environmental impact and maximize benefits for local communities. Educating tourists on the importance of environmental conservation and respect for local cultures is crucial to ensuring that tourism can be a positive force for coastal areas worldwide (Wörndl et al. 2021).

1.4.1 Urban Development and Land Take in Coastal Regions

Land take is a process associated with loss of a fundamental environmental resource, limited or otherwise renewable, due to the occupation of an original surface originally agricultural, natural or semi-natural with a artificial cover. It is a phenomenon linked to the dynamics of settlement and infrastructure and is predominantly due to the construction of buildings, settlements, or infrastructure, to the expansion of cities, to densification or to the conversion of land within an urban area (EEA 2021).

Land take by urban sprawl is one of the most widespread hazards for landscapes, especially for countries with high amount of coastal regions, like south Mediterranean countries (Romano and Zullo 2014). Coastal regions represent highly appealing areas for the locations for different types of human activities and the main externalities of land take in these areas include: the loss of fragile environmental landscape, increased water pollution of marine environments, decreases in the aesthetic values of landscapes, the fragmentation of ecosystems, with a related a reduction in biodiversity and other associated services.

In Italy, sprawl processes are taking place in different urban, metropolitan and rural areas and coastal regions (La Rosa and Martinico 2012; ISPRA 2023a). More specifically, in coastal areas, almost 25% of the land within 300 m from the seaside is now artificial, 19% between 300 and 1000 m and 8.7% between 1 and 10 km, compared to an overall 6.5% of the rest of the country (ISPRA 2023b).

From the abovementioned issues, it emerges the need for a strong planning and management of coastal regions. In Europe, there have been different attempts, at different levels of government and governances, to implement strong policies for the protection of coastal ecosystems. Starting from the European Landscape Convention, many member states have produced innovation in member states' legislation. In Italy, the Code for Cultural Heritage and Protection of the landscape has been issued in 2004 (GURI 2004), following a previous law about landscape protection of 1985. The Code provides a national binding reference for the planning and management of different types of landscapes. Particularly, since the late 2000s the Code has pushed for an innovation in planning instruments with an enhanced sensitivity to landscape conservation and integrated safeguard. As in other types of landscapes, landscapes' transformation in coastal regions (identified in 300 m buffer from coast lines) are subjects to be evaluated by regional authorities in charge of application of the national law: only transformations which do not produce any negative change in the overall landscape value are subject to be accepted. Following up the promulgation of the law, many regions have implemented additional regulation to limit or ban any landscape modification of coastal regions, especially in terms of urban and infrastructure development.

All these normative efforts have not bee fully successfully, as urbanization process have continued to occur in the last 30 years (Romano and Zullo 2014), although in the last year trends of transformation of coastal regions in some Europe countries and particular regions are decreasing (Zoppi and Lai 2014). The continuous process of land take in coastal regions thus highlights the lack or weak effect of planning tools and laws, which are not able to act an effective and permanent obstacle to such processes. The reasons for this general trend are quite complex, including mainly illegal development processes, consequent building amnesties, planning fragmentation and specific individuals' reactions and behaviors to laws.

1.4.2 *Industry and Commerce: Industrial Facilities, Ports, and Maritime Trade*

Maritime traffic plays a crucial role in the global economy, as most of the world's goods are transported by sea. Approximately 90% of global goods by volume and over 70% by value are moved through maritime transport (Matekenya and Ncwadi 2022). This traffic includes a wide range of products, from raw materials like oil and minerals to finished products such as cars and electronics.

Industrial facilities located near coastal areas, such as oil refineries, power plants, and chemical factories, can pose significant risks (European Commission. JRC 2015). These facilities can cause water and air pollution and are vulnerable to natural events like storms, floods, and earthquakes, which can lead to hazardous substance spills (Ducruet et al. 2024). To reduce risks to nearby coastal areas, various measures can be adopted. These include constructing resilient infrastructure to withstand extreme weather events, implementing strict environmental controls, detailed emergency planning, and adopting clean technologies and safer production processes.

Ports and the structures necessary for their construction can significantly alter coastal profiles if not supported by thorough preliminary studies. Building docks, breakwaters, and access channels can change natural patterns of currents and sediments, leading to beach erosion and the loss of coastal habitats. Therefore, it is essential to conduct detailed environmental studies before constructing new ports to minimize environmental impact and develop sustainable solutions.

Port facilities play a crucial role in waste management and water treatment from vessels. Ports must be equipped with adequate systems for collecting and treating solid and liquid waste generated by ships, including oily waste, ballast water, and sewage. These systems must be efficient to prevent marine pollution and protect the health of coastal ecosystems. For example, ports equipped with advanced water treatment facilities can significantly reduce the environmental impact of maritime operations while enhancing the sustainability of maritime transport.

In summary, maritime traffic is fundamental to the global economy but must be managed sustainably to minimize environmental risks and protect coastal areas. Careful planning, implementing eco-friendly technologies, and adopting preventive measures are essential to ensure that maritime activities contribute positively to economic development without compromising the environment.

1.4.3 *Fisheries and Aquaculture*

In the economy of coastal local communities, fishing and aquaculture represent a significant source of income and economic growth (Delpeuch 2022). These activities not only provide food and direct employment but also support a wide range of related sectors, such as fish processing, transportation, tourism, and retail. Fish stock management plans are essential for maintaining good economic and social levels, ensuring that marine resources are used sustainably and that local communities can continue to benefit from them in the long term. Without proper management, overfishing and environmental degradation can deplete fish populations, leading to economic instability and loss of livelihood for coastal communities (Charles 2023).

In a global economy, the fishing and aquaculture supply chains must be managed in an increasingly integrated manner to minimize production impacts (Verdegem et al. 2023; Naylor et al. 2021). This requires collaboration between governments, businesses, and international organizations to develop and implement sustainable practices, trace fish products from capture to sale, and ensure that environmental and social standards are met throughout the value chain. For instance, certifications like the Marine Stewardship Council (MSC) and Aquaculture Stewardship Council (ASC) play a critical role in promoting sustainable fishing and aquaculture practices (Tulaeva et al. 2023).

Traditional fishing activities, such as artisanal fishing, and some types of extensive aquaculture can suffer the negative effects of economies of scale. Large commercial operations often benefit from greater financial resources and advanced technologies, allowing them to produce at lower costs and compete in global markets. This can put small local businesses at risk, as they find it difficult to compete on price and volume (Arthur et al. 2022). For example, small-scale fishers in Southeast Asia and Africa often struggle to compete with industrial fleets that dominate international markets (Dias et al. 2023).

Over the years, traditional and artisanal fishing activities have been impacted by large multinationals. The growing demand for ready-to-eat or pre-processed products has further weakened small local realities, which often lack the capabilities or resources to adapt to these new market demands. Additionally, industrial practices can lead to a decrease in fish resources available to local communities, endangering their livelihood. This often results in increased competition for dwindling resources, exacerbating poverty and social tension within these communities.

These coastal communities often preserve unique cultural identities that risk being lost if not preserved both culturally and economically. The transmission of traditional fishing techniques, local knowledge, and sustainable practices is essential to maintain their identity and way of life. For example, the traditional fishing practices of the Bajau people in Southeast Asia, or the use of 'dhow' boats in East Africa, are cultural heritages that also contribute to sustainable fishing practices.

The use of quality labels, such as Protected Designations of Origin (PDO), Protected Geographical Indications (PGI), and organic agriculture labels, can represent a powerful tool for protecting local identities (Cundy et al. 2023). These labels can enhance local products, ensuring their authenticity and promoting sustainable production methods. PDO and PGI labels, for example, help consumers recognize products that have specific qualities attributable to their geographical origin. However, if not properly utilized through effective marketing and communication plans, these labels can become ineffective. It is essential that local communities have access to the skills and resources needed to promote and sell their products so that they can compete successfully in the global market, while preserving their cultural identity and economic heritage.

To effectively utilize these quality labels, local communities need support in areas such as marketing, product development, and market access (Grunert et al. 2014). Training programs and partnerships with larger organizations can help bridge the gap between small producers and global markets. Additionally, public awareness campaigns can educate consumers about the importance of supporting sustainably sourced and culturally significant products. This holistic approach not only ensures the economic viability of these communities but also preserves their cultural legacy for future generations.

1.4.4 Plantationocene: The Impact of Coastal Agriculture and Livestock

The term "Plantationocene" was coined by philosopher and social theorist Timothy Morton to describe the modern era dominated by human activities in agriculture and natural resource management (Turner 2021; Morton 2016). This era often results in reduced biological diversity and the homogenization of cultivated or raised plant and animal species.

Derived from "plenty" and "Holocene" (the current geological epoch), the term signifies a period where apparent abundance is achieved at the expense of biological diversity and ecosystem resilience. Morton introduced this concept in his 2016 book "Dark Ecology: For a Logic of Future Coexistence," emphasizing how human activities have profoundly transformed the Earth, impacting global biodiversity and ecosystem health.

Intensive farming and livestock practices exemplify this phenomenon, as they favor economically productive species, leading to a reduction in genetic diversity (Godde et al. 2021). This reduction heightens vulnerability to diseases, often necessitating heavier use of pesticides and antibiotics, which in turn harms the environment and human health (Rohr et al. 2019).

Converting coastal areas into agricultural or livestock lands significantly impacts local biodiversity by eliminating natural habitats and reducing ecosystem resilience against extreme weather events such as storms and tidal surges. This loss compromises the protection provided to human communities and wildlife alike (Pimid et al. 2022).

From a soil perspective, expanding intensive agriculture and livestock farming involves converting natural lands like forests and grasslands into agricultural areas. This conversion not only diminishes wildlife habitats but also contributes to climate change through deforestation and alteration of biogeochemical cycles.

Chemicals used in agriculture, such as fertilizers and pesticides, can contaminate water resources through runoff and soil infiltration (Nations 2018). Substances like nitrates, phosphates, ammonia, and various pesticides (including neonicotinoids) contribute to eutrophication in surface waters and aquifers, negatively impacting water quality and marine biodiversity, which in turn affects fishing and tourism.

In conclusion, the plantationocene presents significant challenges to agricultural sustainability and environmental conservation. To address these challenges, it is crucial to enhance crop and livestock genetic diversity, protect coastal habitats, and minimize the environmental impact of agricultural chemicals. Integrated approaches that prioritize biodiversity conservation, sustainable agricultural practices, and ecosystem restoration are essential.

Collaboration among stakeholders—farmers, researchers, policymakers, and civil society—is key to developing and implementing strategies that balance agricultural production with environmental conservation (Melo 2018). By investing in resilient food systems that respect ecological limits and support local communities, we can work towards a sustainable future where both human and ecological well-being thrive.

Ultimately, understanding and mitigating the impacts of the plantationocene are essential for safeguarding biodiversity, ensuring food security, and preserving natural resources for future generations. This requires a paradigm shift towards regenerative practices that promote biodiversity, resilience, and sustainability across agricultural landscapes and beyond.

1.5 The Ecological Transition and Green Deal

The ecological transition represents a fundamental shift in both economy and society towards a sustainable model, where human activities respect the planet's ecological limits. At the heart of this transformation is the European Green Deal, an ambitious strategy launched by the European Commission to guide the EU towards a sustainable economy (Hedberg 2021). Its overarching goal is to achieve climate neutrality by 2050, achieved through reducing greenhouse gas emissions, promoting resource efficiency, and driving green innovation.

Such a strategy is imperative to address climate change, reduce pollution, preserve biodiversity, and secure a sustainable future for generations to come. The reliance on fossil fuels and unsustainable industrial practices has inflicted significant harm on the environment and ecosystems, necessitating an urgent transition towards renewable energy sources and clean technologies.

Benefits of the ecological transition include lower greenhouse gas emissions, the creation of new green jobs, improved air and water quality, and the promotion of a circular economy. Specific targets of the European Green Deal include reducing greenhouse gas emissions by 55% by 2030 and achieving climate neutrality by 2050.

Despite its benefits, implementing green technologies entails risks and concerns, particularly regarding energy storage technologies and the materials used in them. For instance, the production and recycling of batteries can have significant environmental impacts if not managed properly. Raw materials such as lithium and cobalt, essential for batteries, are often extracted under unsustainable conditions, posing risks to both health and the environment.

Careful analysis and improvement of production processes can lead to substantial reductions in polluting emissions. Adoption of advanced technologies for monitoring and controlling emissions, coupled with more sustainable industrial practices, can enhance energy efficiency and reduce the environmental footprint of economic activities.

1.6 Eremocene and Capitalocene: Implications for Coastal Zones

The Eremocene, a term coined by Edward O. Wilson, describes a future epoch characterized by ecological loneliness resulting from mass extinction of species. This concept underscores biodiversity loss as one of the most serious consequences of human impact on the environment. Visible effects of the Eremocene include habitat destruction, declining wildlife populations, and the loss of essential ecosystem services.

The Capitalocene era highlights the geological epoch where capitalist systems have profoundly impacted ecosystems and the environment, where human activities such as overfishing, habitat destruction for coastal development, and pollution have caused significant declines in marine biodiversity (Moore 2017). This period is marked by intensive exploitation of natural resources and severe climatic and environmental disruptions, including climate change, biodiversity loss, and global pollution.

For example, coral reefs, often referred to as the "rainforests of the sea," are facing widespread bleaching and degradation due to rising sea temperatures and ocean acidification, both consequences of human-induced climate change (Voolstra et al. 2023). These reefs support a vast array of marine life and provide crucial ecosystem services, such as shoreline protection and fisheries support. The loss of coral reefs represents a clear manifestation of the Eremocene in coastal ecosystems, where entire ecosystems are at risk of collapse due to human actions.

Furthermore, coastal wetlands, mangrove forests, and estuaries, which serve as nurseries for fish and provide protection against coastal erosion and storm surges, are also under threat (Anu et al. 2024). Their destruction or degradation due to urbanization, agriculture, and aquaculture practices not only diminishes biodiversity but also reduces the resilience of coastal communities to climate impacts, such as sea-level rise and extreme weather events.

The Capitalocene epoch, characterized by intensive resource extraction and exploitation driven by capitalist systems, has profoundly shaped coastal landscapes and ecosystems. Coastal areas are often prime targets for economic development, tourism, and industrial activities due to their accessibility and economic potential. To counteract the negative effects of the Capitalocene, sustainable practices must address the root causes of environmental crises and promote an economic model that prioritizes environmental stewardship and social equity. For example, adopting sustainable agricultural practices, promoting the circular economy, and investing in renewable energies can mitigate environmental impact and foster fairer, sustainable development.

For instance, large-scale tourism developments along coastlines can lead to habitat destruction, pollution from sewage and runoff, and pressure on local resources such as water and energy. The construction of ports and harbors for shipping activities can disrupt natural coastal processes, affecting sediment transport, shoreline stability, and coastal biodiversity.

In addition, the extraction of natural resources such as sand for construction and minerals from coastal sediments can have far-reaching environmental consequences. Sand mining, for example, can lead to erosion, loss of beach habitats, and increased vulnerability to coastal erosion and storm damage. These activities exemplify how capitalist-driven resource extraction in coastal zones contributes to the Capitalocene's legacy of environmental degradation and socio-economic inequality.

Addressing the challenges posed by the Eremocene and Capitalocene in coastal zones requires concerted efforts towards conservation, restoration, and sustainable development (UNESCO 2018). Protecting and restoring critical coastal habitats such as mangroves, salt marshes, and coral reefs is essential for preserving biodiversity and enhancing coastal resilience. Adopting sustainable practices in coastal planning and development, such as incorporating green infrastructure, promoting low-impact tourism, and implementing integrated coastal zone management approaches, can help mitigate environmental impacts while supporting local economies. Involving local communities in decision-making processes and ensuring equitable access to resources can foster stewardship and resilience in coastal areas. By recognizing and addressing the legacies of the Eremocene and Capitalocene in coastal zones, societies can work towards a more sustainable future where coastal ecosystems thrive, biodiversity is preserved, and coastal communities are resilient to environmental change.

Combining principles from the Capitalocene and Eremocene emphasizes the urgent need for adopting sustainable policies and practices to prevent further damage to the environment and ecosystems. Achieving environmental milestones demands global commitment to sustainability, green innovation, and international cooperation.

1.7 Conclusion

Considering concepts like the Entropocene, Eremocene, Capitalocene, Wasteocene, Urbanocene, and Plantationocene, managing coastal zones effectively requires comprehensive strategies that balance development and conservation. Integrated Coastal Zone Management (ICZM) is a crucial approach that seeks to harmonize economic growth with environmental protection through coordinated planning and governance frameworks.

Restoration of degraded habitats, such as dunes and wetlands, plays a crucial role in mitigating the impacts of human activities characteristic of these epochs.

Ecological restoration projects not only enhance biodiversity but also bolster coastal resilience against climate change-induced threats like sea-level rise and storm surges.

Robust regulatory frameworks are essential in safeguarding coastal areas from exploitation and degradation. Developing and enforcing laws and policies tailored to the unique challenges of coastal environments can mitigate unsustainable practices and promote sustainable development practices.

Community engagement and education are vital components of effective coastal management. By fostering awareness and involving local communities in decision-making processes, stakeholders become empowered stewards of their coastal environments. This engagement fosters a sense of responsibility and ownership, crucial for ensuring long-term sustainability and resilience in coastal regions.

In conclusion, as we navigate the complexities of the Antropocene and its various manifestations, from the Capitalocene's exploitation to the Wasteocene's environmental degradation, proactive measures such as ICZM, ecological restoration, stringent regulations, and community involvement are imperative. These strategies not only mitigate current environmental pressures but also pave the way towards a more sustainable future for coastal ecosystems and communities worldwide.

Today, the paradigm of coastal sustainability cannot overlook the social and cultural implications of environmental policies. Sustainable development strategies must consider diverse cultures and the needs of local communities, ensuring that proposed solutions are inclusive and respectful of human rights. Only through a holistic approach that integrates environmental, social, and economic considerations can we build a sustainable future for all.

References

Anu, Henna Parveen K, Sneha VK, Busheera P, Muhammed J, Anu A (2024) Mangroves in environmental engineering: harnessing the multifunctional potential of nature's coastal architects for sustainable ecosystem management. Results Eng 21 (March):101765. https://doi.org/10.1016/j.rineng.2024.101765

Arthur RI, Skerritt DJ, Schuhbauer A, Ebrahim N, Friend RM, Sumaila UR (2022) Small-scale fisheries and local food systems: transformations, threats and opportunities. Fish Fish 23(1):109–124. https://doi.org/10.1111/faf.12602

Barragán JM, de Andrés M (2015) Analysis and trends of the world's coastal cities and agglomerations. Ocean Coast Manag 114(September):11–20. https://doi.org/10.1016/j.ocecoaman.2015.06.004

Charles A (2023) Sustainable fishery systems

Crutzen PJ (2006) The Anthropocene. In: Ehlers E, Krafft T (eds) Earth system science in the Anthropocene. Springer, Berlin, Heidelberg, pp 13–18

Cundy ME, Santana-Garcon J, McLennan AG, Ayad ME, Bayer PE, Cooper M, Corrigan S, Harrison E, Wilcox C (2023) Seafood label quality and mislabelling rates hamper consumer choices for sustainability in Australia. Sci Rep 13(1):10146. https://doi.org/10.1038/s41598-023-37066-4

Delpeuch C (2022) OECD review of fisheries 2022

Dias AC, Esteves DA, Nayak PK, Akintola SL, Arizi EK, Chuenpagdee R, Das BK et al (2023) From vulnerability to viability: a situational analysis of small-scale fisheries in Asia and Africa. Mar Policy 155(September):105731. https://doi.org/10.1016/j.marpol.2023.105731

Ducruet C, Martin BP, Sene MA, Prete ML, Sun L, Itoh H, Pigné Y (2024) Ports and their influence on local air pollution and public health: a global analysis. Sci Total Environ 915(March):170099. https://doi.org/10.1016/j.scitotenv.2024.170099

EEA (2021) Land take and land degradation in functional urban areas, EEA Report n. 17/2021. European Environment Agency, Copenhagen

Environmental indicators: typology and overview (n.d) Publication. European Environment Agency. Accessed 19 June 2024. https://www.eea.europa.eu/publications/TEC25

European Commission. JRC (2015) Best available techniques (BAT) reference document for the refining of mineral oil and gas industrial emissions: industrial emissions directive 2010/75/EU (Integrated pollution prevention and control). Publications Office, LU. https://data.europa.eu/doi/10.2791/010758

Gazzetta Ufficiale della Repubblica Italiana (GURI) (2004) Codice dei beni culturali e del paesaggio, ai sensi dell'articolo 10 della legge 6 luglio 2002, n. 137. Gazzetta Ufficiale n. 45 del 24 febbraio 2004—Supplemento Ordinario n. 28

Glaeser EL, Resseger MG (2010) The complementarity between cities and skills. Reg Sci 50:221–244. https://doi.org/10.1111/j.1467-9787.2009.00635.x

Godde CM, Mason-D'Croz D, Mayberry DE, Thornton PK, Herrero M (2021) Impacts of climate change on the livestock food supply chain; a review of the evidence. Glob Food Secur 28(March):100488. https://doi.org/10.1016/j.gfs.2020.100488

Grunert KG, Hieke S, Wills J (2014) Sustainability labels on food products: consumer motivation, understanding and use. Food Policy 44(February):177–189. https://doi.org/10.1016/j.foodpol.2013.12.001

Hedberg A (2021) The European green deal: how to turn ambition into action. European Policy Centre.

ISPRA (2023a) Indicatori ambientali. Available at: http://groupware.sinanet.isprambiente.it/uso-copertura-e-consumo-di-suolo/library/consumo-di-suolo

ISPRA (2023b) Consumo di suolo, dinamiche territoriali e servizi ecosistemici. Edizione 2023 Report SNPA n. 37/2023–ISBN 978-88-448-1178-5. Available at https://www.snpambiente.it/snpa/consumo-di-suolo-dinamiche-territoriali-e-servizi-ecosistemici-edizione-2023/

Karametou P, Savina G, Agisilaos E (2024) Coastal tourism and sustainable development: case study of Xylokastro (Region of Peloponnese, Greece). In: Katsoni V, Cassar G (eds) Recent advancements in tourism business, technology and social sciences. Springer Nature Switzerland, Cham, pp 557–71. https://doi.org/10.1007/978-3-031-54338-8_32

Kleer J, Nawrot KA (2018) The rise of megacities—an empirical overview of the demographic changes in the world economy and an introduction to an integral theory of megacities. World Scientific Book Chapters. World Scientific Publishing Co. Pte. Ltd. https://econpapers.repec.org/bookchap/wsiwschap/9781786344274_5f0002.htm

Koens JF, Dieperink C, Miranda M (2009) Ecotourism as a development strategy: experiences from Costa Rica. Environ Dev Sustain 11(6):1225–1237. https://doi.org/10.1007/s10668-009-9214-3

La Rosa D, Martinico F, (2012) Assessment of hazards and risks for landscape protection planning in Sicily. J Environ Manage 127, Supplement. https://doi.org/10.1016/j.jenvman.2012.05.030

Magnan AK, Oppenheimer M, Garschagen M, Buchanan MK, Duvat VKE, Forbes DL, Ford JD et al (2022) Sea level rise risks and societal adaptation benefits in low-lying coastal areas. Sci Rep 12(1):10677. https://doi.org/10.1038/s41598-022-14303-w

Malthus T (n.d) An essay on the principle of population

Matekenya W, Ncwadi R (2022) The impact of maritime transport financing on total trade in South Africa. J Shipping Trade 7(1):5. https://doi.org/10.1186/s41072-022-00106-9

Mazzoni F, Marsili V, Alvisi S, Franchini M (2022) Exploring the impacts of tourism and weather on water consumption at different spatiotemporal scales: evidence from a coastal area on the Adriatic Sea (Northern Italy). Environ Res: Infrastruct Sustain 2(2):025005. https://doi.org/10.1088/2634-4505/ac611f

Meadows DH, Meadows DL, Randers J, Behrens WW (1972) The limits to growth. Universe Books, New York

Mejía CV, Brandt S (2015) Managing tourism in the Galapagos Islands through price incentives: a choice experiment approach. Ecol Econ 117(September):1–11. https://doi.org/10.1016/j.ecolecon.2015.05.014

Mejjad N, Rossi A, Pavel AB (2022) The coastal tourism industry in the Mediterranean: a critical review of the socio-economic and environmental pressures and impacts. Tourism Manage Perspect 44(October):101007. https://doi.org/10.1016/j.tmp.2022.101007

Melo V (2018) Collaborative efforts for sustainable development: surveying the literature on multi-stakeholder initiatives to realize the sustainable development goals. https://doi.org/10.13140/RG.2.2.19706.75209

Moore JW (2017) The Capitalocene, part I: on the nature and origins of our ecological crisis. J Peasant Stud 44(3):594–630. https://doi.org/10.1080/03066150.2016.1235036

Morton T (2016) Dark ecology: for a logic of future coexistence. Columbia University Press

Nations, Food and Agriculture Organization of the United (2018) Soil pollution: a hidden reality. Food & Agriculture Org

Naylor RL, Hardy RW, Buschmann AH, Bush SR, Cao L, Klinger DH, Little DC, Lubchenco J, Shumway SE, Troell M (2021) A 20-year retrospective review of global aquaculture. Nature 591(7851):551–563. https://doi.org/10.1038/s41586-021-03308-6

Neumann B, Vafeidis AT, Zimmermann J, Nicholls RJ (2015) Future coastal population growth and exposure to sea-level rise and coastal flooding—a global assessment. PLoS ONE 10(3):e0118571

OECD (2019) OECD regional outlook 2019: leveraging megatrends for cities and rural areas. Organisation for Economic Co-operation and Development, Paris. https://www.oecd-ilibrary.org/urban-rural-and-regional-development/oecd-regional-outlook-2019_9789264312838-en

Olsen SB (2003a) Frameworks and indicators for assessing progress in integrated coastal management initiatives. Ocean Coast Manag 46(3–4):347–361

Olsen SB (2003b) Frameworks and indicators for assessing progress in integrated coastal management initiatives. Ocean Coast Manag 46(3–4):347–361. https://doi.org/10.1016/S0964-5691(03)00012-7

Pimid M, Nasir MRM, Krishnan KT, Chambers GK, Ghafar Ahmad A, Perijin J (2022) Understanding social dimensions in wildlife conservation: multiple stakeholder views. Anim: Open Access J MDPI 12(7):811. https://doi.org/10.3390/ani12070811

PR_WGII_AR6_english.Pdf (n.d). Accessed 19 June 2024. https://www.ipcc.ch/site/assets/uploads/2022/02/PR_WGII_AR6_english.pdf

Rockström J, Steffen W, Noone K, Persson Å, Chapin FS, Lambin EF, Foley JA (2009) A safe operating space for humanity. Nature 461(7263):472–475

Rohr JR, Barrett CB, Civitello DJ, Craft ME, Delius B, DeLeo GA, Hudson PJ et al (2019) Emerging human infectious diseases and the links to global food production. Nat Sustain 2(6):445–456. https://doi.org/10.1038/s41893-019-0293-3

Romano B, Zullo F (2014) The urban transformation of Italy's Adriatic coastal strip: fifty years of unsustainability. Land Use Policy 38:26–36. https://doi.org/10.1016/j.landusepol.2013.10.001

Sachs JD, Mellinger AD, Gallup J (2001) The geography of poverty and wealth. Sci Am 284:70–75. https://doi.org/10.1038/scientificamerican0301-70

Smith TF, Elrick-Barr CE, Thomsen DC, Celliers L, Le Tissier M (2023) Impacts of tourism on coastal areas. Camb Prisms: Coast Futures 1(January):e5. https://doi.org/10.1017/cft.2022.5

Steffen W, Broadgate W, Deutsch L, Gaffney O, Ludwig C (2015) The trajectory of the Anthropocene: the great acceleration. Anthropocene Rev 2(1):81–98

Tulaeva S, Tysiachniouk M, Pappila M, Tynkkynen M (2023) Marine stewardship council certification in Finland and Russia: global standards and local practices. Sustainability 15(5):4063. https://doi.org/10.3390/su15054063

Turner J (2021) Living in the Plantationocene. Bryant Univ J Interdisc Stud 2(1). https://digitalcommons.bryant.edu/isbhs/vol2/iss1/18

UNESCO (2018) Welcome to the Anthropocene!—UNESCO digital library 2018. https://unesdoc.unesco.org/ark:/48223/pf0000261900

Verdegem M, Buschmann AH, Win Latt U, Dalsgaard AJT, Lovatelli A (2023) The contribution of aquaculture systems to global aquaculture production. J World Aquac Soc 54(2):206–250. https://doi.org/10.1111/jwas.12963

Voolstra CR, Peixoto RS, Ferrier-Pagès C (2023) Mitigating the ecological collapse of coral reef ecosystems. EMBO Rep 24(4):e56826. https://doi.org/10.15252/embr.202356826

Wackernagel M, David L (2023) Earth overshoot day. In: Wallenhorst N, Wulf C (eds) Handbook of the Anthropocene. Springer International Publishing, Cham, pp 569–72. https://doi.org/10.1007/978-3-031-25910-4_90.

Williams BA, Watson JEM, Beyer HL, Klein CJ, Montgomery J, Runting RK, Roberson LA et al (2022) Global rarity of intact coastal regions. Conserv Biol 36(4):e13874. https://doi.org/10.1111/cobi.13874

Wörndl W, Koo C, Stienmetz JL (eds) (2021) Information and communication technologies in tourism 2021: proceedings of the ENTER 2021 e-Tourism conference, January 19–22, 2021. Springer International Publishing, Cham. https://doi.org/10.1007/978-3-030-65785-7

Zoppi C, Lai S (2014) Land-taking processes: an interpretive study concerning an Italian region. Land Use Policy 36:369–380. https://doi.org/10.1016/j.landusepol.2013.09.011

Chapter 2
The Effects of Climate Change on Ocean Upwelling and Productivity in Regional Seas

Muhammad Naim Satar, Winfred Marshal, and Mohd Fadzil Akhir

Abstract Climate change is having a big impact on the ocean, affecting its ability to support life. This paper explores how warmer seas, changes in upwelling systems, and shifts in the ocean's ability to produce food are all connected. We reveal the close connection between upwelling, a conveyor belt carrying nutrients, and the thriving biodiversity it supports. We expose the detailed workings of this life-sustaining mechanism, uncovering the physical forces that coordinate its rhythm. Next, we discuss the disruptive impact of climate change, closely examining how it affects regional upwelling patterns and the resulting chain of effects on marine life. We delve into the intricate complexity of biogeochemical models, which act as digital predictors foretelling future scenarios. These models help us understand how they interpret the ocean's symphony into predictions about ecological changes. Armed with these tools, we begin a worldwide journey, tracing the widespread impacts of climate change on primary productivity—the essential force powering marine ecosystems. Ultimately, we focus on specific regional examples, demonstrating how the global symphony of change manifests in localized variations. This paper sheds light on the scientific complexities of this dynamic equation, providing a clear yet optimistic picture of the ocean's ability to withstand an unprecedented challenge.

Keywords Upwelling · Climate change · Productivity · Ocean dynamic

---

The original version of the chapter has been revised. A correction to this chapter can be found at https://doi.org/10.1007/978-3-031-75749-5_15

M. N. Satar · W. Marshal · M. F. Akhir (✉)
Institut Oceanography and Environment, Universiti Malaysia Terengganu, 21030 Kuala Nerus, Terengganu, Malaysia
e-mail: mfadzil@umt.edu.my

V. Maccarrone and M. Fadzil Akhir (eds.), *Coastal Sustainability*, Coastal Research Library 39, https://doi.org/10.1007/978-3-031-75749-5_2

2.1 Introduction

2.1.1 Climate Change Impacts on Ocean

Climate change profoundly affects the Earth's oceans, influencing their physical, chemical, and biological features on a global scale. The ocean plays a crucial role in counteracting climate change by acting as a significant reservoir for heat and carbon. As the Earth heats up, the ocean absorbs and redistributes heat energy, helping to regulate temperatures across the planet. When the ocean absorbs excessive heat and energy due to climate change, the resulting temperature shift triggers a series of unprecedented consequences. These include the melting of ice, rising sea levels, marine heatwaves, and ocean acidification. The ocean acts as a vast carbon sink, absorbing over 35% of the CO_2 released since the Industrial Revolution, primarily from fossil fuel combustion. This human-driven influx of CO_2 is leading to increased ocean acidity, posing a significant threat to marine ecosystems. The global ocean is experiencing a cascade of impacts from climate change, with rising temperatures and increasing carbon dioxide concentrations altering marine ecosystems and impacting coastal communities. Anthropogenic climate change is leaving its indelible mark on the ocean, with documented effects on marine biodiversity, primary productivity, and the socio-economic well-being of coastal populations, both directly and indirectly.

The ocean's vital role in the climate system makes it highly vulnerable to even small alterations in climatic controls, potentially triggering significant shifts in marine ecosystems (Behrenfeld et al. 2006; Mantua et al. 1997). Climate's tight grip on the ocean can lead to dramatic transformations in marine life. In particular, large marine climate modes, such as the El Niño-Southern Oscillation (ENSO) and the Pacific Decadal Oscillation (Mantua and Hare 2002), are susceptible to change. For example, Sea Surface Temperature (SST) in the eastern and equatorial Pacific Oceans are abnormally warming during the warm phases, which further limits the upwelling of high-nutrient water along the eastern Pacific coast and intensifies the stratification. Since the temperature, light, and nutrient availability are all significant elements in phytoplankton development, it is expected that these climate-ocean dynamics might alter the phytoplankton population due to climate change. Typically, large-scale patterns of climatic variability act as a stress on the ecosystems (though this can favorably impact the incidence and development of harmful algal blooms (Moore et al. 2008)).

Climate variabilities are already inducing alterations in marine food webs, leading to changes in the distribution patterns and foraging strategies of marine organisms. Shifts in atmospheric circulation negatively impact coastal salinity, turbidity, and land-based chemical and nutrient inflows. Ocean temperatures are projected to increase by 2–3 °C within the next few decades, following the trend observed since the late twentieth century (Stock et al. 2011). Additionally, rising atmospheric Carbon di-oxide (CO_2) levels will drive ocean acidification, with a projected decrease of 0.14–0.35 in global surface pH by the end of the twenty-first century (Bernstein et al. 2007).

2.1.2 The Vital Link Between Upwelling and Ocean Life

The majority of the world's ocean's surface waters typically have low nutrient levels. New nutrients are needed to maintain primary production. These can originate from rivers that drain into coastal regions and are rich in minerals, from turbulent small-scale mixing in nutrient-rich underlying waters, from the upwelling of deeper nutrient-rich seas, or even from some bacteria fixing nitrogen. Furthermore, intensive marine production is supported by upwelling in some coastal locations and during shelf breaks in several global ocean regions (Kämpf and Chapman 2016). A phenomena known as upwelling occurs in oceanography when prevailing winds drive surface water away from the shore, allowing deeper water to rise and exchange the moving surface water. This movement of cold, nutrient-rich water is vertical and occurs from the deeper layers of the ocean to the top. Productivity is greatly impacted by this process, which is especially important for marine ecosystems and coastal locations. Nitrate, phosphate, and silicate are among the nutrients that are found in enormous amounts in the thermocline, the deepest layers of the ocean. This water full of nutrients is brought to the surface by upwelling, which supplies vital components for phytoplankton growth. More phytoplankton means more food for bigger organisms in the ocean, like fish and whales. When the amount of nutrients available for these phytoplankton changes, it throws off the whole balance and affects productivity in coastal areas where cold, nutrient-rich water wells up from the deep (Wilkerson and Dugdale 2008). Because of increased primary production, upwelling regions hold economic significance, contributing to nearly 20% of the worldwide fish catch. Remarkably, despite constituting less than 1% of the global ocean, these upwelling areas play a crucial role (Pauly and Christensen 1995).

On the flip side, elevated water-column stratification linked to higher surface temperatures hinders the interchange of waters through the pycnocline, leading to a reduction in primary production. The theory that rising surface temperatures brought on by rising greenhouse gas concentrations will reduce the amount of nutrients available to the euphotic zone of ocean ecosystems and harm fisheries has been supported by this dynamical knowledge (Behrenfeld et al. 2006). Low amounts of dissolved oxygen are seen in subsurface and near-bottom waters in coastal seas, which are typically eutrophic and have significant primary productivity at the surface that encourages oxygen consumption through remineralization of sinking organic matter. Surface waters with high productivity are common in coastal upwelling zones and support ecosystems that are significant to the local economy. Thus, deoxygenation can particularly affect upwelling systems (Roegner et al. 2011).

The primary factor influencing hydrographic and biogeochemical variability throughout the upwelling season is wind forcing, which regulates the entry of deep water masses from offshore onto the shelf. The offshore intermediate waters, which are primarily located 100–300 m below the seasonal pycnocline, are the primary source of the upwelled water in the majority of upwelling systems (Kämpf and Chapman 2016). Because wind plays such a significant role in upwelling systems, wind stress affects how much nutrient delivery there is to the surface layer. Weak winds have the potential to lift only modest amounts of water onto the shelf, as was previously said, and this water may not reach the euphotic zone. Because of increased turbulence or being advected out of the coastal zone, phytoplankton may not be able to use the nutrients if the winds are too strong. In that area, upwelling does not always translate into great output. The strongest biological reaction usually occurs when upwelled water meets the surface, but the strength of the reaction varies depending on whether the wind stress is sustained or lessens to permit the development of a phytoplankton bloom. The predominant upwelling process that produces highly productive coastal ecosystems and profitable coastal fisheries is typically wind-driven coastal upwelling. Constantly occurring upwelling-favourable winds provide a steady supply of nutrients to the euphotic zone.

2.1.3 Intention and Objectives of This Paper

The huge ocean, which was formerly thought to be a passive victim of climate change, is now seen as a vital front in the fight against its growing fury. This research examines the complex connection between ocean dynamics and climate change by navigating the many environmental elements affecting the oceans on Earth. By examining the upwelling-productivity relationship via a lens, a fundamental understanding of how climate-induced changes appear in regional upwelling systems is revealed. An in-depth analysis of upwelling mechanisms reveals insights into the intricate mechanisms governing vertical nutrient movement and influencing the biological dynamics of marine habitats. Scientists are closely examining how climate change affects ocean productivity and how it affects phytoplankton, marine food webs, and nutrient cycling.

Our discussion also goes into the domain of biogeochemical models, where basic ideas controlling the dynamics of ecosystems and nutrient cycles are highlighted. These models are very useful for figuring out the complex interactions between the physical, chemical, and biological elements that make up the ocean. By applying these models, we are able to gain a better understanding of how marine ecosystems respond to changing climatic circumstances and develop a prediction framework for evaluating future scenarios.

Then, this scientific story goes beyond simple observation to prediction modelling as we explore the regional and global effects of climate change on primary productivity. This helps us to understand possible changes in species distribution, patterns of biodiversity, and carbon cycling. In the parts that follow, we will examine the intricate workings of upwelling, including its various regional manifestations and how climate change affects its productivity-dancing. Next, we arm ourselves with biogeochemical models' tools and discover their mysteries to understand how regional and global primary output reacts to a changing climate. We want to demonstrate the ocean's susceptibility to climate change with every step we take, emphasising the pressing need for action to protect life as we know it on Earth.

2.2 Upwelling

2.2.1 *How Wind Drives Upwelling?*

Wind plays a major role in the movement of the surface of the seawater. As the wind blows over the ocean, it exerts a force on the ocean surface in the direction of the wind. This force, per unit ocean surface area, is called wind stress. The wind stress creates a surface wave field and drives ocean circulation (Yelland and Taylor 1996). However, since our Earth is rotating, it creates another force called the Coriolis force. Due to this force, the direction of the ocean current does not fully follow the direction of the wind. When the wind blows across the ocean, the water at the surface of the ocean followed the wind direction, but the water under the surface water tends to be deflected to the right (left) in the Northern Hemisphere (Southern Hemisphere) for about 45° from the wind direction. The water below this second layer tends to deflect more than the first one, and it continues until it makes a spiral of moving water 100 to 150 m deep called an Ekman spiral (https://scied.ucar.edu/learning-zone/earth-system/how-ocean-moves-ekman-transport#:~:text=Ocean). The average transport of this deflected water is around 90° from the wind direction, called Ekman transport.

This Ekman transport plays a crucial role in upwelling formation. For example, the wind blows parallel to the coastline in the Northern Hemisphere (Fig. 2.1; (Satar et al. 2023: p. 2)), the Ekman transport deflects the water to the right, making the water to move offshore. The water from the deeper layer then moves upwards, replacing the surface water that moves offshore, called coastal upwelling. Apart from Ekman transport, there is also another process called Ekman pumping. It formed due to the wind stress curl, which creates surface divergence and upward movement of the seawater. This upwelled water from the deeper layer is usually colder and rich in nutrients, thus enhancing the area's primary productivity in the upwelling region. This is why the upwelling locations usually serve as high fishing grounds worldwide. Upwelling only covers around $< 1\%$ of the total surface area of the ocean, but it provides 8% of the global primary production and more than 20% of the world's capture fisheries (Pauly and Christensen 1995).

2.2.2 *Upwelling Around Our World*

Our world has four major upwelling systems, called the Eastern Boundary Upwelling Systems (EBUS), located in the eastern part of the Pacific Ocean and the Atlantic Ocean for both hemispheres (Fig. 2.2; (Satar et al. 2023: p. 2)). EBUS includes the California, Humboldt, Canary, and Benguela current systems. They are driven by the continuously blowing equatorward winds, which blow parallel to the eastern borders of the ocean basins (western shores of landmasses in the Pacific Ocean and Atlantic Ocean), induce the Ekman transport, and lead to the upwelling event (Bakun et al. 2015). These regions provide a great deal of profit in terms of economy (Costanza

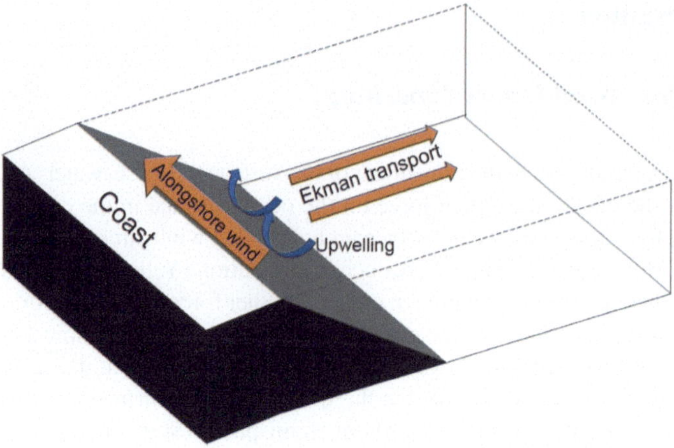

Fig. 2.1 Schematic figure on how coastal upwelling is generated. The figure is sourced from Satar et al. (2023)

et al. 1998), as well as significant recreational services to people living along these coastlines, linking indirectly to the rest of the world (Wang et al. 2015).

Apart from EBUS, many small scales of upwelling in many regions occur seasonally, or what we call seasonal upwelling (Fig. 2.2). These upwelling usually depend on the seasonal blowing winds parallel to the coastline. For example, the South China Sea region, where many seasonal upwelling locations can be found. The South China Sea is mainly influenced by the East Asian Monsoon, which can be divided into two monsoon seasons, the southwest and the northeast monsoon (Fang et al. 2006). These monsoon seasons are mainly caused by different seasons between the northern and southern hemispheres. In boreal summer (southwest monsoon), the southwesterly

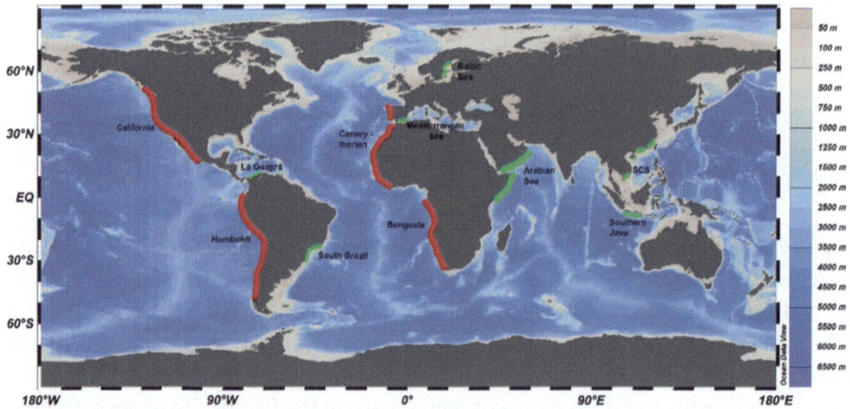

Fig. 2.2 Upwelling locations in our world. The red line indicates the EBUS regions, and the green line indicates the seasonal upwelling regions

wind blows in the South China Sea from May to September, while in boreal winter (northeast monsoon), the direction of the wind is reversed to northeasterly, and it prevails from November to March. On the other hand, the period between the northeast and southwest monsoon is dominated by the inter-monsoon, where the wind prevails in a variable direction and usually occurs in April and October (Morgan et al. 1983).

Depending on the upwelling locations, these two monsoonal winds are responsible for generating the upwelling in the South China Sea region. Most of the upwelling in the South China Sea occurred during southwest monsoon such as on the East Coast of Peninsular Malaysia (Akhir et al. 2015; Kok et al. 2017), southern Vietnam (Kuo et al. 2004), Taiwan Strait (Hong et al. 2009) and Hainan Island (Lü et al. 2008). Only two occur during northeast monsoon, which are in the Luzon Strait (Wang et al. 2010) and northwest Sabah (Yan et al. 2015; Satar et al. 2020). Apart from the South China Sea, many other seasonal upwelling locations can be found throughout our world (Fig. 2.2).

2.2.3 How Climate is Changing the Wind Pattern?

Since upwelling is majorly dependent on the wind, any wind speed and direction changes can impact the upwelling. A climate event like ENSO, which alters the atmospheric process, can affect the upwelling event depending on the upwelling location. ENSO is a periodic fluctuation in SST and the air pressure of the overlying atmosphere (Southern Oscillation) across the equatorial Pacific Ocean. Usually, it occurs between 2 to 7 years. ENSO can be divided into two phases, El Niño, and La Niña. The warm ENSO (El Niño) occurs when the SST in the central and eastern tropical Pacific Ocean is warming or above the average, and the convection shifts to the central or eastern Pacific (Macias et al. 2012). In contrast, the cool ENSO is called La Niña, the opposite phase of El Niño.

ENSO can influence upwelling strength as this event impacts the wind intensity. For example, the upwelling in Peru usually weakens during El Niño and strengthens during La Nina. When the El Niño event takes place, it will weaken the trade winds, thus reducing equatorial upwelling, and causing an anomalous increase in the coastal SST and an anomalous deepening of the thermocline off the Peruvian coast as well as a decrease in nutrients (Barber and Chavez 1983). Nevertheless, some areas experienced an intensified upwelling during El Niño, such as in the north South China Sea and northwest Sabah (Satar et al. 2020; Jing et al. 2011). This occurred due to the presence of an anticyclonic atmospheric circulation anomaly, which intensifies the monsoonal wind during El Niño.

That is one example of one climatic event. Our world has several other climatic events such as Indian Ocean Dipole, Pacific Decadal Oscillation, Madden Julian Oscillation, Atlantic Multidecadal Oscillation, and others. However, one of the biggest concerns today is not these climatic events, but climate change or global warming. Climate change can cause a lot of consequences to land, atmosphere, and ocean. Upwelling is also one of the processes that will be affected by climate change. Bakun (1990) came with a theory that later known as Bakun Hypothesis. He said that climate change that increase our global temperature will heat our world much more. And land is surely will be heated and warm much higher and faster than the ocean. This will increase the land and ocean thermal contrast, leading to stronger pressure gradients and producing stronger alongshore winds. With stronger winds, it will finally lead to stronger upwelling. Thus, Bakun Hypothesis conclude that upwelling will become stronger under the climate change scenario. Apart from this theory, (Rykaczewski et al. 2015) said that rather than deepening the low continental pressure by warming of land, climate change will cause the shifting or movement of high-pressure systems towards the poleward region and cause the upwelling to become more intensified in the poleward region and thus weakened in the equator part.

Changes in upwelling intensity can cause many consequences. Weakened upwelling will cause less nutrient input will be pumped to the surface, which mean less primary productivity, while strengthened upwelling will cause more nutrients will be pumped to the surface. Even though more nutrients input means more primary productivity, but it also can lead to harmful algal bloom event. Furthermore, deeper water also contains less oxygen concentration, which can also lead to the hypoxic event at the surface water in the upwelling locations (Chan et al. 2008). Moreover, the stronger (weakened) upwelling also lead to stronger (weakened) offshore transport of the surface water, and this can affect the plankton composition (Maranón and Fernández 1995; Bakun and Weeks 2004).

Other than changes in upwelling intensity, climate change also indirectly affects phytoplankton growth by increasing upper-ocean stratification, thus limiting nutrient supply to the sunlit layer. Another effect of warmer ocean is it can increase the metabolic rates for plankton, especially the zooplankton, and can cause higher grazing rates by zooplankton which can decrease the amount of phytoplankton.

2.2.4 Climate Change Versus Upwelling

2.2.4.1 Unmasking Climate Change: A Global Modeling Mission

Realizing the impacts of climate change on our world, the Intergovernmental Panel on Climate Change (IPCC) was formed under the United Nations (UN) in 1988. Thousands of people from all over the world contribute to the work of the IPCC. Their objective is to provide governments at all levels with scientific information that they can use to develop climate policies. IPCC reports are also a key input into international climate change negotiations. They divide their team into three working groups. Working Group I deal with The Physical Science Basis of Climate Change, Working Group II with Climate Change Impacts, Adaptation and Vulnerability, and Working Group III with Mitigation of Climate Change.

Table 2.1 Five different types of SSP and its narratives

SSP	Narratives
1	Sustainability—taking the green road (Low challenges to mitigation and adaptation)
2	Middle of the road (Medium challenges to mitigation and adaptation)
3	Regional rivalry—a rocky road (High challenges to mitigation and adaptation)
4	Inequality—a road divided (Low challenges to mitigation, high challenges to adaptation)
5	Fossil-fueled development—taking the highway (High challenges to mitigation, low challenges to adaptation)

Apart from the IPCC, another important group, the Working Group on Coupled Modelling (WGCM) of the World Climate Research Programme (WCRP), worked on a climate model. Realizing how climate change can threaten our world, and modeling can help recreate these multifaceted phenomena using computers to approximate reality, this group has organized a collaborative framework designed to improve knowledge of climate change, called Coupled Model Intercomparison Project (CMIP). They compared models with real-world observations and inter-compared model simulations of the Earth's future climate. Their objective is to use multi-model to understand better past, present, and future climate changes brought on by natural, unforced variability or in response to changes in radiative forcing. CMIP began in 1995, and their first set of common experiments compared the model response to an idealized forcing—a constant rate of increase, which was accomplished using a CO_2 increase of 1% per year compounded. CMIP kept growing after that, with various experiments being conducted and developed. The Earth System Grid Federation gathered all the data from CMIP output, giving open access to all research communities. With CMIP, it helps the researchers in their estimation of the consequences of further changes in our climate, and the IPCC is using their results to develop suitable policies. The latest CMIP that has been produced is the 6th phase (CMIP6), with more than 50 modeling centers participating in this project. One of the important components in their models is the projection based on the Shared Socioeconomic Pathways (SSP). This SSP (previously in the 5th Assessment Report known as Representative Concentration Pathways, RCP) is the scenario of projected socioeconomic global changes up to 2100. They are used to derive greenhouse gas emissions scenarios with different climate policies. Thus, they are divided into five different projections based on 5 different climate policies (Table 2.1). The higher number means weaker policies and implementations are being made to counter climate change.

2.2.4.2 How Climate Change is Altering Ocean's Upwelling Symphony?

Many researchers worldwide try to analyze changes that are already happening in upwelling. Some of them also used the global climate model such as CMIP to make the future upwelling projection based on a different scenario. Most of this research focused more on the EBUS areas, as they are large-scale upwelling ecosystems. Below here is several upwelling research conducted in the EBUS area.

In 2014, (Sydeman et al. 2014) conducted a meta-analysis by synthesizing 22 studies from 1990 and 2012 on upwelling and climate change in the EBUS area. They tested whether the evidence from these studies was consistent (strengthening wind) or inconsistent (weakening wind) with Bakun Hypothesis. They found that intensification wind present in California, Humboldt, and Benguela but not in the Canary-Iberian as it showed weakening wind strength for the whole latitude in Iberian and southern half of Canary system. Later, (Varela et al. 2015) used global wind data obtained from NCEP CFSR from 1982–2010 to investigate the temporal and spatial trend of upwelling regime worldwide. The results showed different trends for each region. For both Benguela and Canary, most of the region showed positive (increasing) trend of upwelling especially northern part of Benguela except for a small region in southern part for both regions. For Humboldt, upwelling in Chile (southern part of Humboldt) showed dominant negative (decreasing) trend except for the middle part of the upwelling region, while in the northern part of Humboldt (Peru upwelling), positive trend of upwelling was found except for the southernmost part of this region. Meanwhile for the California region, negative trend also been found in the southern region, but only significant in southernmost part, while northern region of California showed positive trend of upwelling except for the southernmost part as well.

If we go specifically in every part of the EBUS, there are much research published related to this upwelling and climate change with the usage of CMIP data in their research. Usually, researchers used two types of data, historical and future projection. The future projection is basically from the SSP scenario. Then, they will find the difference between future and historical to predict the future changes in upwelling. For instance, in the California region, (Xiu et al. 2018) use the CMIP5 data, under the RCP8.5 (worst case scenario) to analyze the historical (1990–2009) and future (2030–2049) trend of upwelling by coupling the Regional Ocean Modeling System (ROMS) and the Carbon, Silicate, and Nitrogen Ecosystem model (CoSiNE-31) that provide physical, biological, and chemical data. The results showed an increasing trend in SST data especially at the offshore region, which increase in the upper-ocean stratification. However, the nitrates and silicates concentration also showed an increment, especially in the coastal region and the upwelling intensity also showed an increasing trend due to stronger alongshore wind, wind stress curl, and formation of stronger eddies in the study area.

Later, a regional ocean model of Geophysical Fluid Dynamics Laboratory (GFDL)-CM3 was used by Arellano and Rivas (2019) to explore how primary production in California's upwelling region would react to the future climate change scenarios under the RCP6.0 and RCP8.5. The SST projection of RCP6.0 and RCP8.5 showed an increase of SST between 1.0 to 2.5 °C and 1.0 to 4.0 °C that could lead to an increase in stratification. However, the wind also intensified two times and four times stronger during spring and summer, respectively. Further, under the RCP8.5, a decrease in thermocline depth was observed in the coastal area, and it can be related to the increase in upwelling activities. Moreover, except for small regions, these upwelling activities under RCP8.5 could counteract the increase in stratification that decreased the chlorophyll-a concentration.

In the case of the Canary-Iberian region, specifically in Senegalo-Mauritanian (the southern part of the Canary Upwelling System), the upwelling intensity has been analyzed using 47 climate models from the CMIP5 database under the RCP8.5 scenario (Sylla et al. 2019). Even though each model showed diverse responses to the RCP8.5 scenario in terms of amplitude, the general picture demonstrated a reduction of the upwelling toward the end of the twenty-first century. This was evidenced by decrements in SST and the SST upwelling index (offshore SST minus coastal SST) for most models in the northern and southern parts of the Senegalo-Mauritanian upwelling area. The main cause of this weakening of upwelling is the reduction of the wind forcing linked to a northward shift of the Azores anticyclone and a more regional modulation of the low pressures found over Northwest Africa.

In the northwest of the Iberian Peninsula (the north part of the Canary-Iberian Upwelling System), (Sousa et al. 2020) used the Delft3D-Flow model to observe CMIP5 data for historical upwelling status from the year 1976 to 2005 and predict the future upwelling in the years 2070–2099 under the condition of RCP 8.5. The results showed an increment of the upwelling index for both historical and future projections. However, the stratification of the seawater also increased, and the thermocline deepened from 5–15 m in historical data and is expected to be deepened to 5–25 m in the future. This increment in ocean stratification is the main factor that will decrease the upwelling intensity in the future. High stratification will limit the amount of deeper water reaching the sea surface even though the wind stress is increased based on Bakun Hypothesis.

However, using GCM from the CMIP6 project recently, future SST changes were evaluated for both the coast and the offshore regions under SSP 5–8.5 in the Canary-Iberian upwelling region (Varela et al. 2022). Results showed that the upwelling will intensify in the future for the northern and middle regions of the Canary-Iberian Upwelling system and weaken in the southern region, which implies that upwelling changes depending on the latitudinal displacement as suggested by Rykaczewski et al. (2015).

2.2.5 Future of Upwelling in Regional Seas: Case Studies and Current Insights

Unlike EBUS areas, research on upwelling and climate change in the seasonal upwelling region is less focused and only several research used the CMIP data to predict the future upwelling event such as in Arabian Sea and in South Brazil.

In the Arabian Sea, specifically in Somalia and Oman, they used ROMS with CMIP5 model outputs (under RCP 8.5) to analyze the impact of monsoon low-level jet (MLJJ) shift in terms of upwelling and productivity in Omanian and Somali coasts under the global warming scenario (Praveen et al. 2016). The results showed intensifying wind stress, zonal mass transport, and vertical transport over Oman and weakening in Somali, thus leading to a stronger upwelling in Oman and weakened upwelling in Somali due to the poleward shift in MLJJ. Previously, analysis of wind from National Centers for Environmental Prediction—Climate Forecast System Reanalysis (NCEP-CFSR) by Varela et al. (2015) also found the weakening of the Somalian upwelling system, which supported the findings from Praveen et al. (2016).

However, in the same year, the analysis of upwelling by using both the Global Climate Model (GCM) and Regional Climate Model (RCM) found that upwelling in Somali is projected to increase with latitude, even higher than the EBUS under the RCP 4.5 and RCP 8.5 with a robust value at 8.5° N (Castro et al. 2016). The intensification of upwelling here is more affected by the Ekman transport relative to Ekman pumping, as the projected Ekman pumping did not show clear trends for most latitudes. A significant decrease in Ekman pumping is found in the southernmost latitudes. Due to global warming, air pressure and temperature variations between land and sea are also projected to increase significantly throughout the twenty-first century. This intensification significantly impacts the coastal upwelling, supporting and strengthening Bakun's theory.

Meanwhile, by downscaling Earth System Models (ESM) using the Hamburg Shelf Ocean Model (HAMSOM), de (Souza et al. 2020) managed to project the future upwelling projection under RCP8.5 in the South Brazil Bight. Even though the Ekman forcing showed an increment trend, the vertical velocities at the bottom of the mixed layer showed a decrement trend at the end of the century. An increase in water column stability due to the surface warming and a slight reduction of the Brazil Current transport south of 25° latitude tend to decrease the upwelling intensity in this area.

As for the South China Sea, several research regarding upwelling and climate change by using historical data from various sources. For instance, a sediment core was taken at an upwelling area in Vietnam, providing 3200 years of historical data (Xu et al. 2021). The core provides historical data that shows the rise and fall of the selected terrestrial element's concentration, such as aluminum, titanium, and potassium. These fluctuations were then correlated with the upwelling intensity to know whether the upwelling was stronger or weakened over the past year. Increased upwelling enhances marine production, leading to decreased concentrations of terrestrial elements. The result showed a rapid decrease (average 25%) in all selected elements since 1950, suggesting an abrupt intensification of upwelling from the same period. The record of summer monsoon wind data over the past 1810 year taken by Zhang et al. (2008) was also well matched with the data from the sediment core, where the wind showed an intensification during the summer monsoon, which are the primary driver of upwelling in Vietnam.

Meanwhile in the north South China Sea region, they used coral core at the eastern Hainan Island to provide SST record from 1876–1996, which been used to reveal the upwelling fluctuation (Liu et al. 2013). From the SST anomaly data, they divide the data into three periods based on their trend. The latest period (1961–1996) showed a signature weakening in upwelling intensity based on the high SST anomaly during that period. This weakened upwelling has also been discovered by Xie et al. (2016), as the upwelling index from 1982–2012 showed a decreasing trend with a rate of 0.01 °C per year. Later, (Zhu et al. 2023) revealed the warming trend of SST trend in the Hainan Island region by using data obtained from the MODIS-Aqua from 2003–2021. However, based on the Empirical Orthogonal Function (EOF) analysis, the upwelling intensity around Hainan Island showed an increasing trend from 2003–2021, especially in the period after 2013. But this positive trend could not compete with the overall warming trends of SST in the northwest area of the SCS.

2.3 The Engine of Marine Life: Ocean Productivity

Ocean productivity is an important mechanism that supports marine life and exerts an impact on the global climate. It relates to the conversion of sunlight and carbon dioxide into organic substances through photosynthesis, carried out by autotrophic elements within the marine plankton community. This synthesis of organic matter forms the basis of the marine food chain, promoting a diverse range of aquatic organisms spanning from minute zooplankton to immense whales. Phytoplankton, minute self-sustaining marine organisms, flourish in the sunlit portions of the ocean's upper layer. They create fresh biomass from inorganic sources such as carbon dioxide and mineral nutrients, coupling with light energy. Accounting for approximately half of the planet's photosynthetic processes, they play a pivotal role in governing the Earth's carbon cycle and maintaining global climate equilibrium (Field et al. 1998). Various elements, including nutrient availability, light exposure, temperature and water currents, together to shape the dispersion and quantity of phytoplankton. Nutrients like nitrogen, phosphorus, and iron are crucial for their development, are furnished to the ocean's upper layer through diverse mechanisms like upwelling, mixing, and the deposition of atmospheric dust. The ocean productivity cycles involve several key terms and concepts:

i. **Gross Primary Production (GPP)**: Cumulative organic carbon synthesis by autotrophic organisms. The overall quantity of carbon compounds generated by autotrophic populations within the marine environment during a specific timeframe.
ii. **Respiration (RP)**: Energy is expended during the conversion of organic carbon back into carbon dioxide through oxidation. This occurs during nighttime when photosynthesis is not possible due to the absence of light (Bowers and Brett 2008).
iii. **Net Primary Production (NPP)**: Outcomes arise from the conversion of atmospheric carbon into organic substances by marine phytoplankton, initiating the conveyance of organic materials and energy to marine ecosystems. This can be distilled as the pace at which autotrophic metabolism generates biomass over a specific timeframe. Both gross and net primary production are usually denoted in terms of mass per area per unit time (g C m − 2 yr − 1).

iv. **Secondary Production (SP)**: Secondary production entails the creation of living biomass within a heterotrophic individual or a collective of organisms within a specific timeframe. It can also signify the build-up of mass across a whole trophic level. The unit of measurement for secondary production is similar to NPP includes a time component (Benke and Huryn 2007).
v. **Net Ecosystem Production (NEP)**: The rate of net ecosystem production (NEP) can be described as NEP = GPP + RP, with GPP representing positive values and RP indicating negative values. In essence, it captures the equilibrium between the organic carbon gains through GPP and the losses through RP, offering an indication of surplus production influencing shifts in internal storage within the ecosystem's boundaries (Sato et al. 2022).
vi. **Export Production or New Production**: Export production constitutes a portion of the overall primary production, involving the descent of non-recycled organic materials, generated by autotrophs within the well-lit euphotic zone, to the deeper aphotic zone (Palmer and Totterdell 2001). These collectively offer a more efficient mechanism for conveying carbon to the deep sea compared to individual particles.

These concepts support in understanding the carbon dynamics within the ocean, the development of diverse life forms, and the equilibrium of carbon within marine ecosystems. The effectiveness of carbon transference and its implications for fisheries are subjected on these mechanisms. Consequently, they facilitate an enhanced grasp of carbon's important role in the oceanic biological pump. This biological pump elucidates the comprehensive journey of carbon, starting from atmospheric sources and river influx to its descent into the depths of the ocean basin. Various physical and chemical factors make the ocean a best repository for the greenhouse gas CO_2, enabling it to assimilate more CO_2 than the atmosphere. Encompassing approximately 71% of the Earth's surface, the oceans furnish an extensive arena for gas interchange. This expansive surface area facilitates substantial interaction between CO_2 and the ocean's aqueous environment. CO_2 exhibits higher solubility in water compared to air, resulting in a greater capacity for a specific volume of water to absorb CO_2 than an equivalent volume of air. The ocean's considerable volume, coupled with CO_2's solubility, synergistically enables it to effectively take in substantial quantities of this gas.

Oceanic currents and vertical mixing mechanisms facilitate the dispersion of CO_2 across the water column, permitting a more profound integration of dissolved CO_2. This mixing dynamic empowers the ocean to stock CO_2 not just at its surface, but also within its deeper strata. Consequently, the lower ocean layers exhibit significantly higher density than the upper layers, causing an absence of intermixing between these zones. This phenomenon leads to a substantial portion of anthropogenic CO_2 being absorbed by the surface ocean, with only a marginal fraction infiltrating the deeper ocean layers. Once integrated into the ocean, CO_2 can remain dissolved for extended spans, ranging from decades to centuries. This prolonged retention period transforms the ocean into an extended ground for overabundance of atmospheric CO_2. While the ocean possesses the capability to assimilate CO_2, there exist constraints on its optimal storage capacity. The growing absorption of CO_2 has put up the ocean acidification, bearing significance for marine ecosystems and life forms. Moreover, the ocean's capacity to absorb CO_2 is not boundless, and persistent elevated emissions of CO_2 stemming from human actions can destabilize equilibrium, potentially resulting in detrimental outcomes for both the oceanic realm and the to the whole planet. Phytoplankton takes in dissolved $_{CO2}$ from the seawater and utilizing it as a foundational element for the process of photosynthesis. This CO_2 undergoes a sequence of biochemical transformations, culminating in the creation of organic carbon. While engaging in photosynthesis and harnessing available nutrients, phytoplankton undergo growth and replication. Consequently, the phytoplankton population within the euphotic zone experiences expansion.

With the progress in time or upon their demise, phytoplankton cells gain density and descend towards the depths under the influence of gravity. Furthermore, a portion of the organic carbon synthesized through photosynthesis can coalesce into more sizable entities. These descending entities, collectively known as "marine snow," comprise the aggregation of sinking particles. Oceanic currents, turbulence, and mixing mechanisms facilitate the conveyance of sinking particles and organic carbon from the upper ocean layers to more deep strata. This vertical transportation is supported by physical phenomena such as Ekman transport and thermohaline circulation, which entail deep ocean currents propelled by variances in temperature and salinity. Some sinking particles may also serve as feed for zooplankton and other marine life forms, becoming a nutritional source. This carbon can subsequently traverse the marine food chain, possibly culminating in the deposition of carbon into seafloor sediments upon the demise of these organisms. Carbon stored in the depths of the ocean or embedded within sediments can persist for extended periods ranging from centuries to millennia, playing a role in the management of worldwide carbon circulation and climate regulation. It's crucial to acknowledge that although marine biological pumping functions as a natural mechanism upholding equilibrium in the carbon cycle, human actions like the combustion of fossil fuels have disrupted this equilibrium by discharging substantial quantities of CO_2 into the atmosphere.

Oceanic productivity establishes the fundamental basis for marine food chains and sustains a wide array of aquatic organisms. The presence of productive oceans bears paramount importance for the worldwide fishing and aquaculture sectors, playing a pivotal role in supplying a substantial fraction of global protein consumption and underpinning the livelihoods of countless individuals. Flourishing marine ecosystems featuring ample food supplies contribute to elevated fish populations, thereby advancing the cause of sustainable seafood provision. Beyond this, the productivity of oceans with the flow of marine currents, yields an influence over climate dynamics. The efficacy of productive oceans extends to the regulation of global climate factors such as temperature, precipitation and atmospheric conditions through the redistribution of heat and energy across the Earth. Marine productivity, primarily steered by photosynthesis-driven primary productivity, depends on a spectrum of input factors that furnish the vital conditions for the growth of marine autotrophs. Nutrients stand as important chemical elements for the feed and development of organisms. Macronutrients, demanded in major measures, encompass carbon, hydrogen, oxygen, nitrogen, phosphorus, potassium, sulphur, magnesium, and calcium, whereas micronutrients like iron, copper, and zinc are requisite in smaller quantities (Smith et al. 1998). Human-induced origins of nutrients within the marine realm primarily origin from terrestrial activities, encompassing agricultural runoff, sewage and industrial effluents, as well as emissions from fossil fuel combustion in the atmosphere.

2.3.1 Unraveling Ocean Productivity: The Power of Biogeochemical Models

The Biogeochemistry (BGC) model serves as a scientific structure designed to replicate and explain the interplays and mechanisms encompassing biological, geological, and chemical constituents within either an ecosystem or a distinct environmental system. These models strive to grasp and measure the circulation of nutrients, elements, and compounds across diverse Earth system compartments, including the atmosphere, oceans, soils, and living entities. BGC models facilitate researchers in examining intricate associations and reciprocal processes that oversee the motion and alteration of substances within the innate surroundings. Biological activities involve the functions carried out by living organisms, including phytoplankton, zooplankton, and detritus. These processes encompass photosynthesis, respiration, decomposition, nutrient absorption, growth, and other metabolic functions that impact the circulation of elements and compounds. Geological aspects pertain to the physical elements of the Earth system, encompassing weathering, erosion, sedimentation, and mineral creation. These mechanisms contribute to the release, conveyance, and alteration of nutrients and elements within the environment. Chemical constituents assume a central role in biogeochemistry models, engaging in the alteration of elements and compounds through processes such as oxidation, reduction, precipitation, dissolution, and complexation. Within biogeochemistry models, interconnected loops of the elements are integrated, wherein alterations within one part of the system ripple through to influence other part, giving rise to intricate interplays and reactions. As an illustration, shifts in temperature can influence the accessibility of nutrients, subsequently affecting biological productivity. Environmental drivers encompass external elements that impact biogeochemical processes, including climate parameters such as temperature, precipitation, and solar radiation.

Nutrient- Phytoplankton- Zooplankton-Detritus (NPZD) models, coined by Fasham et al. (1990), are simple, straightforward, dynamic models reliant on concentrations to depict the fundamental dynamics of a planktonic ecosystem. These models focus on alterations in state variables, represented as concentrations (e.g., 3 μ mol of phytoplankton nitrogen per litre), rather than individual counts. The linkage between state variables is established through transfer functions, which explains the interplays among these variables. These functions essentially act as verbs, illustrating the

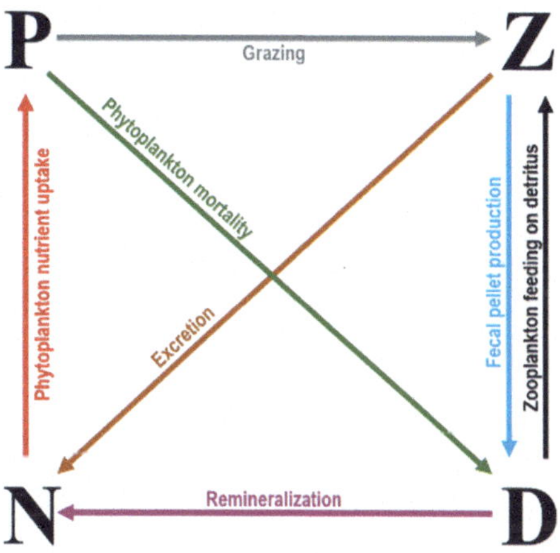

Fig. 2.3 State variables and their transfer functions of BGC model

conveyance and alteration of energy or matter within the ecosystem. Distinct transfer functions explain particular ecosystem processes (as portrayed in the Fig. 2.3).

Mathematical representations commonly depict biogeochemistry models, encapsulating a collection of equations that elucidate the velocities of diverse processes and interconnections. These equations are based on scientific principles, empirical data, and observations from field studies. The model is subsequently transcribed into a sequence of differential equations (as shown in Eqs. (2.1–2.4), which articulate the general formulation of the authentic differential equation).

$$\frac{dN}{dt} = -\text{uptake} + \text{remineralization} \tag{2.1}$$

$$\frac{dP}{dt} = \text{uptake} - \text{grazing} \tag{2.2}$$

$$\frac{dZ}{dt} = \text{grazing} - \text{excretion} \tag{2.3}$$

$$\frac{dD}{dt} = \text{excretion} - \text{remineralization} \tag{2.4}$$

Modeling can encompass a spectrum spanning uncomplicated conceptual frameworks to complex numerical simulations. These models can function across various spatial extents, whether local, regional, or global, and temporal spans ranging from hours to centuries. A basic NPZD model might not invariably capture important system dynamics. In such case, the inclusion of supplementary state variables and transfer functions (Fig. 2.4) becomes essential. There are practically as many models as there are modellers, however it gets harder to parameterize and falsify the

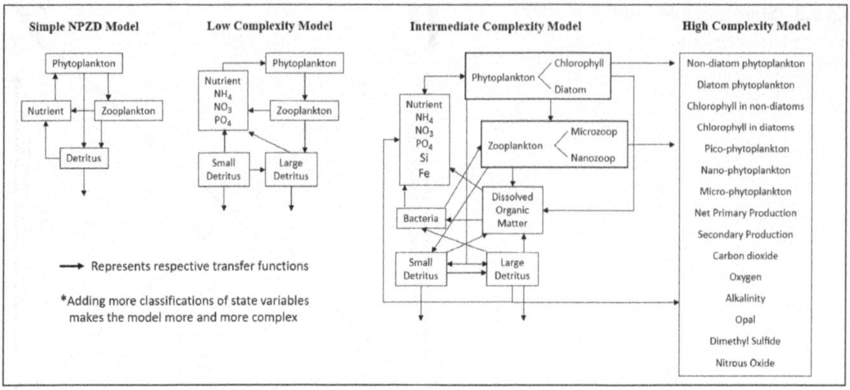

Fig. 2.4 Model's complexity integration and their possible variable sub-classes

model as more state variables (and transfer functions… and parameters) are introduced. Models can be brought closer to real-world observations by introducing additional complexity, particularly when refining model performance through parameter adjustments.

Depending on the complexity of the ecosystem and the governing equations, applications of biogeochemical models to various ecosystems will display varying performances (Ismail and Al-Shehhi 2022). Typically, these models are coupled with physical models via the advection–diffusion equation, characterized by horizontal resolutions ranging from a few kilometres to over 200 kms. The concentration dynamics of NPZD are simulated within distinct compartments representing varying depths, locations, and more, throughout a "biological" time interval (Fig. 2.5). The spectrum of these models encompasses straightforward one-dimensional (1D) models, where biological dynamics are averaged across the mixed layer, to comprehensive three-dimensional (3D) models featuring advanced turbulence-closure submodels. Through the integration of the NPZD ecosystem model with diverse physical models, an array of physical-biological interactions within the oceanic realm has been explored. Diverse categories of biogeochemistry modeling methodologies have been devised to understand and explain complex interplays, including Ecological Models, Process-Based Models, Biophysical Models, Nutrient Cycling Models, Global Biogeochemical Models, Integrated Earth System Models, and more. These frameworks empower scientists to simulate and see how distinct factors and processes collaborate to shape the flow of nutrients, elements and compounds within marine ecosystems. Each modeling approach boasts its pros and cons, prompting researchers to often employ an ensemble of models to attain a more understanding of biogeochemical shifts in the ocean. These modeling endeavours enhance our capacity to anticipate and manage the effects of human activities and climate fluctuations on marine ecosystems and their corresponding services.

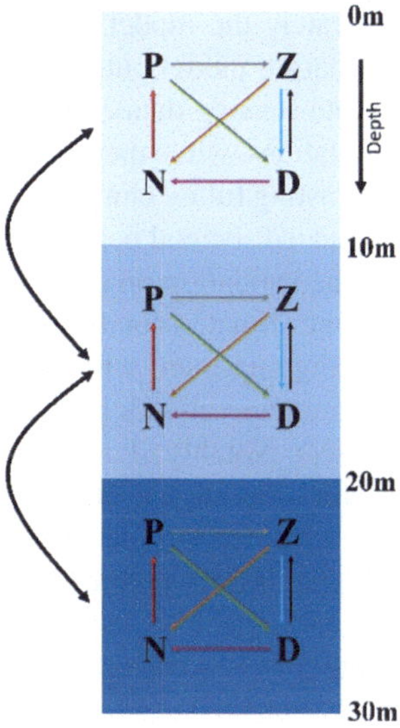

Fig. 2.5 Vertical individual compartment of BGC model function

The process of identifying biogeochemical alterations in climate models entails the examination of model outputs in order to discern changes in crucial variables associated with biogeochemical cycles. (i) The process of Change Detection Analysis entails the comparison of model outputs across distinct timeframes encompassing the past, present, and future, with the aim of pinpointing alterations in biogeochemical factors. Statistical methodologies like t-tests, Analysis of Variance and regression analysis are applicable to find noteworthy imbalance between these timeframes. For example, (Ilori and Ajayi 2020) utilized methods such as the Standard Normal Homogeneity Test, Buishand's range, and Pettitt's test in climate-oriented domains to achieve this objective. (ii) Spatial and Temporal trend analysis centers on the recognition of extended shifts or patterns in biogeochemical factors throughout time. Utilizing methods of time series analysis, like linear regression, assists in quantifying these trends and evaluating their significance. To assess the corresponding consequences and vulnerabilities within a specific region, visualizing current and future climate conditions spatially plays a pivotal role. An appropriate statistical tools and visualization strategies facilitates the generation of outputs that help in understanding the complex climate analyses to a broader audience, as demonstrated by Brett et al. (2021). (iii) Anomaly detection encompasses the recognition of deviations from anticipated or historical norms. Aberrations in biogeochemical factors can offer insights into shifts in ecosystem operations. Statistical techniques such as z-scores or standard deviations can be employed for anomaly identification. When climate patterns diverge from established norms in a specific area and timeframe, manifesting an unusual climate or weather scenario, fluctuations and incongruities emerge. Calculating anomalies holds immense significance in climate science, constituting a foundational aspect as many climate data analyses rely on initial anomaly computations (Wahyono et al. 2020) (e.g., (Mogen et al. 2022)). (iv) It is possible to evaluate how accurately the model simulates actual biogeochemical processes and changes by comparing model outputs to real-world observations (such as field measurements, remote sensing data). The outputs of a climate model are generally referred to as simulations, while they take on the designation of projections when directed towards forecasting future climates based on predefined sets of anthropogenic influences and associated societal behaviours termed scenarios.

The validation process for climate models primarily concentrates on simulating current climatic conditions (spanning at least 20 to 30 years from the present) and aligning them with meteorological observations. It's crucial to ensure a truly autonomous validation process by considering the calibration of models with specific datasets. Notably, during the developmental phase, climate models are tailored to existing conditions, introducing empirical values and assumptions that might evolve over time due to ongoing human emissions and land use shifts. Consequently, a comprehensive approach, tailored to the particular research inquiry and accessible data, is frequently employed to systematically uncover and analyse biogeochemical changes within climate models. Thorough deliberation of the merits and limitations of each method, as well as the interpretation of outcomes within the wider scientific comprehension of biogeochemical procedures and climate alteration, remains imperative.

2.3.2 Global Productivity in the Grip of Climate Change

Temperature plays an important role in exerting significant impact on marine productivity (Falkowski and Oliver 2007). The effects of temperature extend directly to the rates of photosynthesis and respiration in marine organisms, especially phytoplankton. Elevated temperatures typically induce heightened metabolic activity, resulting in accelerated rates of photosynthesis and growth. Consequently, primary production escalates, forming the foundational component of the marine food chain. However, excessive warmth can impose stress on phytoplankton, adversely affecting their development and overall well-being. The influence of temperature extends to the realm of nutrient availability and cycling within marine ecosystems. Elevated temperatures can increase recycling of nutrients through heightened microbial activity and nutrient remineralization. This in turn rise nutrient accessibility for phytoplankton, thereby stimulating their growth rapidly. Furthermore, the temperature impact is also felt by zooplankton, which graze on phytoplankton. Ocean temperature fluctuations give rise to distinct layers of vertical stratification. In locations where elevated surface temperatures arrest the vertical mixing termed as thermal stratification, the nutrient-enriched waters from deeper layers might not ascend to the surface, potentially constraining productivity. On the other hand, events of upwelling, influenced by temperature and wind patterns, provide nutrient-rich deep waters upward, elevating productivity in those regions. The impact of human-induced climate change is yielding sustained increments in oceanic temperatures. These transformations can introduce complex uncertain impacts for oceanic productivity. The warming seas might contribute to alterations in species dispersion, shifts in food chain dynamics, and potential disturbances to ecosystem functions.

The influence of climate change on organisms and ecosystems extends beyond persistent, multi-decadal patterns, also encompassing the increased occurrence and severity of extreme occurrences. Instances like marine heatwaves and tropical cyclones can induce swift alterations in SST and vertical mixing, influencing the accessibility of nutrients and light exposure for phytoplankton. As oceanic temperatures evolve, specific phytoplankton species may flourish while others wane. This dynamic can trigger modifications in the makeup of phytoplankton communities, subsequently inducing a cascade of effects throughout the marine food chain. The warming of ocean waters can result in reduced solubility of oxygen, giving rise to zones with depleted oxygen levels commonly referred to as "ocean dead zones." These areas can exert adverse impacts on marine ecosystems and disrupt the productivity of the affected regions. Climate change has the capacity to trigger feedback loops that make worse the influence on oceanic productivity. A case in point is the melting of Arctic ice, which reveals darker ocean surfaces and makes it to start absorb more sunlight, consequently instigating further warming and potentially inducing alterations within marine ecosystems.

From NASA satellite data, (Sarmiento et al. 2004) have underscored the intimate correlations between ocean productivity and worldwide climate patterns. The rise in surface temperatures enhances the imbalance in density, thereby intensifying the vertical "stratification" of oceanic waters. Consequently, there is diminished interaction between the upper layers of water, where phytoplankton thrive, and the deeper layers, which stocks the essential nutrients for their growth. This circumstance poses a challenge for tropical phytoplankton populations, as their nutrient supply stands to diminish due to reduced mixing (Siegel et al. 2013). The research conducted by Behrenfeld et al. (2006) revealed that on a global scale, a warmer climate might prompt a rapid and widespread decline in marine life. To assess the robustness of ENSO cycles, the study combined sea-level pressure, surface winds, SST, surface air temperature, and cloudiness to formulate a Multivariate ENSO Index (MEI). These physical factors were regarded as pivotal climate variables twisted with ocean circulation and productivity. The study uncovered a conspicuous and robust link between the variability of MEI and anomalies in NPP and Chlorophyll based on SeaWiFS data. Warmer MEI conditions correlated with decreased NPP and Chl levels, and the inverse was true as well. This connection underscores the paramount influence of climate fluctuations on trends in ocean productivity. Their findings distinctly establish a connection between changes in Chl and Net NPP with SST, stratification, and fluctuations in climate conditions. These outcomes present an avenue for comparing with simulated ocean responses to a warming climate. Coupling atmosphere–ocean general circulation models with ocean ecosystem models becomes important for understanding the interplay between the global carbon cycle and climate feedbacks. When driven by increased atmospheric CO_2 concentrations, these predictive models consistently project elevated SSTs and overall reductions in NPP within stratified oceanic regions (Bopp et al. 2001; Boyd and Doney 2002).

The most pronounced seasonality in productivity is observed in high latitudes, primarily propelled by light availability (as portrayed in the Fig. 2.6). Summertime boasts more intense and lengthier sunlight exposure, directly benefiting photosynthesis. Furthermore, the upper ocean's mixed layer becomes shallower during this period, preventing phytoplankton from descending into darkness as they grow (Siegel et al. 2013). The spring season sees the mixed layer undergoing increased mixing, largely due to heightened sunlight that induces warming and freshening, both elevating surface water buoyancy. This process is sometimes amplified by generally calmer weather in spring and summer, leading to reduced wind-driven turbulence. During the spring bloom, NPP surpasses the loss of phytoplankton biomass caused by grazing and mortality, which leads to a temporary high in export production and net biomass accumulation. In certain subpolar areas, productivity attains its peak during spring when phytoplankton experience a shift from light to nutrient restriction. At the utmost latitudes, although the key nutrients Nitrogen and Phosphorus are present in noteworthy amounts, the trace element Iron can become a limiting factor as summer progresses. In a few of these polar ecosystems, there's evidence suggesting that light and iron can jointly impede photosynthesis during the summer months (Boyd et al. 2000, 2007).

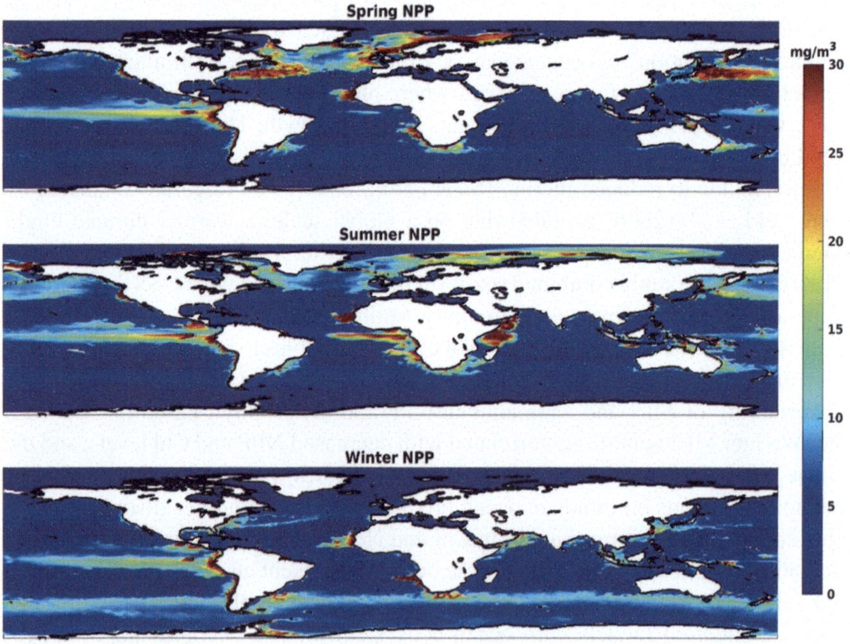

Fig. 2.6 Global surface NPP from (*Data*: CMEMS ocean biogeochemistry hindcast)

2.3.3 The Productivity of Regional Seas: A Closer Look

Oceanic primary productivity exhibits non-uniform dispersion throughout the seas. Coastal zones and locations with upwelling activities typically manifest higher levels of productivity, owing to the continuous influx of nutrient-enriched waters from deeper ocean strata. In contrast, the open ocean frequently encounters reduced primary productivity due to constraints in nutrient availability. Regional seas constitute zones where human society interfaces most closely with the marine realm, thus holding significant socio-economic significance. Predicted climate change scenarios are characterized by anthropogenic emissions of greenhouse gases, which induce alterations in radiative forcing, thereby modifying the climate system. These modifications reverberate through regional seas via three distinct channels: atmospheric, oceanic, and terrestrial. Two robust shifts in atmospheric conditions associated with global warming are increased air temperature and bottom troposphere water vapour. These shifts are projected to lead to a reduction in both sensible and latent heat flux emanating from the ocean, until the ocean attains a state of dynamic thermal equilibrium. For shallow shelf seas, this equilibrium is swiftly achieved on a seasonal basis, while for deeper basins, the process is notably prolonged, predominantly dictated by the over-turning circulation (Held and Soden 2006).

Predicted alterations in haline stratification are projected to hold significant move on a basin-wide scale, broadly aligning with the intensification of the hydrological cycle (Held and Soden 2006). For instance, in the context of the study by Capotondi et al. (2012) contend that increased haline stratification gains prominence in the North Atlantic and Arctic regions. The progression towards elevated permanent stratification leads to a reduction in winter mixing, consequentially diminishing the depths of winter mixed layers. Consequently, the overall quantity of nutrients drawn into the winter mixed layer decreases. This decline in nutrient availability curtails the potential for phytoplankton growth, a well-documented consequence of climate change in the open ocean. Climate change has the potential to influence a majority of the physical mechanisms within regional seas, consequently exerting downstream effects on ecosystem biogeochemistry and the lower trophic levels. The reaction of the interconnected physical-lower trophic level framework to climatic stimuli depends upon transport processes originating from the open ocean and deeper waters. These processes interplay with factors like benthic efflux, river inflow, atmospheric deposition, and biogeochemical losses, collectively establishing the elemental and chemical energy budget accessible for the biological activity specific to a given region (Holt et al. 2016).

The rise in ocean temperature is driven by increased heat absorption, precipitating changes in oceanic circulation patterns that facilitate the distribution of warm and cold waters worldwide. These alterations in circulation have a substantial impact on SST, which in turn influences the rate of evaporation, serving as a moisture source for the atmosphere (Sadhukhan and Chakraborty 2023). Marine heatwaves represent anomalous SST conditions, where temperatures surpass the 90th percentile threshold, leading to notable repercussions on marine biodiversity. The research by Chakraborty et al. (2023) delves into marine heatwave events in the Arabian Sea, revealing that an average of two such events occurred annually, displaying an ascending linear trend in the northern Arabian Sea sector. Their findings also indicate a decline in several biogeochemical parameters during marine heatwave occurrences, with a discernible diminishing trend aligning with a rise in marine heatwave days. The interaction between the atmosphere and the ocean governs the exchange of mass, momentum, and heat across their interface. Significantly, these interactions play a pivotal role in shaping the evolution of atmospheric and oceanic systems across scales ranging from sub-mesoscale to global dimensions. On extensive spatial scales, air-sea interactions are predominantly steered by atmospheric dynamics, with the ocean adjusting to the variability in the atmosphere. Conversely, at oceanic mesoscales, the ocean drives the variations, prompting the atmosphere to respond to shifts in SST anomalies, which arise from rapid oceanic advection (Perlin et al. 2020). This mesoscale interplay between the atmosphere and the ocean can also wield influence over biogeochemical settings and primary productivity. As demonstrated by Gaube et al. (2013) through satellite observations, the responses of wind stress to mesoscale SST and currents introduce disturbances in the form of Ekman upwelling and downwelling, thereby promoting striking mesoscale plankton blooms in the middle of the ocean.

Newly-formed cyclonic eddies play a role in stimulating productivity by augmenting the influx of nutrients to surface waters. Cyclonic eddies are characterized by their distinctive traits: they exhibit pronounced stratification, lower sea surface height, and colder SST compared to their surrounding waters. Conversely, anti-cyclonic eddies display the opposite characteristics to cyclonic eddies. In-situ investigations focusing on cyclonic eddies have documented a surge in primary productivity and biomass within specific regions, including the Sargasso Sea, the subtropical Pacific, the Peru upwelling system, and the California upwelling system (Hayward and Mantyla 1990; Sweeney et al. 2003; Benitez-Nelson et al. 2007; Stramma et al. 2013), respectively). EBUS rank among the most productive biomes in the ocean, contributing to a substantial portion of the world's wild marine fish catch. These ecosystems are defined by prevailing ocean currents that convey water rich in nutrients but low in oxygen to coastlines flanking the eastern fringes of global ocean basins. The intensification of upwelling in the poleward sectors of EBUSs might serve as a counterbalance to the warming tendency near the coastal surface, while simultaneously enhancing the gradients of SST across the shoreline. This phenomenon has been observed in the context of the northern California Current System (Li et al. 2014) and the northern Canary Current System (Varela et al. 2022).

2.4 Conclusion

To sum up, our thorough discussion has aimed to unravel the complex relationship between oceans and climate change, with a particular focus on the vital interaction between upwelling dynamics and marine productivity. By carefully dissecting upwelling mechanisms, we have shed light on how regional upwelling systems are impacted by climate change and how this influences the ecological dynamics of marine environments. The analysis of how ocean productivity is affected by climate change has brought to light the significant consequences for marine food webs, phytoplankton, and nutrient cycling, highlighting the pressing need to address human influences on Earth's climate.

The exploration of biogeochemical models has shown these instruments as essential tools for understanding the intricate linkages found in maritime settings. We can more accurately forecast and lessen the effects of climate change on marine ecosystems by comprehending the basic ideas guiding nutrient cycles and ecosystem dynamics. Our investigation into the regional and global effects on primary productivity has highlighted how susceptible these systems are to environmental shifts, highlighting the necessity of taking preventative action to preserve the quality of our oceans.

The knowledge gained from this paper lays the groundwork for future research projects as the scientific community works to better comprehend these complex processes. Through establishing a connection between theoretical models and empirical facts, we may work together to develop sustainable practices and policies that protect marine ecosystems from the effects of a changing climate.

References

Akhir MF, Daryabor F, Husain ML, Tangang F, Qiao F (2015) Evidence of upwelling along Peninsular Malaysia during southwest monsoon. Open J Mar Sci 05(03):273–279. https://doi.org/10.4236/ojms.2015.53022

Arellano B, Rivas D (2019) Coastal upwelling will intensify along the Baja California coast under climate change by mid-21st century: Insights from a GCM-nested physical-NPZD coupled numerical ocean model. J Mar Syst 199:103207. https://doi.org/10.1016/j.jmarsys.2019.103207

Bakun A (1990) Global climate change and intensification of coastal ocean upwelling. Science 247:198–201.https://doi.org/10.1126/science.247.4939.198

Bakun A, Weeks SJ (2004) Greenhouse gas buildup, sardines, submarine eruptions and the possibility of abrupt degradation of intense marine upwelling ecosystems. Ecol Lett 7(11):1015–1023. https://doi.org/10.1111/j.1461-0248.2004.00665.x

Bakun A, Black BA, Bograd SJ, García-Reyes M, Miller AJ, Rykaczewski RR et al (2015) Anticipated effects of climate change on coastal upwelling ecosystems. Curr Clim Change Rep 1(2):85–93. https://doi.org/10.1007/s40641-015-0008-4

Barber RT, Chavez FP (1983) Biological consequences of El Niño. Science 222(4629):1203–1210. https://doi.org/10.1126/science.222.4629.1203

Behrenfeld MJ, O'Malley RT, Siegel DA, McClain CR, Sarmiento JL, Feldman GC et al (2006) Climate-driven trends in contemporary ocean productivity. Nature 444(7120):752–755. https://doi.org/10.1038/nature05317

Benitez-Nelson CR, Bidigare RR, Dickey TD, Landry MR, Leonard CL, Brown SL et al (2007) Mesoscale eddies drive increased silica export in the subtropical Pacific ocean. Science 316(5827):1017–1021. https://doi.org/10.1126/science.1136221

Benke AC, Huryn AD (2007) Secondary production of macroinvertebrates. In: Methods in stream ecology. Elsevier, pp 691–710. https://doi.org/10.1016/B978-012332908-0.50041-3

Bernstein L, Bosch P, Canziani O, Chen Z, Christ R, Davidson O et al (2007) Contribution of working groups I, II and III to the fourth assessment report of the intergovernmental panel on climate change

Bopp L, Monfray P, Aumont O, Dufresne J, Le Treut H, Madec G et al (2001) Potential impact of climate change on marine export production. Global Biogeochem Cycles 15(1):81–99. https://doi.org/10.1029/1999GB001256

Bowers DG, Brett HL (2008) The relationship between CDOM and salinity in estuaries: an analytical and graphical solution. J Mar Syst 73(1–2):1–7. https://doi.org/10.1016/j.jmarsys.2007.07.001

Boyd PW, Watson AJ, Law CS, Abraham ER, Trull T, Murdoch R et al (2000) A mesoscale phytoplankton bloom in the polar Southern ocean stimulated by iron fertilization. Nature 407(6805):695–702. https://doi.org/10.1038/35037500

Boyd PW, Jickells T, Law CS, Blain S, Boyle EA, Buesseler KO et al (2007) Mesoscale iron enrichment experiments 1993–2005: synthesis and future directions. Science 315(5812):612–617. https://doi.org/10.1126/science.1131669

Boyd PW, Doney SC (2002) Modelling regional responses by marine pelagic ecosystems to global climate change. Geophys Res Lett 29(16). https://doi.org/10.1029/2001GL014130

Brett GJ, Whitt DB, Long MC, Bryan F, Feloy K, Richards KJ (2021) Sensitivity of 21st-century projected ocean new production changes to idealized biogeochemical model structure. Biogeosciences 18(10):3123–3145. https://doi.org/10.5194/bg-18-3123-2021

Capotondi A, Alexander MA, Bond NA, Curchitser EN, Scott JD (2012) Enhanced upper ocean stratification with climate change in the CMIP3 models. J Geophys Res: Oceans 117(C4). https://doi.org/10.1029/2011JC007409

Chakraborty A, Chandrakar R, Kumar S, Sadhukhan B, Kumar A (2023) Analysis of marine heat-waves and biogeochemistry in the Northern Arabian Sea. Reg Stud Mar Sci 63:103019. https://doi.org/10.1016/j.rsma.2023.103019

Chan F, Barth JA, Lubchenco J, Kirincich A, Weeks H, Peterson WT et al (2008) Emergence of anoxia in the California current large marine ecosystem. Science 319(5865):920–920. https://doi.org/10.1126/science.1149016

Costanza R, d'Arge R, de Groot R, Farber S, Grasso M, Hannon B et al (1998) The value of the world's ecosystem services and natural capital. Ecol Econ 25(1):3–15. https://doi.org/10.1016/S0921-8009(98)00020-2

de Castro M, Sousa MC, Santos F, Dias JM, Gómez-Gesteira M (2016) How will Somali coastal upwelling evolve under future warming scenarios? Sci Rep 6(1):30137. https://doi.org/10.1038/srep30137

de Souza MM, Mathis M, Mayer B, Noernberg MA, Pohlmann T (2020) Possible impacts of anthropogenic climate change to the upwelling in the South Brazil Bight. Clim Dyn 55(3–4):651–664. https://doi.org/10.1007/s00382-020-05289-0

Falkowski PG, Oliver MJ (2007) Erratum: mix and match: how climate selects phytoplankton. Nat Rev Microbiol 5(12):966–966. https://doi.org/10.1038/nrmicro1792

Fang G, Chen H, Wei Z, Wang Y, Wang X, Li C (2006) Trends and interannual variability of the South China sea surface winds, surface height, and surface temperature in the recent decade. J Geophys Res: Oceans 111(C11). https://doi.org/10.1029/2005JC003276

Fasham MJR, Ducklow HW, McKelvie SM (1990) A nitrogen-based model of plankton dynamics in the oceanic mixed layer. J Mar Res 48(3):591–639. https://doi.org/10.1357/002224090784984678

Field CB, Behrenfeld MJ, Randerson JT, Falkowski P (1998) Primary production of the biosphere: integrating terrestrial and oceanic components. Science 281(5374):237–240. https://doi.org/10.1126/science.281.5374.237

Gaube P, Chelton DB, Strutton PG, Behrenfeld MJ (2013) Satellite observations of chlorophyll, phytoplankton biomass, and Ekman pumping in nonlinear mesoscale eddies. J Geophys Res: Oceans 118(12):6349–6370. https://doi.org/10.1002/2013JC009027

Hayward TL, Mantyla AW (1990) Physical, chemical and biological structure of a coastal eddy near Cape Mendocino. J Mar Res 48(4):825–850

Held IM, Soden BJ (2006) Robust responses of the hydrological cycle to global warming. J Clim 19(21):5686–5699. https://doi.org/10.1175/JCLI3990.1

Holt J, Schrum C, Cannaby H, Daewel U, Allen I, Artioli Y et al (2016) Potential impacts of climate change on the primary production of regional seas: a comparative analysis of five European seas. Prog Oceanogr 140:91–115. https://doi.org/10.1016/j.pocean.2015.11.004

Hong H, Zhang C, Shang S, Huang B, Li Y, Li X et al (2009) Interannual variability of summer coastal upwelling in the Taiwan Strait. Cont Shelf Res 29(2):479–484. https://doi.org/10.1016/j.csr.2008.11.007

How seawater moves: Ekman transport. https://scied.ucar.edu/learning-zone/earth-system/how-ocean-moves-ekman-transport#:~:text=Ocean water at the surface, deep called an Ekman spiral. Accessed 26th January 2024

Ilori OW, Ajayi VO (2020) Change detection and trend analysis of future temperature and rainfall over west Africa. Earth Syst Environ 4(3):493–512. https://doi.org/10.1007/s41748-020-00174-6

Ismail K, Al-Shehhi MR (2022) Reviews and syntheses: assessment of biogeochemical models in the marine environment. Biogeosciences Discuss

Jing Z, Qi Y, Du Y (2011) Upwelling in the continental shelf of northern South China Sea associated with 1997–1998 El Niño. J Geophys Res 116(C2):C02033. https://doi.org/10.1029/2010JC006598

Kämpf J, Chapman P (2016) Upwelling systems of the world. Springer International Publishing, Cham. https://doi.org/10.1007/978-3-319-42524-5

Kok PH, Mohd Akhir MF, Tangang F, Husain ML (2017) Spatiotemporal trends in the southwest monsoon wind-driven upwelling in the southwestern part of the South China Sea. PLoS ONE 12(2):e0171979. https://doi.org/10.1371/journal.pone.0171979

Kuo NJ, Zheng Q, Ho CR (2004) Response of Vietnam coastal upwelling to the 1997–1998 ENSO event observed by multisensor data. Remote Sens Environ 89(1):106–115. https://doi.org/10.1016/j.rse.2003.10.009

Li H, Kanamitsu M, Hong SY, Yoshimura K, Cayan DR, Misra V et al (2014) Projected climate change scenario over California by a regional ocean–atmosphere coupled model system. Clim Change 122(4):609–619. https://doi.org/10.1007/s10584-013-1025-8

Liu Y, Peng Z, Shen C, Zhou R, Song S, Shi Z et al (2013) Recent 121-year variability of western boundary upwelling in the northern South China Sea. Geophys Res Lett 40(12):3180–3183. https://doi.org/10.1002/grl.50381

Lü X, Qiao F, Wang G, Xia C, Yuan Y (2008) Upwelling off the west coast of Hainan Island in summer: its detection and mechanisms. Geophys Res Lett 35(2). https://doi.org/10.1029/2007GL032440

Macias D, Landry MR, Gershunov A, Miller AJ, Franks PJS (2012) Climatic control of upwelling variability along the western North-American coast. PLoS ONE 7(1):e30436. https://doi.org/10.1371/journal.pone.0030436

Mantua NJ, Hare SR (2002) The Pacific decadal oscillation. J Oceanogr 58(1):35–44. https://doi.org/10.1023/A:1015820616384

Mantua NJ, Hare SR, Zhang Y, Wallace JM, Francis RC (1997) A pacific interdecadal climate oscillation with impacts on salmon production. Bull Am Meteor Soc 78(6):1069–1079. https://doi.org/10.1175/1520-0477(1997)078%3c1069:APICOW%3e2.0.CO;2

Maranón E, Fernández E (1995) Changes in phytoplankton ecophysiology across a coastal upwelling front. J Plankton Res 17(10):1999–2008. https://doi.org/10.1093/plankt/17.10.1999

Mogen SC, Lovenduski NS, Dallmann AR, Gregor L, Sutton AJ, Bograd SJ et al (2022) Ocean biogeochemical signatures of the North Pacific Blob. Geophys Res Lett 49(9). https://doi.org/10.1029/2021GL096938

Moore SK, Trainer VL, Mantua NJ, Parker MS, Laws EA, Backer LC et al (2008) Impacts of climate variability and future climate change on harmful algal blooms and human health. Environ Health 7(Suppl 2):S4. https://doi.org/10.1186/1476-069X-7-S2-S4

Morgan JR, Valencia MJ (1983) The natural environmental setting. In: Morgan JR, Valencia MJ (eds) Atlas for marine policy in Southeast Asian seas. University of California Press, pp 4–17
Palmer JR, Totterdell IJ (2001) Production and export in a global ocean ecosystem model. Deep Sea Res Part I 48(5):1169–1198. https://doi.org/10.1016/S0967-0637(00)00080-7
Pauly D, Christensen V (1995) Primary production required to sustain global fisheries. Nature 374(6519):255–257. https://doi.org/10.1038/374255a0
Perlin N, Kamenkovich I, Gao Y, Kirtman BP (2020) A study of mesoscale air–sea interaction in the Southern ocean with a regional coupled model. Ocean Model 153:101660. https://doi.org/10.1016/j.ocemod.2020.101660
Praveen V, Ajayamohan RS, Valsala V, Sandeep S (2016) Intensification of upwelling along Oman coast in a warming scenario. Geophys Res Lett 43(14):7581–7589. https://doi.org/10.1002/2016GL069638
Roegner GC, Needoba JA, Baptista AM (2011) Coastal upwelling supplies oxygen-depleted water to the Columbia river estuary. PLoS ONE 6(4):e18672. https://doi.org/10.1371/journal.pone.0018672
Rykaczewski RR, Dunne JP, Sydeman WJ, García-Reyes M, Black BA, Bograd SJ (2015) Poleward displacement of coastal upwelling-favorable winds in the ocean's eastern boundary currents through the 21st century. Geophys Res Lett 42(15):6424–6431. https://doi.org/10.1002/2015GL064694
Sadhukhan B, Chakraborty A (2023) Role of local and remote forcing on the decadal variability of mixed layer depth in the Bay of Bengal. Dyn Atmos Oceans 102:101349. https://doi.org/10.1016/j.dynatmoce.2022.101349
Sarmiento JL, Slater R, Barber R, Bopp L, Doney SC, Hirst AC et al (2004) Response of ocean ecosystems to climate warming. Global Biogeochem Cycles 18(3):6–7. https://doi.org/10.1029/2003GB002134
Satar MN, Akhir MF, Kok PH, Daud NR (2020) Upwelling in the northwest Sabah during the northeast monsoon and its relation with El-Niño. Res Mar Sci 5(2):681–698
Satar MN, Akhir MF, Zainol Z, Chung JX (2023) Upwelling in marginal seas and its association with climate change scenario—a comparative review. Climate 11(7):151. https://doi.org/10.3390/cli11070151
Sato Y, Nishihara GN, Tanaka A, Belleza DFC, Kawate A, Inoue Y et al (2022) Variability in the net ecosystem productivity (NEP) of seaweed farms. Front Mar Sci 9. https://doi.org/10.3389/fmars.2022.861932
Siegel DA, Behrenfeld MJ, Maritorena S, McClain CR, Antoine D, Bailey SW et al (2013) Regional to global assessments of phytoplankton dynamics from the SeaWiFS mission. Remote Sens Environ 135:77–91. https://doi.org/10.1016/j.rse.2013.03.025
Siegel DA, Maritorena S, Nelson NB, Hansell DA, Lorenzi-Kayser M (2002) Global distribution and dynamics of colored dissolved and detrital organic materials. J Geophys Res: Oceans 107(C12). https://doi.org/10.1029/2001JC000965
Smith RL, Smith TM, Hickman GC, Hickman SM (1998) Elements of ecology. San Francisco USA
Sousa MC, Ribeiro A, Des M, Gomez-Gesteira M, de Castro M, Dias JM (2020) NW Iberian Peninsula coastal upwelling future weakening: competition between wind intensification and surface heating. Sci Total Environ 703:134808. https://doi.org/10.1016/j.scitotenv.2019.134808
Stock CA, Alexander MA, Bond NA, Brander KM, Cheung WWL, Curchitser EN et al (2011) On the use of IPCC-class models to assess the impact of climate on living marine resources. Prog Oceanogr 88(1–4):1–27. https://doi.org/10.1016/j.pocean.2010.09.001
Stramma L, Bange HW, Czeschel R, Lorenzo A, Frank M (2013) On the role of mesoscale eddies for the biological productivity and biogeochemistry in the eastern tropical Pacific Ocean off Peru. Biogeosciences 10(11):7293–7306. https://doi.org/10.5194/bg-10-7293-2013
Sweeney EN, McGillicuddy DJ, Buesseler KO (2003) Biogeochemical impacts due to mesoscale eddy activity in the Sargasso sea as measured at the Bermuda Atlantic time-series study (BATS). Deep Sea Res Part II 50(22–26):3017–3039. https://doi.org/10.1016/j.dsr2.2003.07.008

Sydeman WJ, García-Reyes M, Schoeman DS, Rykaczewski RR, Thompson SA, Black BA et al (2014) Climate change and wind intensification in coastal upwelling ecosystems. Science 345(6192):77–80. https://doi.org/10.1126/science.1251635

Sylla A, Mignot J, Capet X, Gaye AT (2019) Weakening of the Senegalo-Mauritanian upwelling system under climate change. Clim Dyn 53(7–8):4447–4473. https://doi.org/10.1007/s00382-019-04797-y

Varela R, Álvarez I, Santos F, deCastro M, Gómez-Gesteira M (2015) Has upwelling strengthened along worldwide coasts over 1982–2010? Sci Rep 5(1):10016. https://doi.org/10.1038/srep10016

Varela R, Rodríguez-Díaz L, de Castro M, Gómez-Gesteira M (2022) Influence of Canary upwelling system on coastal SST warming along the 21st century using CMIP6 GCMs. Global Planet Change 208:103692. https://doi.org/10.1016/j.gloplacha.2021.103692

Wahyono T, Heryadi Y, Soeparno H, Abbas BS (2020) Anomaly detection in climate data using stacked and densely connected long short-term memory model. J Comput 31(4):42–53

Wang J, Tang D, Sui Y (2010) Winter phytoplankton bloom induced by subsurface upwelling and mixed layer entrainment southwest of Luzon Strait. J Mar Syst 83(3–4):141–149. https://doi.org/10.1016/j.jmarsys.2010.05.006

Wang D, Gouhier TC, Menge BA, Ganguly AR (2015) Intensification and spatial homogenization of coastal upwelling under climate change. Nature 518(7539):390–394. https://doi.org/10.1038/nature14235

Wilkerson F, Dugdale RC (2008) Coastal upwelling. In: Nitrogen in the marine environment. Elsevier, pp 771–807. https://doi.org/10.1016/B978-0-12-372522-6.00017-7

Xie LL, Zong X, Yi X, Li M (2016) The interannual variation and long-term trend of Qiongdong upwelling. Oceanol Limnol Sin 47:43–51

Xiu P, Chai F, Curchitser EN, Castruccio FS (2018) Future changes in coastal upwelling ecosystems with global warming: the case of the California current system. Sci Rep 8(1):2866. https://doi.org/10.1038/s41598-018-21247-7

Xu L, Ji C, Kong D, Guo M (2021) Abrupt change in Vietnam coastal upwelling as a response to global warming. J Quat Sci 36(3):488–495. https://doi.org/10.1002/jqs.3292

Yan Y, Ling Z, Chen C (2015) Winter coastal upwelling off northwest Borneo in the South China Sea. Acta Oceanol Sin 34(1):3–10. https://doi.org/10.1007/s13131-015-0590-2

Yelland M, Taylor PK (1996) Wind stress measurements from the open ocean. J Phys Oceanogr 26(4):541–558. https://doi.org/10.1175/1520-0485(1996)026%3c0541:WSMFTO%3e2.0.CO;2

Zhang P, Cheng H, Edwards RL, Chen F, Wang Y, Yang X et al (2008) A test of climate, sun, and culture relationships from an 1810-year Chinese cave record. Science 322(5903):940–942. https://doi.org/10.1126/science.1163965

Zhu J, Zhou Q, Zhou Q, Geng X, Shi J, Guo X et al (2023) Interannual variation of coastal upwelling around Hainan Island. Front Mar Sci 10. https://doi.org/10.3389/fmars.2023.1054669

Chapter 3
Sea Level Rise

Fatin Izzati Minhat, Nur Hidayah Roseli, Muhammad Hafeez Jeofry, and Muhammad Noah Irfan Azran

Abstract Global sea levels have gone through a series of fluctuations since millions of years ago. The rising and falling sea level observed through previous sedimentology, and fossil records indicated that this process occurred slowly over a long period. A global consensus on the recent sea-level rise rate has caught the community's interest. The main concern is that human-induced climate change may accelerate the relative sea-level rise in the next coming century. Most of the countries in Southeast Asia are maritime, surrounded by the ocean. Therefore, understanding past, current and future sea level changes is crucial to draw the picture of future sea levels in this region.

Keywords Holocene · Future projection · Sea level dynamics · Ice sheets · Global mean sea level

3.1 Introduction

Over millions of years, global sea levels have experienced a series of fluctuations that are evident in sedimentology and fossil records. This gradual process of rising and falling took place over an extended period. Recently, the community has become increasingly interested in reaching a global consensus on the rate of sea-level rise. The primary concern is that human-induced climate change could hasten the relative rise in sea levels in the next century. Since many Southeast Asian countries are maritime and bordered by the ocean, it is vital to comprehend the historical, current, and future sea level changes to anticipate the future sea levels in this region. This chapter will explore the history of sea level changes during the last glacial period, particularly in Southeast Asia. In addition, this chapter will discuss the present-day changes in sea level based on the IPCC 6th Assessment Report, which indicate that climate change has led to the rise in sea levels and its potential impacts on coastal communities.

F. I. Minhat (✉) · N. H. Roseli · M. H. Jeofry · M. N. I. Azran
Faculty of Science and Marine Environment, Universiti Malaysia Terengganu, 21030 Kuala Nerus, Terengganu, Malaysia
e-mail: fatinminhat@umt.edu.my

V. Maccarrone and M. Fadzil Akhir (eds.), *Coastal Sustainability*, Coastal Research Library 39, https://doi.org/10.1007/978-3-031-75749-5_3

Finally, a summary of the future projections of sea level in these regions will be presented.

3.2 Past Sea Level Changes in Southeast Asia

Sea level rise is one of the most pressing challenges facing the planet today, with the potential to impact millions of people living in coastal regions around the world. But while sea level changes may seem like a modern phenomenon, they have actually been occurring for millions of years. Global sea levels have experienced significant fluctuations due to a range of factors, such as changes in climate and tectonic activity, during the late Quaternary period that began approximately 2.6 million years ago.

During the last 30 ka, the Southeast Asia has experienced prominent palaeogeographic changes due to the glaciations and deglaciations events (Voris 2000; Hanebuth et al. 2009). The global cooling during the Last Glacial Maximum (LGM) contributed to the lowering of sea-level to ~ 116 m below present level (Voris 2000; Sathiamurthy and Voris 2006). During this period Southeast Asia formed a large land area that connects the Peninsular Malaysia, Sumatra, Java and all the islands of Borneo together (Sathiamurthy and Voris 2006). This massive landform is known as the Sundaland (Tjia 1996).

Based on geophysical records, Hanebuth et al. (2009) suggested that the Last Glacial Maximum has come to an end at ~ 21 ka BP. The sea level then rise slowly after 21 ka BP with rate of 0.10 m/100 year (Hanebuth et al. 2000). Later in ~ 19 ka BP, Clark and Mix (2002) suggest that a rapid sea-level rise occurred due to rapid glacial melting at higher latitude. During this period the sea level around the Sunda Shelf rises at a rate of 10 m within 800 years (Hanebuth et al. 2009). By 12.7 ka BP a long extreme global cooling event (Younger Dryas) took place (Woodroffe and Horton 2005) which slowed down the sea level rise (Bard et al. 2010). Thus, between 12.7 and 10.8 ka BP, the sea level was − 60 m relative to present-day level (Hanebuth et al. 2011). During this period, the central shelf regions of Southeast Asia experienced inundation (Hanebuth et al. 2011). After the Younger Dryas another significant global sea level rise was recorded in the beginning of Holocene (~ 10.8 ka BP) when the sea level rise from − 60 to − 40 m below present level (Hanebuth et al. 2011; Bard et al. 2010). By then, the inner shelf of Sunda Shelf regions are inundated and the land bridges that connect Borneo and Sumatra became narrower (Voris 2000; Hanebuth et al. 2011). By the next few hundred years, the sea level reached − 20 m below present level in ~ 9.5 ka BP. This led to the disappearance of land bridges that connect most of the southeast Asian islands and consequently the opening of ocean passages (Voris 2000). The sea level continued to rise after 6.07 ka BP until it reached the Holocene highstand (+ 5 m above present level) by 4.2 ka BP. During the Mid-Holocene highstand, the Southeast Asia experienced delta progradation where low coastal plains and deltas (e.g. the Chao Phraya delta and the Mekong delta) are submerged (Sathiamurthy and Voris 2006; Hanebuth et al. 2011). Then after 4.2 ka BP, the sea level fell gradually to present level at about 1 ka BP (Sathiamurthy and

Voris 2006). However, according to Tjia (1996) between 2.7 and 2.5 ka the sea level actually rose again before regressing back to present level around 1.5 ka BP.

3.3 Summary of Present Sea Level Changes Based on IPCC 6th Assessment Report

Climate change is a concerning issue going into the twenty-first century. Sea level rise being one of the impacts of climate change is a concerning issue as it would have the potential of impacting countries especially maritime countries. Contributions of global mean sea level rise include thermal expansion, Antarctic and Greenland ice sheet, glaciers and land water storage (Church et al. 2013). Factors contributing to sea level rise are thermal expansion of ocean water and discharging of water bodies from ice melt (i.e., Antarctica and Greenland) as a consequence of atmospheric and ocean warming (Nicholls and Cazenave 2010). Supplemented with storm surge events, sea level rise would have the potential to negatively impact coastal communities from erosion, damage to infrastructure and inundation of lowland coastal regions (Lyu et al. 2000; Oppenheimer et al. 2019; Overpeck et al. 2006). Significant impacts on coastal areas would include coastal area floods, coastline erosion, subsidence of ecosystems such as wetlands and mangroves (Sarkar et al. 2014). Based on IPCC AR6, heating of the climate system has resulted in global mean sea level rise as a combination of ice loss on land and thermal expansion from ocean warming; thermal expansion accounted for 50% of sea level rise between 1971 and 2018 (Arias et al. 2021). While thermal expansion was the primary driver of sea level rise from 1971 to 2018, the acceleration of ice sheet loss means that ice melt is currently the dominant contributor of sea level rise (Siegert et al. 2020). The global mean sea level rise is estimated at approximately 1.3 mm/year (1901–1971), 1.9 mm/year (1971–2006) and 3.7 mm/year (2006–2018), giving an average sea level rise of 0.2 m between 1901 and 2018 (IPCC 2021).

3.4 Future Projections of Future Sea Level Changes in Southeast Asia

Divided into two regions; Mainland Southeast Asia and Maritime Southeast Asia, Malaysia is one of the maritime countries in Southeast Asia and is likely to be impacted by sea level rise. Southeast Asia is a warm region located near the equator; heat movement via thermohaline circulation is critical in cooling the region. Döös et al. (2012) describe the thermohaline circulation as a conveyor belt that carries heat and freshwater from the Indo-Pacific to the Atlantic. Freshwater mixing slows the thermohaline circulation, allowing the sea to absorb excess heat trapped in the region. As the sea warms, its temperature rises, causing the water to expand. As a result of thermal expansion, sea levels in the South China Sea are rising. The altering thermohaline circulation may have an impact on sea level rise (Anthoff et al. 2016).

3.4.1 Steric Sea Level Dynamic

Steric sea level is referred to as the variation in ocean volume caused by changes in density (expansion and contraction of water masses) as well as changes in sea water temperature (thermosteric) and salinity (halosteric) (Raj et al. 2020). Calculating the sum of steric sea level change can be done through identifying the known quantity of the ocean which is integrated with the temperature and salinity of the sea (McKay et al. 2011). One of the factors underlying climate change is that when global temperatures rise, seawater expands, making more room in the ocean basin and raising the water level. The thermosteric and halosteric change combined equals to the stereodynamic change that influences the density and volume of the ocean, are both indicative of local changes in ocean level. Thermosteric sea level change, one of the components of sea level rise caused by thermal expansion due to ocean heat uptake (Jevrejeva et al. 2020).

3.4.2 Antarctic and Greenland Ice-Sheet

Located in the southern hemisphere, the ice sheets of Antarctica can be divided into two major parts, the smaller marine-based West Antarctic Ice Sheet (WAIS) and the larger East Antarctic Ice Sheet (EAIS). The Antarctic ice sheet has a total area of up to 13.92 million km^2 (Fretwell et al. 2013) and covers 98% of the Antarctic continent (Cornelius 1991). The Antarctic ice sheet has enough ice volume (~ 27 million km^3), potentially increasing global sea level by 58 m if the ice were to melt (Fretwell et al. 2013). Antarctica saw an increase in recent years of total mass loss, 40 ± 9 Gt/year in 1979–1990 to 50 ± 14 Gt/year in 1989–2000 with an increase of 166 ± 18 Gt/year in 1999–2009 and up to 252 ± 26 Gt/year in 2009–2017 (Rignot et al. 2019).

Whereas, in the northern hemisphere, the ice sheets of Greenland is the world's largest island which contributes most of the ice loss rate (McKay et al. 2011). The Greenland ice sheet would potentially increase global sea level by 6 ± 7 m (Cuffey and Marshall 2000). It stretches up to 1250 km east to west and 2675 km north to south and spans nearly 24° of latitude. Satellite monitoring of the GrIS revealed an increase in mass loss since the late 1990s, with 263 ± 30 Gt/year between 2005 and 2010, corresponding to a 0.73 mm/year rise in sea level (Csathó et al. 2014).

3.4.3 Glaciers

Sea level rise will likely increase in the future, glaciers (not including Antarctic and Greenland ice sheet) would play a role in increasing the global sea levels. Glaciers are enormous ice formations formed over land in the poles and on mountain peaks by snowfall that has turned into ice and compressed over many years. Glaciers span roughly 706,000 km^2 globally, with an estimated total volume of 170,000 km^3, or 0.4 m of potential sea level rise equivalent (Zemp et al. 2019). This means that glaciers could raise global sea levels if they melt, glacier melting since has increased since 2003 resulting in a 1.4 $\pm$ 0.25 mm/year equivalent sea level rise in 2006 (Cazenave and Llovel 2010; Cogley 2009; Meier et al. 2007). Throughout observation period of 1961–2016, global glacier mass changes totalled to − 9625 $\pm$ 7975 Gt, when a linear rate is assumed, glaciers contribute 27 $\pm$ 22 mm to global sea level or a contribution of 0.5 $\pm$ 0.4 mm/year (Zemp et al. 2019).

3.4.4 Glacial Isostatic Adjustment

Glacial Isostatic Adjustment (GIA) or vertical land motion, is not caused by current glacier melt but by the rebound of the Earth from thick ice sheets that covered the Earth around 20,000 years ago. GIA describes the response of the solid Earth, the gravitational field, and the oceans to the growth and decay of the global ice sheets, a commonly studied component of GIA is "postglacial rebound", which specifically relates to uplift of the land surface following ice melt (Whitehouse 2018). Given that the adjusting process takes thousands of years (i.e., the response since the last glaciation) (Harrison et al. 2021), most studies estimate the effects of GIA as minimal and constant.

3.4.5 Groundwater

Land water changes are defined as the sum of water changes in soil, wetlands, groundwater, lakes, and other reservoirs (Karabil et al. 2021) which contributes to the global mean sea level. Groundwater does not increase the global mean sea level significantly as thermal expansion and ice melting, but it is present when calculating the global mean sea level. Over the periods 1971–2010 and 1993–2010, land water storage driven sea level rise has been estimated to be 0.12 ($\pm$ 0.09) mm/year and 0.38 ($\pm$ 0.12) mm/year, respectively (Church et al. 2013; Karabil et al. 2021; Wada et al. 2012; Konikow 2011). Recent estimation using GRACE Satellite, between 2002 and 2014 there is a positive contribution from groundwater depletion, wetland loss, and water loss from endorheic basins offset by an increase in land water storage due to precipitation increase, resulting in a negative land water storage contribution to GMSL of 0.33 mm/year (Karabil et al. 2021; Reager et al. 2016).

3.5 Conclusion

The historical perspective on sea level changes in Southeast Asia underscores the dynamic nature of this phenomenon over geological time scales. The region has witnessed significant shifts, from the formation of the Sundaland landmass during the Last Glacial Maximum to the subsequent inundation and opening of ocean passages during periods of rapid sea-level rise. These past events, influenced by global climatic and glacial dynamics, highlight the region's susceptibility to environmental transformations.

Looking forward, the projections for future sea level changes in Southeast Asia are shaped by a combination of factors, including steric sea level dynamics, the contribution of Antarctic and Greenland ice sheets, glacier melting, glacial isostatic adjustment, and land water changes. The intricate interplay of these elements, as discussed in the IPCC's 6th Assessment Report, reveals the ongoing impact of climate change on global mean sea levels. The accelerated contribution of ice sheet loss emphasizes the urgency of addressing climate change to mitigate the potential consequences for coastal communities, including erosion, infrastructure damage, and inundation.

As Malaysia, situated in Maritime Southeast Asia, grapples with the implications of rising sea levels, it becomes crucial to consider adaptive measures and sustainable policies. Understanding the multifaceted drivers of sea level change provides a foundation for informed decision-making to safeguard coastal regions and mitigate the broader impacts of climate change on our planet. In the face of this complex and evolving challenge, a collaborative and forward-thinking approach is essential to ensure a resilient and sustainable future for Southeast Asia and beyond.

Acknowledgements The editors and authors of this book wish to thank the research grant co-founded by the Erasmus+ Programme of the European Union and MARE, Marine Coastal and Delta Sustainability for the Southeast Asia to provide funding to complete this book. Authors also wish to acknowledge the reviewers for their time, effort and constructive recommendations that enhanced the value of this book.

References

Anthoff D, Estrada F, Tol RS (2016) Shutting down the thermohaline circulation. Am Econ Rev 106(5):602–606. https://doi.org/10.1257/aer.p20161102

Arias PA, Bellouin N, Coppola E, Jones RG, Krinner G, Marotzke J et al (2021) Technical summary. In: Masson-Delmotte V, Zhai P, Pirani A, Connors SL, Péan C, Berger S et al (eds) Climate change 2021: the physical science basis. Contribution of working group I to the sixth assessment report

of the Intergovernmental Panel on Climate Change. Cambridge University Press, Cambridge, United Kingdom, New York, NY, pp 33–144. https://doi.org/10.1017/9781009157896.002
Bard E, Hamelin B, Delanghe-Sabatier D (2010) Deglacial meltwater pulse 1B and Younger Dryas sea levels revisited with boreholes at Tahiti. Science 327(5970):1235–1237
Cazenave A, Llovel W (2010) Contemporary sea level rise. Annu Rev Mar Sci 2:145–173
Church JA, Clark PU, Cazenave A, Gregory JM, Jevrejevac S, Levermann A et al (2013) Sea level change. In: Stocker TF, Qin D, Plattner GK, Tignor M, Allen SK, Boschung J et al (eds) Climate change 2013: the physical science basis. Contribution of working group I to the fifth assessment report of the Intergovernmental Panel on Climate Change. Cambridge University Press, Cambridge, United Kingdom, New York, NY
Clark PU, Mix AC (2002) Ice sheets and sea level of the Last Glacial Maximum. Quatern Sci Rev 21(1–3):1–7
Cogley JG (2009) A more complete version of the world glacier inventory. Ann Glaciol 50:32–38
Cornelius PE (1991) Life in Antarctica. In: Springer eBooks, pp 9–14. https://doi.org/10.1007/978-1-4612-3012-0_2
Csathó B, Schenk AF, van der Veen CJ, Babonis GS, Duncan K, Rezvanbehbahani S et al (2014) Laser altimetry reveals complex pattern of Greenland ice sheet dynamics. Proc Natl Acad Sci USA 111(52):18478–18483. https://doi.org/10.1073/pnas.1411680112
Cuffey KM, Marshall SJ (2000) Substantial contribution to sea-level rise during the last interglacial from the Greenland ice sheet. Nature 404(6778):591–594. https://doi.org/10.1038/35007053
Döös K, Nilsson J, Nycander J, Brodeau L, Ballarotta M (2012) The world ocean thermohaline circulation. J Phys Oceanogr 42(9):1445–1460. https://doi.org/10.1175/jpo-d-11-0163.1
Fretwell PT, Pritchard HD, Vaughan DG, Bamber J, Barrand NE, Bell RE et al (2013) Bedmap2: improved ice bed, surface and thickness datasets for Antarctica. Cryosphere 7(1):375–393. https://doi.org/10.5194/tc-7-375-2013
Hanebuth T, Stattegger K, Grootes PM (2000) Rapid flooding of the Sunda Shelf: a late-glacial sea-level record. Science 288(5468):1033–1035
Hanebuth TJ, Stattegger K, Bojanowski A (2009) Termination of the Last Glacial Maximum sea-level lowstand: the Sunda-Shelf data revisited. Glob Planet Change 66(1–2):76–84
Hanebuth TJ, Voris HK, Yokoyama Y, Saito Y, Okuno JI (2011) Formation and fate of sedimentary depocentres on Southeast Asia's Sunda Shelf over the past sea-level cycle and biogeographic implications. Earth Sci Rev 104(1–3):92–110
Harrison BJ, Daron J, Palmer MD, Weeks J (2021) Future sea-level rise projections for tide gauge locations in South Asia. Environ Res Commun 3(11):115003. https://doi.org/10.1088/2515-7620/ac2e6e
IPCC (2021) Summary for policymakers. In: Masson Delmotte V, Zhai P, Pirani A, Connors SL, Péan C, Berger S et al (eds) Climate change 2021: the physical science basis. Contribution of working group I to the sixth assessment report of the Intergovernmental Panel on Climate Change
Jevrejeva S, Bricheno L, Brown JM, Byrne D, de Dominicis M, Matthews A et al (2020) Quantifying processes contributing to marine hazards to inform coastal climate resilience assessments, demonstrated for the Caribbean Sea. Nat Hazards Earth Syst Sci 20(10):2609–2626. https://doi.org/10.5194/nhess-20-2609-2020
Karabil S, Sutanudjaja EH, Lambert E, Bierkens MF, van de Wal RSW (2021) Contribution of land water storage change to regional sea-level rise over the twenty-first century. Front Earth Sci 9. https://doi.org/10.3389/feart.2021.627648
Konikow LF (2011) Contribution of global groundwater depletion since 1900 to sea-level rise. Geophys Res Lett 38:1–5. https://doi.org/10.1029/2011GL048604
Lyu K, Zhang X, Church JA, Wu Q (2000) Processes responsible for the southern hemisphere ocean heat uptake and redistribution under anthropogenic warming. J Clim 33(9):3787–3807. https://doi.org/10.1175/jcli-d-19-0478.1
McKay NP, Overpeck JT, Otto-Bliesner BL (2011) The role of ocean thermal expansion in last interglacial sea level rise. Geophys Res Lett 38:4–9

Meier MF, Dyurgerov MB, Rick UK, O'Neel S, Pfeffer WT, Anderson RS, Anderson SP, Glazovsky AF (2007) Glaciers dominate eustatic sea-level rise in the 21st century. Science 317(5841):1064–1067. https://doi.org/10.1126/science.1143906

Nicholls RJ, Cazenave A (2010) Sea-level rise and its impact on coastal zones. Science 328:1517–1520. https://doi.org/10.1126/science.1185782

Oppenheimer M, Glavovic BC, Hinkel J, van de Wal R, Magnan AK, Abd-Elgawad A et al (2019) Sea level rise and implications for low-lying islands, coasts and communities. In: Pörtner HO, Roberts DC, Masson-Delmotte V, Zhai P, Tignor M, Poloczanska E et al (eds) IPCC special report on the ocean and cryosphere in a changing climate. Cambridge University Press, Cambridge, UK, New York, NY, pp 321–445. https://doi.org/10.1017/9781009157964.006

Overpeck JT, Otto-Bliesner BL, Miller GH, Muhs DR, Alley RB, Kiehl JT (2006) Paleoclimatic evidence for future ice-sheet instability and rapid sea-level rise. Science 311(5768):1747–1750. https://doi.org/10.1126/science.1115159

Raj RP, Andersen OB, Johannessen JA, Gutknecht BD, Chatterjee S, Rose SK et al (2020) Arctic sea level budget assessment during the GRACE/Argo time period. Remote Sens 12:2837. https://doi.org/10.3390/rs12172837

Reager JT, Gardner AS, Famiglietti JS, Wiese DN, Eicker A, Lo MH et al (2016) A decade of sea level rise slowed by climate-driven hydrology. Science 351:699–703. https://doi.org/10.1126/science.aad8386

Rignot E, Mouginot J, Scheuchl B, van den Broeke M, van Wessem JM, Morlighem M (2019) Four decades of Antarctic ice sheet mass balance from 1979–2017. Proc Natl Acad Sci USA 116(4):1095–1103. https://doi.org/10.1073/pnas.1812883116

Sarkar MSK, Begum RA, Pereira JJ, Jaafar A, Saari MY (2014) Impacts of and adaptations to sea level rise in Malaysia. Asian J Water Environ Pollut 11(2):29–36

Sathiamurthy E, Voris HK (2006) Maps of Holocene sea level transgression and submerged lakes on the Sunda Shelf. Trop Nat Hist 2:1–44

Siegert M, Alley RB, Rignot E, Englander J, Corell R (2020) Twenty-first century sea-level rise could exceed IPCC projections for strong-warming futures. One Earth 3(6):691–703. https://doi.org/10.1016/j.oneear.2020.11.002

Tjia HD (1996) Sea-level changes in the tectonically stable Malay-Thai Peninsula. Quatern Int 31:95–101

Voris HK (2000) Maps of Pleistocene sea levels in Southeast Asia: shorelines, river systems and time durations. J Biogeogr 27(5):1153–1167

Wada Y, van Beek LPH, Weiland FCS, Chao BF, Wu YH, Bierkens MFP et al (2012) Past and future contribution of global groundwater depletion to sea-level rise. Geophys Res Lett 39:1–6. https://doi.org/10.1029/2012GL051230

Whitehouse PL (2018) Glacial isostatic adjustment modelling: historical perspectives, recent advances, and future directions. Earth Surf Dyn 6(2):401–429. https://doi.org/10.5194/esurf-6-401-2018

Woodroffe SA, Horton BP (2005) Holocene sea-level changes in the Indo-Pacific. J Asian Earth Sci 25(1):29–43

Zemp M, Huss M, Thibert E, Eckert N, McNabb R, Huber J et al (2019) Global glacier mass changes and their contributions to sea-level rise from 1961 to 2016. Nature 568(7752):382–386. https://doi.org/10.1038/s41586-019-1071-0

Chapter 4
The Flow Near the Seabed: The Boundary Layer at the Bottom of Surface Waves

Pietro Scandura, Daniele La Rosa, Vincenzo Maccarrone, and Hee Min Teh

Abstract The wave-bottom boundary layer is a thin region near the seabed where vorticity generated by the adhesion of the fluid to the bottom is significant, while the flow outside this region is essentially irrotational in comparison. The importance of the boundary layer is due to several reasons: it is the place where most of the wave energy dissipation occurs outside the surf zone; the bottom friction is strongly dependent on the flow in the boundary layer; the sediment transport, both as bed load and suspended load, is most dependent on the flow in the boundary layer. This review presents the basic mathematical modelling of the boundary layer and describes the most relevant results available in the literature, obtained by experimental and numerical approaches. The mechanism leading to the generation of steady streaming is also illustrated, and the solution for the steady velocity component is worked out in analytical form for the case of linear waves propagating on a horizontal bottom and a laminar boundary layer.

Keywords Sea wave · Bottom boundary layer · Steady streaming · Rough bed

4.1 Introduction

Sea waves propagating from the open sea to the coast undergo transformations that affect both their direction of propagation and their shape. These phenomena occur

P. Scandura (✉) · D. La Rosa
Department of Civil Engineering and Architecture, University of Catania, Catania, Italy
e-mail: pietro.scandura@unict.it

D. La Rosa
e-mail: d.larosa@unict.it

V. Maccarrone
Consiglio Nazionale delle Ricerche IRBIM, Mazara del Vallo, Italy
e-mail: vincenzo.maccarrone@cnr.it

H. M. Teh
Universiti Teknologi PETRONAS, Seri Iskandar, Malaysia
e-mail: heemin.teh@utp.edu.my

V. Maccarrone and M. Fadzil Akhir (eds.), *Coastal Sustainability*, Coastal Research Library 39, https://doi.org/10.1007/978-3-031-75749-5_4

when the waves enter the intermediate water depth region ($0.05 < h/L < 0.5$; h is the water depth and L is the wavelength), where the interaction between the flow generated by the surface waves and the seabed becomes important.

One of the most obvious phenomena caused by decreasing water depth is the steepening of the waves due to the decrease in wavelength and increase in wave height. This phenomenon is accompanied by the generation of higher order harmonic components of the free surface displacement, so that while waves in deep water ($h/L > 0.5$) are almost sinusoidal, in intermediate water they become skewed, with sharp crests and shallower, flatter troughs. Such waves are called 'skewed waves' and an example is the second order Stokes wave, which can be thought of as the superposition of two cosine waves, the second of which has twice the angular frequency of the first. As the wave moves into shallower water, the non-linearities distort the wave further and it takes on an asymmetric shape, with a steep front and a gently sloping rear face. This type of waves are called 'asymmetric waves'. These phenomena have a strong influence on other important processes related to the interaction of the wave with the seabed and sediment transport.

The propagation of water waves is generally modelled by assuming that the flow is irrotational, i.e. the fluid particles move without rotating, so that vorticity vanishes everywhere and the velocity can be expressed as the gradient of a potential function Φ. By imposing the incompressibility condition, it follows that Φ satisfies the Laplace equation, so that the mathematical problem of determining the fluid motion produced by a sea wave consists in solving the Laplace equation with the appropriate boundary conditions. In general, this approach still involves insurmountable complications due to the nonlinearities of the boundary conditions at the free surface, so further simplifications are introduced according to the Airy wave theory, which assumes that the steepness of the wave is much less than one, thus allowing the governing equations to be linearised. In what follows, we assume that the reader is familiar with most of the concepts involved in this theory (Dean and Dalrymple 2000).

The results of this theory provide a solid basis for analysing most of the phenomena involved in wave propagation, but it does not allow all the physical conditions involved in fluid motion to be satisfied. For example, irrotational flow does not satisfy the physical condition of fluid adhesion to the seabed, so the fluid slides unnaturally over the bottom. The correct fulfilment of this no-slip condition involves the generation of seafloor vorticity, which diffuses into the water column due to fluid viscosity. The region affected by this vorticity is known as 'boundary layer'. The concept of boundary layer is relevant in several fields ranging from aeronautical engineering to geophysical sciences. The classic example of a boundary layer, illustrated in many fluid mechanics textbooks, is that of a flat plate moving in a stationary fluid, such as an aircraft wing. In this case, a boundary layer is observed to grow along the plate. The thickness of this layer is limited by the finite length of the plate, whereas in the case of wave propagation, the thickness is limited by the finite wavelength L or, equivalently, by the finite period T. Indeed, since the flow reverses direction every half wave period and the viscosity of the water is rather low, the vorticity does not spread at a large distance but it remain confined within a thin layer of fluid.

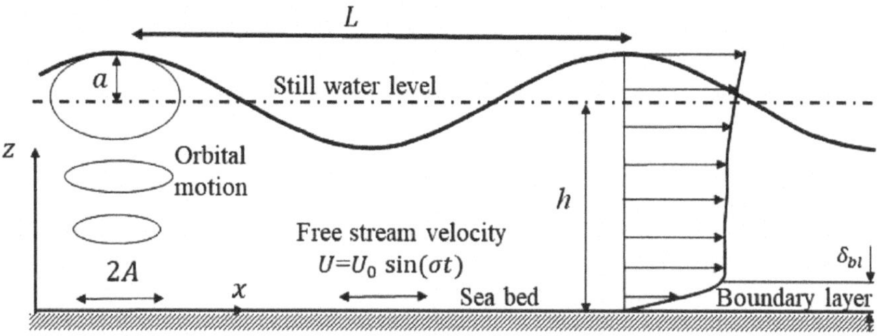

Fig. 4.1 Sketch of a sinusoidal wave propagating from left to right over a horizontal bed

The order of magnitude of the thickness of this layer can be expressed by the relation $\delta = \sqrt{2\nu/\sigma}$, where ν is the kinematic viscosity of water (approximately $\nu = 10^{-6}$ m^2/s) and $\sigma = 2\pi/T$ is the angular frequency of the surface wave. It is worth noting that for periods T of the order of 10 s, typical of sea waves, this thickness is 1.8 mm. The actual thickness δ_{bl} (see Fig. 4.1) is (5–50)δ, the exact value depends on the current hydrodynamic conditions and the nature of the seabed (smooth, rough, rippled).

Although the boundary layer covers a small area compared to the water depth, it plays a key role in several nearshore phenomena. For example, energy dissipation in non-breaking waves occurs mainly in the bottom boundary layer, and sediment transport, both as bedload and suspended load, is influenced by boundary layer hydrodynamics.

The nonlinear dynamic interactions occurring in the boundary layer generate a steady streaming with the direction of wave propagation (Longuet-Higgins 1953). The mechanism that induces this streaming results from nonlinearities in the boundary layer due to the spatial variability of the flow and occurs even for linear waves.

In the case of wave nonlinearities that induce skewness and/or asymmetry in the wave shape so that the wave is no longer sinusoidal, the turbulence generated in the boundary layer differs in intensity between two half cycles, and even between the accelerating and decelerating parts of the flow cycle if the wave is asymmetric. In these cases, an additional steady streaming is generated in the boundary layer, generally in the opposite direction to the wave propagation. There are therefore two competing effects that create a steady streaming in opposite directions between them. Which of the two prevails depends on the flow parameters.

In this paper we review the main aspects of boundary layer hydrodynamics and its mathematical treatment, also in the light of the most important recent results reported in the literature.

In Sect. 4.2 we recall some aspects of classical mathematical modelling of wave-bottom boundary layer flow. Section 4.3 illustrates literature results obtained by experimental and numerical modelling of the flow for both smooth and rough beds. Section 4.4 illustrates the mathematical treatment of the governing equation to derive the expression of the steady streaming for the case of linear waves and laminar boundary layer. Discussion and bibliographic references are given for the case of non-linear waves and turbulent boundary layer.

4.2 Basic Concepts of Wave Bottom Boundary Layer Modelling in Laminar Regime

Figure 4.1 shows a sketch of a sinusoidal sea wave propagating over a horizontal bottom. The wave has amplitude a, wavelength L, angular frequency σ and propagates over a water depth h. Assuming that the ratio a/L is much smaller than 1, the flow outside the boundary layer can be described by the Airy theory of linear irrotational waves. The boundary layer is affected by the viscosity of the fluid, so the Navier–Stokes equations must be used to study the flow in this region.

The equations of conservation of mass and momentum can be written as follows:

$$\frac{\partial u}{\partial x} + \frac{\partial w}{\partial z} = 0 \tag{4.1}$$

$$\frac{\partial u}{\partial t} + u\frac{\partial u}{\partial x} + w\frac{\partial u}{\partial z} = -\frac{1}{\varrho}\frac{\partial p}{\partial x} + \nu\left(\frac{\partial^2 u}{\partial x^2} + \frac{\partial^2 u}{\partial z^2}\right) \tag{4.2}$$

$$\frac{\partial w}{\partial t} + u\frac{\partial w}{\partial x} + w\frac{\partial w}{\partial z} = -\frac{1}{\varrho}\frac{\partial p}{\partial z} + \nu\left(\frac{\partial^2 w}{\partial x^2} + \frac{\partial^2 w}{\partial z^2}\right) \tag{4.3}$$

where u and w are the velocity components along the x and z directions, p is the pressure and ϱ is the density of the fluid. Since the thickness δ is much smaller than L, important simplifications can be introduced into Eqs. (4.1)–(4.3), but first it is necessary to gain insight into the order of magnitude of the quantities involved in these equations. The linear wave theory (Dean and Dalrymple 2000) shows that at the bed $u = a\sigma\cos(kx - \sigma t)/\sinh(kh)$, where k is the wavenumber $2\pi/L$, so that we can say that just outside the boundary layer $u \sim U_0 = a\sigma/\sin(kh)$. Moreover, according to the same theory $p \sim \varrho g a \approx \varrho a\sigma^2 L = \varrho U_0^2 L/a$, where $\sigma^2 L$ is of the same order of the acceleration of gravity g. To see the order of magnitude of the vertical velocity component W_0, considering that $x \sim L$ and $z \sim \delta$ from the continuity equation, we get $W_0/U_0 \sim \delta/L << 1$, so W_0 is much smaller than U_0.

Table 4.1 shows the order of magnitude of all the terms of Eqs. (4.2) and (4.3). The numbering shown in the first column starts from the left hand side with the time derivative and continues on the right hand side. In order to show the relative importance of each term, the ratio between each order of magnitude and the order of magnitude of the time derivative is shown in the third and sixth columns for Eqs. (4.2) and (4.3) respectively.

For the N-S equations in the x direction the time derivative, the pressure term and the second viscous term are all of the same order of magnitude (see third column

Table 4.1 Summary of the order of magnitude of the different terms of Eqs. (4.2) and (4.3)

Term	Order magn. Eq. (4.2)	Ratio	Order magn. Eq. (4.3)	Ratio
1	$U_0\sigma$	1	$W_0\sigma$	1
2	U_0^2/L	A/L	W_0^2/δ	A/L
3	U_0^2/L	A/L	W_0^2/δ	A/L
4	U_0^2/a	1	$U_0^2 L/(a\delta)$	L^2/δ^2
5	$\nu U_0/L^2$	$\delta^2/(2L^2)$	$\nu W_0/L^2$	$\delta^2/(2L^2)$
6	$\nu U_0/\delta^2$	1/2	$\nu W_0/\delta^2$	1/2

of Table 4.1). For the convective term the ratio is equal to A/L, where A is the amplitude of the fluid particle oscillation given by U_0/σ, and for the first viscous term the ratio is δ^2/L^2. Both A/L and δ^2/L^2 are much smaller than 1, therefore, in a first approximation, the boundary layer is described by the following simplified equation:

$$\frac{\partial u_1}{\partial t} = -\frac{1}{\varrho}\frac{\partial p_1}{\partial x} + \nu\frac{\partial u_1}{\partial z^2}, \tag{4.4}$$

where the subscript 1 is introduced to indicate that the quantities in Eq. (4.4) are approximations of the real quantities. For the N-S equation in the z direction (see sixth column of Table 4.1), we observe that the pressure term is much larger than all other terms, so the equation along z is simplified to:

$$\frac{\partial p_1}{\partial z} = 0, \tag{4.5}$$

that is, the pressure is constant throughout the thickness of the boundary layer, hence it is equal to the pressure given in the irrotational region. Assuming that the flow just outside the boundary layer oscillates like $u_1 = U_0 \cos(kx - \sigma t)$, considering that the viscous term in Eq. (4.4) vanishes at the outer edge of the boundary layer, we obtain that the pressure gradient term is given by $-U_0\varrho\sigma \sin(kx - \sigma t)$. Then Eq. (4.4) becomes:

$$\frac{\partial u_1}{\partial t} = U_0\sigma \sin(kx - \sigma t) + \nu\frac{\partial u_1}{\partial z^2}. \tag{4.6}$$

The solution to this equation with the condition $u = 0$ at $z = 0$ is the following:

$$u_1 = U_0\left[\cos(kx - \sigma t) - e^{-\frac{z}{\delta}} \cos\left(kx - \sigma t + \frac{z}{\delta}\right)\right]. \tag{4.7}$$

The second term of Eq. (4.7) describes an oscillation propagating in the fluid with velocity $\sigma\delta$ but with an amplitude that decreases exponentially, so that at a distance from the seafloor approximately equal to 6δ these oscillations essentially disappear.

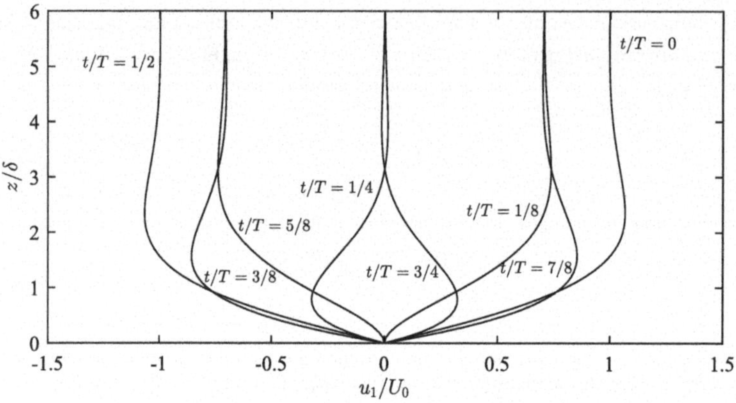

Fig. 4.2 Velocity distribution in the boundary layer at different phases of the wave period

Figure 4.2 shows the time variation of the velocity profile in the boundary layer during the wave period at $x = 0$. It is observed that although the velocity far from the bed is zero in certain phases, the velocity in the boundary layer is different from zero (see $t/T = 1/4$ and $t/T = 3/4$). In addition, due to the viscous effect, the maximum velocity exceeds the free stream velocity at $t/T = 0$ and $t/T = 1/2$.

The tangential stress τ applied to the bottom is given by:

$$\tau = \mu \left(\frac{\partial u_1}{\partial z} \right)_{z=0} = \varrho \sqrt{\sigma \nu} U_0 \cos \left(kx - \sigma t - \frac{\pi}{4} \right) \tag{4.8}$$

which shows that the bed shear stress has a phase lead of $\pi/4$ with respect to the free stream velocity, where $\mu = \varrho\nu$ is the dynamic viscosity. We remark that this result is valid when the flow is laminar and the seabed is smooth and flat. As shown in Sect. 4.3, this phase lead decreases when the flow is turbulent. It has been shown that in the case of a smooth bed, the departure from laminar conditions and the transition to the turbulent regime occurs when the Reynolds number $R_\delta = U_0\delta/\nu$ is approximately equal to 550 (Hino et al. 1976). It should be noted that in the literature there is another way of expressing the Reynolds number as $Re = U_0 A/\nu = U_0^2/(\nu\sigma)$, which is related to R_δ by the equation $Re = R_\delta^2/2$.

In the turbulent regime, all velocity components are non-zero and fluctuate with large amplitudes in both time and space, making the non-linear terms in the N-S equations non-negligible. More specifically, the order of magnitude analysis given in Table 4.1, which shows that the convective terms are negligible to a first approximation, is no longer valid, since in turbulent flow vortex structures are generated whose size scales with the thickness of the boundary layer, making the gradients very large even in the x direction. This precludes any possibility of obtaining an analytical solution to the governing equations, so only experimental or numerical approaches can be used to obtain information about the flow.

Several studies have addressed the case of a turbulent boundary layer through experimental measurements, turbulence modelling and direct numerical simulation (DNS). The next section is devoted to the description of the main results obtained from experimental laboratory measurements and numerical simulations for the case of a turbulent boundary layer.

4.3 Experimental Measurements and Numerical Modelling Results on the Bottom Boundary Layer Under Sea Waves

Experimental measurements and numerical simulations in the field of wave boundary layer studies aim to determine several quantities, including the bed friction, the thickness of the boundary layer δ_{bl}, the phase lead of the wall shear stress and the statistics of the turbulent velocity fluctuations. Knowledge of these statistics is important for a number of reasons, in particular for the development of turbulence models, which are essential when studying the flow through the numerical solution of the Reynolds Averaged Navier–Stokes Equations (RANS).

The statistical quantities are usually defined by a phase average. Taking the velocity component u as an example, the phase average $< u >$ is defined as the arithmetic mean of the instantaneous values of u taken at the same phase for several oscillation periods. The fluctuation u', is given by the difference between the instantaneous velocities u and the phase-averaged velocity $< u >$. Other statistical quantities of interest are: the Reynolds stress $\varrho < u'v' >$, the turbulent kinetic energy per unit of volume $1/2\varrho < u'^2 + v'^2 + w'^2 >$, and several others too numerous to mention here as this would go beyond the scope of this article.

Experimental measurements in the wave-bottom boundary layer require the use of facilities capable of generating an oscillating flow similar to that induced by sea waves. The most obvious facility for this would be a flume equipped with a wave generator, where surface waves similar to those observed in the sea can be generated and propagated. In nature, wave boundary layer flow is mostly turbulent, making it the most interesting flow regime for practical applications. However, turbulence can only occur under relatively large waves, which require large wave flumes to reproduce. Unfortunately, such facilities require significant financial investment to build and maintain, which is why they are only available in a few laboratories around the world.

More economically affordable facilities are the oscillatory flow tunnels (OFT). There are different types of these units, but the most common is a horizontal tube of rectangular cross section with two risers open to the atmosphere at each end. The oscillatory flow inside the tunnel is generated by the alternating motion of a piston driven by an electrohydraulic system and can use water or even air as working fluid. An important difference between the boundary layer under sea waves and that in an oscillatory flow tunnel is that the latter is spatially constant along the flow direction, whereas the former varies due to the finite wavelength of the waves. From a mathematical point of view, the difference between the two types of flow is the same as that induced by the simplifications that lead to Eqs. (4.4) and (4.5).

Due to the constancy of the velocity along the flow direction, in an OFT there is no steady streaming with the same direction of wave propagation. However, this streaming is generally small and does not alter the main turbulence properties of the boundary layer, so it can be neglected in the first instance, making the OFT suitable for studying the relevant properties of the wave-bottom boundary layer.

Experimental measurements in wall bounded turbulent flows have always detected a region where the velocity is described by a logarithmic relationship,

$$\frac{u}{u^*} = \frac{1}{\kappa} \ln \left(\frac{zu^*}{\nu} \right) + C + \Delta \tag{4.9}$$

where $u^* = \sqrt{\tau/\varrho}$ is the friction velocity, τ is the tangential stress applied to the bottom, κ is the von Kármán constant which takes a value of 0.4 and C is a constant equal to 5. The value of Δ is generally negative, differing from zero only in the presence of a rough bed, and depends on the roughness Reynolds number $Re_k = u^* k_s/\nu$, where k_s is the Nikuradse sand roughness size. This relationship was first discovered for steady flow in pipes (Nikuradse 1933) but it has been observed even in oscillatory flows when the Reynolds number is sufficiently high (Jensen et al. 1989). One of the important aspects of this relationship is that it allows the bottom shear stress to be related to the velocity profile so that it can be used to determine the bottom shear stress by finding its best fit to the experimental velocity data. To determine the maximum shear stress τ_0 on the bottom a relationship similar to the Darcy–Weisbach formula is often used, which is written as $\tau_0 = 1/2 f_w U_0^2$ where f_w is the friction factor, which depends on the Reynolds number and the roughness. Using Eq. (4.8), it can be shown that in the laminar regime the friction factor is equal to $2\sqrt{2}/R_\delta$. For the turbulent regime, the determination of f_w is based on experimental measurements or numerical simulations and will be presented in the following.

When the roughness Reynolds number Re_k is sufficiently large (the threshold value depends on the type of roughness), the effect of the viscosity becomes negligible and the flow is denoted as fully rough. Under these circumstances Eq. (4.9) becomes:

$$\frac{u}{u^*} = \frac{1}{\kappa} \ln \left(\frac{z}{k_s} \right) + 8.5. \tag{4.10}$$

In general, the form of roughness is different from that of sand. However, for a given surface roughness, it is possible to define its equivalent Nikuradse sand roughness as the sand grain roughness which, in the fully rough regime, gives the same friction factor as the actual roughness. Therefore the value of k_s in Eqs. (4.9) and (4.10) can be determined for any rough bed.

After this premise, we describe below the main studies on the wave bottom boundary layer in the turbulent regime and report some of the main results.

Bagnold (1946) carried out experiments in which a rippled plate was made to oscillate in still water. He observed that for amplitude of oscillations smaller than the ripple wavelength the friction factor remained constant while for oscillations larger than the wavelength it decreased. Kamphuis (1975) performed experiments in an OFT and reported the trend of the friction factor as a function of the Reynolds numbers for different values of A/k_s ranging from 1 to 2800. Jonsson and Carlsen (1976) carried out oscillatory flow experiments on a wall characterised by a roughness consisting of large triangular elements. The authors reported data on the velocity profiles and on the bottom shear stress during the flow cycle for Reynolds numbers of 2.74×10^6 and 6.01×10^6.

Sleath (1987) performed velocity measurements in an OFT near a bed made rough by a single layer of sand, gravel or pebbles using an anemometer Laser Doppler for velocity measurements. This author showed that at large values of Re the statistics of the velocity fluctuations, such as $< u'^2 >$ and $< w'^2 >$, depend only on A/k_s. Jensen et al. (1989) extended the study of Sleath (1987) to $Re = 10^6$ and found that the transition to turbulence starts at $Re = 10^5$ and occurs during the wave cycle just after the reversal of the near-bed flow. The presence of the roughness causes a significant increase of the turbulent intensity by 50% and of the Reynolds stress by 100%. The logarithmic layer appears first after the flow reversal and covers an increasing portion of the boundary layer as the flow progresses through the half cycle.

van der A et al. (2011) carried out an experimental study of the hydrodynamics of acceleration-skewed oscillating boundary layers on a rough wall using a large oscillating water tunnel. This type of boundary layer occurs under asymmetric waves and is characterized by positive acceleration greater than negative acceleration. Under these conditions, the intensity of turbulence during positive acceleration phases becomes greater than during negative acceleration phases, and this difference provides a mechanism capable of driving an offshore steady streaming in the boundary layer. Another important aspect of asymmetric waves is that the bottom shear stress also takes on an asymmetric trend during the wave cycle.

Yuan and Madsen (2014) performed several tests in an oscillatory water tunnel using a Particle Image Velocimeter for velocity measurements and considering sinusoidal, skewed and asymmetric waves. Three types of bottom were considered: smooth, sandpaper roughness and ceramic-marble bottoms. The Reynolds number Re ranged from 10^6 up to 3.1×10^6. In all cases, the velocity near the bed was found to follow a logarithmic law and this was found to be the best option for estimating the bottom shear stress using a best fit procedure. A boundary layer streaming was found for both second order Stokes waves and asymmetric waves with a magnitude of 1–3% of the maximum free flow velocity. This result is in agreement with Scandura (2007), van der A et al. (2011), Scandura et al. (2016) and Dunbar et al. (2023). In particular, Scandura (2007) reproduced numerically the steady streaming, on a smooth wall, generated in an OFT when the free stream fluid oscillations follows the second order Stokes waves theory. The streaming was found to be due to a non-vanishing time-averaged Reynolds stress, the gradient of which is balanced by the gradient of viscous stresses generated by the steady streaming. Scandura et al. (2016) carried out DNS simulations of the boundary layer flow at the bottom of asymmetric waves,

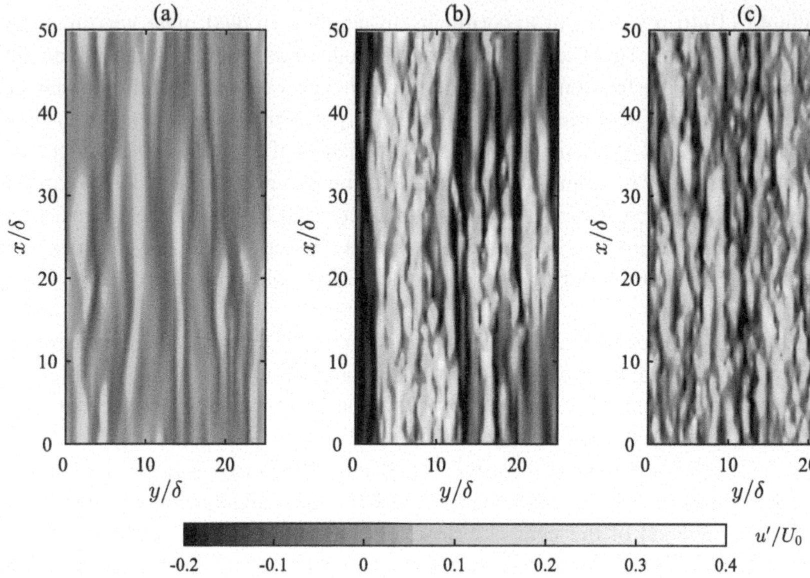

Fig. 4.3 Low and high speed streaks in an acceleration skewed oscillatory flow undergoing breaking during the accelerating phase (Scandura et al. 2016). In this flow the free stream velocity is zero at $2\pi t/T = 0$ and attains its peak at $2\pi t/T = 1$. **a** $2\pi t/T = \pi/8$, **b** $2\pi t/T = \pi/4$, **c** $2\pi t/T = 3\pi/8$. The main flow is in the x direction and u' denotes the velocity fluctuations with respect to the horizontal plane average

still for the case of a smooth wall, and found a steady streaming even for this type of wave. The study showed that during the accelerating phase low and high speed streaks of fluid develops near the wall. Figure 4.3 shows an example of this phenomenon for an acceleration skewed oscillatory flow characterized by the skewness parameter $\beta = \dot{U}(t=0)/[\dot{U}(t=0) - \dot{U}(t=T/2)] = 0.75$ and Reynolds number $Re = U_0\delta/\nu = 1100$ where a dot denotes a time derivative. As long as these streaks remain approximately straight, the wall shear stress is not much different from that in laminar flow. Before the end of the accelerating phase, the streaks become unstable and break into smaller structures. During this breaking process, the bottom shear stress increases rapidly to values much higher than those in laminar flow. This can be seen in Fig. 4.4, where at the beginning of the flow cycle the wall shear stress in the laminar regime takes values close to those in the turbulent regime. However, at a certain stage the wall shear stress in the turbulent regime starts to increase and reaches values much higher than in the laminar regime. Figure 4.4 also shows that in the laminar regime there is a phase lead of the bottom shear stress with respect to the free stream velocity, whereas in the turbulent regime the bottom shear stress tends to be in phase with the free stream velocity.

Mazzuoli et al. (2018) carried out numerical simulations at $R_\delta = 246.6$ on a bed of spheres distributed in order to reproduce the sand bed obtained in a laboratory experiments. The transition to turbulence occurred at lower Reynolds numbers com-

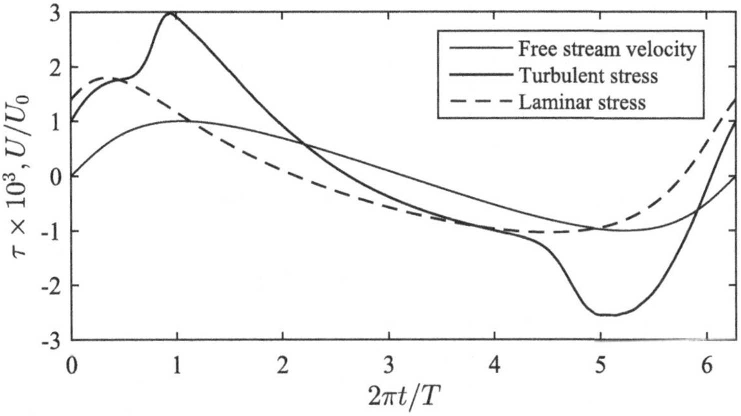

Fig. 4.4 Free stream velocity and trend of the bottom shear stress in the turbulent and laminar regime for an acceleration-skewed boundary layer flow

pared to the flat bed case. The fluctuation of the forces acting on the individual grains was found to be large compared to the average value, suggesting that the threshold for incipient motion based on average values may not be fully appropriate. The case of a rough wall has also recently been addressed by Dunbar et al. (2023), who carried out both experiments and numerical simulations for different flow conditions: sinusoidal waves, asymmetric waves, skewed waves and combined asymmetric-skewed waves. The shape of the rough bottom was obtained by scanning a real gravel bed. The Reynolds number Re ranged from 3.21×10^5 to 6.63×10^5 for the numerical simulations and from 3.21×10^5 to 2.82×10^6 for the experiments. The study included the analysis of various flow characteristics including the boundary layer thickness δ_{bl}, the phase lead ϕ of the bottom shear stress, the friction factor, and the turbulent statistics of velocity fluctuations up to the fourth order. In particular, a steady streaming was determined similar to that previously observed on a smooth wall. The study also reports non-Gaussian turbulence statistics in the near-bottom layer, which become more Gaussian as one moves away from the bottom. The characteristics of the rough bed oscillatory flow have also been studied numerically by Ghodke and Apte (2016) who performed DNS over a rough bed consisting of hexagonally arranged spheres and reported that the statistics of velocity and pressure fluctuations were well described by a fourth-order Gram–Charlier distribution.

Measurement in the Stokes layer have been also carried out in wave flume facilities. Sleath (1970) carried out measurements in a wave tank and noted that vortex formation due to the bed roughness causes an increase of the boundary layer thickness and a reduction of the phase lead of the velocity close to the bed. Cox et al. (1996) studied the characteristics of the bottom boundary layer under breaking waves determining the presence of a logarithmic layer. Dixen et al. (2008) performed experimental measurements on the boundary layer of relatively large roughness (ratio A/k_s of order 1) consisting of stones of diameter 1.4 and 3.85 cm and regular spherical

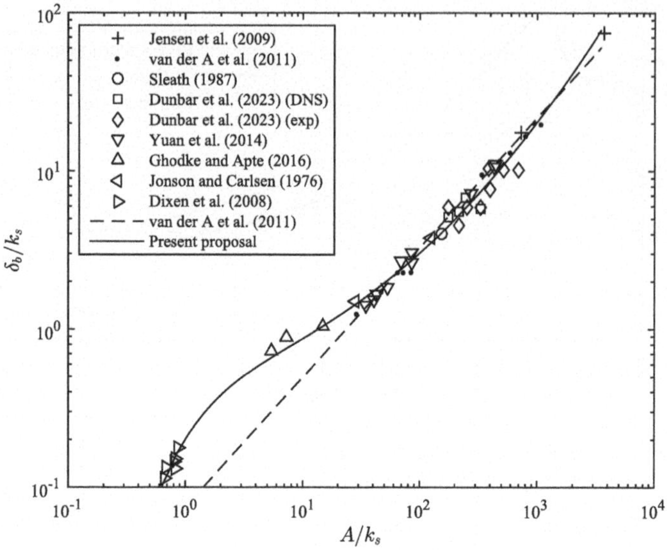

Fig. 4.5 Dimensionless thickness of the boundary layer versus the ratio between the amplitude of fluid particles oscillations A and the Nikuradse sand roughness k_s. The current proposed interpolation formula is expressed by Eq. (4.11)

elements of 3.6 cm in size giving rise to a ratio A/k_s in the range 0.6–3. The study shows that the Nikuradse equivalent sand roughness is $k_s = 2.5D$ for the regular spherical elements and $k_s = (2.0 - 2.3)D$ for the stones.

Most of the studies mentioned above have provided information on the thickness of the boundary layer, the phase lead and the friction factor that will be discussed below. In general the thickness of the boundary layer δ_{bl}/A is a function of the Reynolds number and of A/k_s, unless the Reynolds number is sufficiently high. In that case δ_{bl}/A is a function of A/k_s only. Figure 4.5 shows the dimensionless boundary layer thickness δ_{bl}/k_s versus the ratio A/k_s for Reynolds numbers large enough to assume independence on the Reynolds number itself. For large values of A/k_s, in logarithmic scale the trend is close to be linear and is well described by the equation proposed by van der A et al. (2011): $\delta_{bl} = 0.075(A/k_s)^{0.82}$. For small values of A/k_s the trend of the experimental data seems to deviate from the linear trend, probably because for values of A/k_s of order 1 the phenomenon loses the characteristics of a flow over a rough bed and acquires that of a flow over obstacles.

An interpolation formula that reproduces the trend of δ_{bl}/k_s for both small and large A/k_s is proposed here and is given by Eq. (4.11).

$$\frac{\delta_{bl}}{k_s} = \exp\left[-0.394 - 1.21\left(\frac{A}{k_s}\right)^{-1}\right]\left(\frac{A}{k_s}\right)^{0.0713\log\left(\frac{A}{k_s}\right)}. \tag{4.11}$$

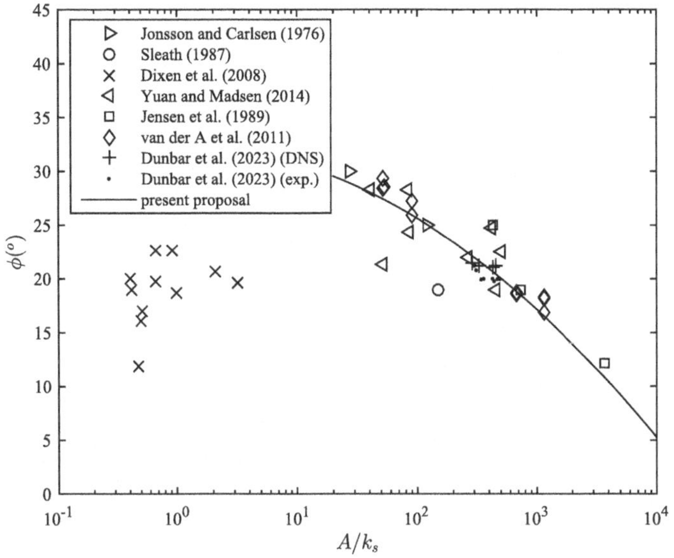

Fig. 4.6 Phase lead of the bottom stress versus the ratio between the amplitude of fluid particles oscillations A and the Nikuradse sand roughness k_s. The continuous line is given by Eq. (4.12)

Figure 4.6 shows the phase lead of the bottom shear stress versus the logarithm of A/k_s. It can be seen that for large A/k_s the phase lead decreases with A/k_s, but for small values of A/k_s the trend is unclear. Further experiments would be needed to clarify the trend of ϕ in the range A/k_s = 3–20 between the data of Dixen et al. (2008) and the rest of the data.

Therefore, when determining the interpolation equation (4.12), the data from Dixen et al. (2008) were excluded.

$$\phi = -0.32 \log\left(\frac{A}{k_s}\right) \tag{4.12}$$

It can be observed that Eq. (4.12) reasonably reproduces the trend of the data for A/k_s greater than 20.

As mentioned above, the maximum bed shear stress during the wave period is usually estimated using the formula $\tau_0 = 1/2 f_w \varrho U_0^2$, where f_w is the friction factor. Generally f_w is a function of the Reynolds number and of the relative roughness k_s/A. When the Reynolds number is large its effect becomes negligible and the friction factor becomes a function of k_s/A only. This regime occurs approximately for Re larger than 10^4 and A/k_s smaller than 10. For A/k_s larger than 10 the flow can be rough turbulent if $Re > 10^3 A/k_s$.

Figure 4.7 shows the values of the friction factor measured in some of the studies mentioned above. For cases where the flow was not sinusoidal, because of skewed and/or asymmetric waves, the orbital excursion A was adjusted according to $A \rightarrow$

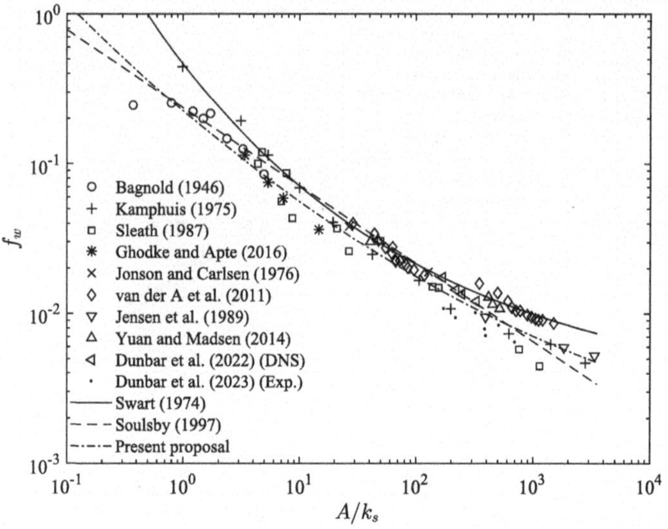

Fig. 4.7 Friction factor versus the ratio between the amplitude of fluid particles oscillations A and the Nikuradse sand roughness k_s. The current proposed interpolation formula is given by Eq. (4.13)

$2AT_{ac}/T_c$ (van der A et al. 2011), where T_{ac} is the time to get the maximum free stream velocity starting from the beginning of the flow cycle and T_c is the duration of the positive part of the flow cycle. The relationships proposed by Swart (1974) and by Soulsby (1997) are shown along with the present proposal expressed by the relationship:

$$f_w = \left(\frac{A}{k_s}\right)^{[0.022\log(A/k_s)-0.657]} \tag{4.13}$$

4.4 The Steady Streaming

As discussed in Sect. 4.3, in addition to the oscillatory velocity component, the flow near the seafloor has a steady streaming component, which can arise from two different mechanisms. In this section, the steady streaming due to the streamwise variability of the flow in the boundary layer is determined analytically for the case of laminar flow.

In order to obtain information on the steady streaming, we need to improve the approximation in the description of the boundary layer flow made in Sect. 4.2. To achieve this goal, we determine u_2 and p_2 which are the contribution to the velocity and pressure fields due to the two largest neglected terms in Eq. (4.2), which are both of order A/L when compared to the largest terms (see Table 4.1). Therefore, even u_2

and p_2 are of order A/L compared to u_1, that is $u_2/u_1 \sim A/L$. These terms can be approximated using the solutions obtained at the first order of approximation u_1 and w_1, where the latter can be calculated from the continuity equation (see Eq. (4.1)) as follows:

$$w_1 = -\int_0^z \frac{\partial u_1}{\partial x} dz \tag{4.14}$$

Therefore, at the order of approximation A/L, the x momentum equation can be written as follows:

$$\frac{\partial u_2}{\partial t} + u_1 \frac{\partial u_1}{\partial x} + w_1 \frac{\partial u_1}{\partial z} = -\frac{1}{\varrho}\frac{\partial p_2}{\partial x} + \nu \frac{\partial^2 u_2}{\partial z^2} \tag{4.15}$$

It can be observed that all the terms in Eq. (4.15) have the same order of magnitude. The viscous term $\partial^2 u_2/\partial x^2$ has been neglected because much smaller than the other terms (see fifth terms in the third column of Table 4.1). To determine an equation for p_2, we consider Eq. (4.3). As can be seen in Table 4.1, the pressure term in this equation is much larger than all the others, and this remains true even when considering that p_2 is of order A/L. Therefore, Eq. (4.3) gives $\partial p_2/\partial z = 0$, showing that p_2, like p_1, is constant along z and equal to the value taken by the pressure just outside the boundary layer, where the flow is irrotational.

Since we are interested in the steady velocity component, we take the time average of Eq. (4.15) during one wave period. The time average of the first term on the left hand side vanishes because it is the time derivative of a periodic quantity. The second term can be written as $1/2\partial u_1^2/\partial x$. Then we note that the time average of this term can be related to the spatial average in the x direction over a wavelength, and since u_1^2 is spatially periodic of period L, this average is also zero. The same reasoning applies to the pressure term, for which the time average is also zero. Therefore, the time average of Eq. (4.15) can be written as follows:

$$\overline{w_1 \frac{\partial u_1}{\partial z}} = \nu \frac{\partial^2 \overline{u_2}}{\partial z^2}, \tag{4.16}$$

where a bar denotes the time average. The integration of Eq. (4.16), with the boundary condition $\overline{u_2} = 0$ at $z = 0$, provides the time-averaged velocity $\overline{u_2}$. Skipping the mathematical elaboration, the final result is given by Eq. (4.17).

$$\begin{aligned}\overline{u_2} = &-\frac{U_0^2 k}{2\sigma} e^{-z/\delta}\left[\cos\left(\frac{z}{\delta}\right) + \sin\left(\frac{z}{\delta}\right)\right]\frac{z}{\delta} + \frac{U_0^2 k}{\sigma} e^{-z/\delta}\left[\frac{1}{2}\sin\left(\frac{z}{\delta}\right) - \cos\left(\frac{z}{\delta}\right)\right] \\ &+ \frac{U_0^2 k}{4\sigma} e^{-2z/\delta} + \frac{3}{4} U_0^2 \frac{k}{\sigma}\end{aligned} \tag{4.17}$$

In Eq. (4.17) we observe that $\overline{u_2}$ is proportional to the ratio U_0^2/c where $c = \sigma/k$ is the wave celerity. We also note that all terms of Eq. (4.17) decay exponentially

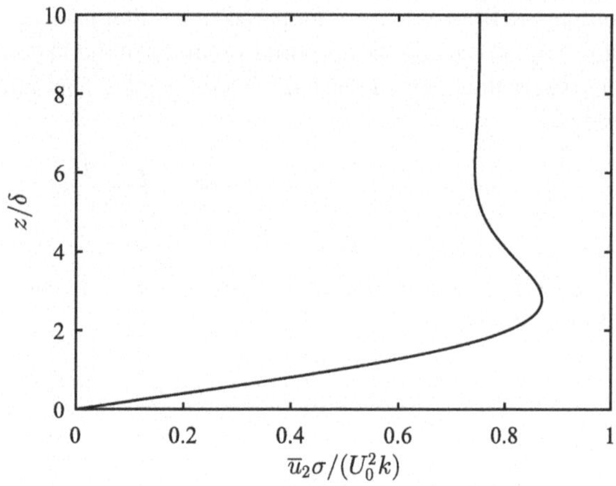

Fig. 4.8 Dimensionless steady streaming versus z/δ in laminar flow

with z/δ except the last one, which is constant, so it persists to the outer edge of the boundary layer thus affecting the flow along the entire water depth. It should also be noted that this term does not depend on the viscosity (δ does not appear), although it is essential to consider the viscosity in order to obtain the steady streaming.

Figure 4.8 shows the trend of $\overline{u_2}\sigma/(U_0^2 k)$ versus z/δ. The maximum occurs at $z/\delta \approx 2.79$ and it is equal to 0.87 approximately. To give a numerical example, consider a wave of amplitude $a = 0.4$ m propagating in a water depth of 3.5 m, with a wavelength of 20 m. The angular frequency σ calculated from the dispersion relation ($\sigma^2 = gk \tanh(kh)$) is 1.57 rad/s, $U_0 = a\sigma/\sinh(kh) = 0.47$ m/s and finally the constant term in Eq. (4.17) gives $\overline{u_2} = 0.033$ m/s, which is the 7% of U_0.

The results described above are valid for laminar flow; if the flow is turbulent, the steady streaming can be modelled numerically, allowing the deviation of the wave shape from the sinusoidal to be taken into account. The boundary layer flow including the presence of steady streaming in the turbulent regime has been studied by Holmedal and Myrhaugh (2009) for second order Stokes waves. Therefore, in this study the steady streaming is due to both the nonlinearities in the boundary layer and the skewness of the wave. Under these circumstances the steady streaming can be onshore or offshore depending on water depth and wavelength. The study showed that the maximum value of the steady streaming increases with the bottom roughness but its value at the outer edge of the boundary layer is not much influenced by the roughness. Blondeaux et al. (2012) developed a model capable of reproducing the steady streaming and the sediment transport in the boundary layer. The results have shown that the steady streaming can be either onshore or offshore, while the sediment transport is always in the onshore direction over the range of parameters studied. This may be due to the fact that for non-linear waves the maximum velocity in the onshore direction is greater than the maximum velocity in the offshore direction, so a greater amount of sediment is mobilised when the flow is in the onshore direction, resulting in a net sediment transport in that direction in all cases.

4.5 Conclusions

This review describes the main results of studies on the boundary layer at the bottom of sea waves reported in the past and more recent literature. The topics covered are by no means exhaustive of all possible aspects of bottom boundary layer hydrodynamics, but the literature cited may help the interested reader to find further sources of information on details not sufficiently explored in this review.

Under the assumption of linear waves and laminar flow, the solution for the velocity field has been worked out. When the flow is turbulent, the terms of the governing equation that are neglected in the case of laminar flow, allowing an analytical solution to be obtained, are no longer negligible. In this case, the analysis of the boundary layer flow must be based on experimental measurements or numerical simulations.

Several experiments and numerical simulations have been carried out in the past decades to explore the characteristics of the flow in the turbulent regime. In the case of a smooth seabed, it now seems established that the flow becomes three-dimensional and turbulent when the Reynolds number R_δ exceeds 550. In the field, the seabed is predominantly rough, so the rough surface case is the most interesting for engineering applications. Studies have shown that in the case of a rough surface, the transition to turbulence occurs for values of R_δ well below 550 and the stress applied to the bed is greater than on a smooth bed with a friction factor f_w that can increase by an order of magnitude.

At large Reynolds numbers the flow on a rough bed enters the fully rough regime where the stress on the bed is mainly due to the pressure jump between the windward and leeward sides of the roughness elements and the Reynolds number no longer affects the flow variables which remain under the influence of the roughness alone. For this regime, quantities such as the boundary layer thickness, the phase lead of the bottom shear stress and the friction factor have been presented as a function of the inverse of the dimensionless roughness A/k_s. New interpolation functions of experimental and numerical data have been included.

It has been shown that a steady streaming develops in the boundary layer, the intensity and direction of which depends on flow parameters such as wavelength, water depth and the degree of nonlinearity. Details have been provided for the steady streaming generated by the nonlinearities in the boundary layer, which has the direction of wave propagation. Wave bottom boundary layer research has already addressed several aspects of this flow, but there is still room for improvement in the case of shoaling waves, which are characterised by a high degree of non-linearity and irregular waves that are ubiquitous in the near-shore region.

Competing Interest The authors have no conflicts of interest to declare that are relevant to the content of this chapter.

References

Bagnold R (1946) Motion of waves in shallow water. Interaction between waves and sand bottoms. Proc R Soc A 187:18

Blondeaux P, Vittori G, Bruschi A, Lalli F, Pesarino V (2012) Steady streaming and sediment transport at the bottom of sea waves. J Fluid Mech 697:115–149

Cox D, Kobayashi N, Okayasu A (1996) Bottom shear stress in the surf zone. J Geophys Res 101(C6):14337–14348

Dean R, Dalrymple R (2000) Water wave mechanics for engineers and scientists. World Scientific

Dixen M, Hatipoglu F, Sumer BM, Fredsøe EJ (2008) Wave boundary layer over a stone-covered bed. Coast Eng 55:1–20

Dunbar D, van der A DA, Scandura P, O'Donoghue T (2023) An experimental and numerical study of turbulent oscillatory flow over an irregular rough wall. J Fluid Mech 955

Ghodke CD, Apte S (2016) DNS study of particle-bed-turbulence interactions in an oscillatory wall-bounded flow. J Fluid Mech 792:232–251

Hino M, Sawamoto M, Takasu S (1976) Experiments on transition to turbulence in an oscillatory pipe flow. J Fluid Mech 75(part 2):193–207

Holmedal L, Myrhaugh D (2009) Wave-induced steady streaming, mass transport and net sediment transport in rough turbulent ocean bottom boundary layers. Cont Shelf Res 29:911–926

Jensen B, Sumer B, Fredsøe J (1989) Experimental and theoretical investigation in an oscillatory turbulent boundary layer. J Fluid Mech 206:265–297

Jonsson I, Carlsen N (1976) Experimental and theoretical investigations in an oscillatory turbulent boundary layer. J Hydraul Res 14(1):45–60

Kamphuis J (1975) Friction factor under oscillatory waves. J Waterw Harbors Coast Eng Div 101(2):135–144

Longuet-Higgins MS (1953) Mass transport in water waves. Phil Trans R Soc Lond A 345:535–581

Mazzuoli M, Blondeaux P, Simeonov J, Calantoni J (2018) Direct numerical simulation of oscillatory flow over a wavy, rough, and permeable bottom. J Geophys Res Oceans 123(3):1595–1611

Nikuradse J (1933) Stromungsgesetze in rauhen rohren. In: Forschung auf dem Gebiete des Ingenieurwesens, Forschungsheft 361. VDI Verlag, Berlin, Germany (in German). English translation: Laws of flow in rough pipes, NACA TM 1292, 1950

Scandura P (2007) A numerical investigation of the Stokes boundary layer in the turbulent regime. J Fluid Mech 571:265–280

Scandura P, Faraci C, Foti E (2016) A numerical investigation of acceleration-skewed oscillatory flows. J Fluid Mech 808:576–613

Sleath J (1970) Velocity measurements close to the bed in a wave tank. J Fluid Mech 42:111–123

Sleath J (1987) Turbulent oscillatory flow over rough beds. J Fluid Mech 182:369–409

Soulsby RL (1997) Dynamics of marine sands: a manual for practical applications. Technical report. Thomas Telford, London

Swart DH (1974) Offshore sediment transport and equilibrium beach profiles. Technical report. Delft Hydraulics Laboratory

van der A DA, O'Donoghue T, Davies AG, Ribberink JS (2011) Experimental study of the turbulent boundary layer in acceleration-skewed oscillatory flow. J Fluid Mech 684:251–283

Yuan J, Madsen OS (2014) Experimental study of turbulent oscillatory boundary layers in an oscillating water tunnel. Coast Eng 89:63–84

Chapter 5
Coastal Erosion and Accretion in Vietnam: Some Case Studies

Thoa Thi Kim Le, Kim Thi Tran, Diem Thi My Phung, Huy Ngoc Dinh, and Thinh Nam Ngo

Abstract This chapter provides an in-depth analysis of coastal erosion and accretion in Vietnam, spanning from the 1990s onward. Through exemplary studies, it aims to comprehend the complexity of these natural disasters, which pose significant threats to coastal communities and ecosystems. The discussion explores the contemporary understanding of erosion and accretion, delving into the intricate correlations with human activities and the rapid changes in coastal morphology. The chapter thoroughly examines the causes, influencing factors, and challenges associated with coastal erosion and accretion in Vietnam, particularly in the context of sea reclamation works and the integrated coastal management strategy. Additionally, it offers an overview of the current state of these phenomena, exploring mechanisms, policies, and laws aimed at disaster prevention and control. The chapter concludes with illustrative case studies, shedding light on the complexity of erosion and accretion in various areas and providing insights into the causes and solutions to mitigate past natural disasters. Overall, it presents a comprehensive perspective on erosion and coastal accretion in Vietnam, emphasizing the threats and challenges posed to biodiversity, marine resources, and the livelihoods of coastal communities.

Keywords Coastal area · Estuary · Erosion · Accretion · Coastline change · Coastal protection

5.1 Introduction

Coastal erosion and accretion are the results of marine geodynamics or marine geodynamics combined with river geodynamics. They frequently occur along coastlines worldwide with varying degrees, intensities and different spheres of influence (Cambers et al. 2002). Coastal and estuarine erosion are severe natural disaster, exhibiting highly complex developments that result in significant damages to human lives, properties, infrastructure, and socio-economic and ecological environments

T. T. K. Le (✉) · K. T. Tran · D. T. M. Phung · H. N. Dinh · T. N. Ngo
Ho Chi Minh City University of Natural Resources and Environment, Ho Chi Minh City, Vietnam
e-mail: thoa.ltk@hcmunre.edu.vn

V. Maccarrone and M. Fadzil Akhir (eds.), *Coastal Sustainability*, Coastal Research Library 39, https://doi.org/10.1007/978-3-031-75749-5_5

(Tho et al. 2020, Nicholls et al. 2007). Meanwhile, accretion along the coast and estuaries forms valuable alluvial grounds for many regions. However, in numerous places. It also leads to serious disasters, causing saltation in shipping lanes, impacting port operations, obstructing river channels, hindering transportation, reducing flood drainage capacity, leading to widespread flooding, and altering the characteristics of lagoons and bays, especially in terms of freshwater intrusion... In Vietnam, coast and estuarine erosion are formidable natural disaster occurring in all three regions. The situations are exceedingly intricate, resulting in substantial losses in terms of human lives and property, leaving long-term consequences on the socio-economic and ecological environment (Van Cu and Shibayama 2014; Hoanh 2014; Mai et al. 2009).

5.2 Coastal Erosion and Accretion in Vietnam

5.2.1 Coastal Erosion and Accretion in the Northern Region

The coastal area from Mong Cai to Cua Ong is currently experiencing the impact of high tidal ranges and the development of mangrove forest. Due to the small size of estuaries in this region and limited sediment, the formation of accretion banks progresses slowly, and the coastal strip undergoes minimal changes. Studies have documented a decreasing trend in the current state of erosion and accretion (Duc et al. 2008). Erosion mainly occurs due to ebb currents and involves two mechanisms: erosion and abrasion of the beach surface, along with deep erosion of the tidal creek system. In the coastal area from Cua Ong to Yen Lap, the land is predominantly stable with human activities playing a significant role in shaping the coastline. The expansion of Ha Long urban area is leading to localized changes in landscape. Coastal areas like Cam Pha, Cai Dam and Tuan Chau are facing land reclamation for industrial and commercial purposes, as well as the development of new urban areas. Next, from Yen Lap to Do Son, this coastal stretch exhibits a funnel-shaped morphology of the Bach Dang estuary, accompanied by a densely developed tidal creek system and mangrove forest on high tidal flats. This region is undergoing transformations related to both marine and terrestrial environments. The area from Do Son to Lach Truong is characterized by the presence of numerous large rivers discharging into the sea, carrying substantial sediment loads. As a result, the coastline in this region undergoes frequent changes, with accretion processes prevailing (Cu et al., 2017). In the Red River—Thai Binh River Delta, erosion occurs along relatively long stretches of the coastline, a crucial characteristic determining the complex and prolonged nature of erosion and accretion phenomena in the Red River Delta. Strongly eroded bank sections such as Dong Chau, Hai Hau, Xuan Thuy are interspersed between estuaries with strong accretion processes like Thai Binh, Tra Ly, Ba Lat, and Day. Depending on the degree of accretion—erosion, the Red River Delta coastline can be divided into stages with different intensities and scales of accretion—erosion, including weak, moderate, strong and very strong.

According to research by Tran Duc Thanh and colleagues in 2002, coastal erosion in the Northern region has been divided into 51 segments, with a total length of up to 113,930 m, accounting for 34.2% of the total length of the coastline, with an average rate of approximately 6.0 m/year. Each year, an area of 68 ha is lost due to erosion. Erosion intensity is categorized into 4 levels: weak (0–2.5 m/year) account for 22.40%, moderate (2.5–5 m/year) account for 34.2%, strong (5–10 m/year) accounts for 16.6% and very strong ($\geq$10 m/year) accounts for 26.8% of the total length of erosion. The phenomenon of coastal erosion occurs on a large scale, exhibiting complex variations, with many erosion segments having strong and very strong intensity, and the trend is increasing over time. From Mong Cai to Do Son, erosion occurs on a large scale and prolonged scale, predominantly with moderate and weak intensity. From Do Son to Ba Lat, erosion was previously strong but has recently decreased. From Ba Lat to Lach Truong, erosion is particularly strong and increasing in Hai Hau and Hau Loc. Some new erosion points have appeared with small-scale but with strong and very strong intensity, although they are less affected by storms. The main concentration points of erosion on the Northern coastal region are Cat Hai (Hai Phong) with a length of 6.4 km, Hai Hau (Nam Dinh) with a length of 17.2 km and Hau Loc (Thanh Hoa) with a length of 5.0 km long. These areas have large scales, strong and very strong intensity, and a trend of increasing over time (Duc et al. 2002). The coastal area from Hai Phong to Ninh Binh is a distinctive region with many large rivers flowing into the sea, resulting in the presence of large sediment loads. Due to the substantial sediment input, the coastline in this area undergoes significant changes. The accretion process still takes place but unevenly in space and time, exhibiting complexity. This area belongs to the modern Red River Delta, with varying morphological characteristics, structures and developmental trends of coastline. The coastline in this area developed on a modern tectonic subsidence, with the capable of compensating accretion exceeding the inundation capacity. However, the uneven distribution of sediment between estuaries may lead to heterogeneity in the accretion process. This area is often affected by large waves from both the Northeast and Southwest monsoon seasons. As a result, local erosion and accretion processes frequently alternate. Nevertheless, the accretion process tends to dominate in this overall process (Cu et al. 2017).

5.2.2 *Coastal Erosion and Accretion in the Central Region*

A study by Pham Huy Tien in 2001 identified the situation of coastal erosion in the Central region, extending from Thanh Hoa (Lach Truong estuary) to Ninh Thuan (Son Hai, Ninh Phuoc), covering 275 segments with a total length of 328.16 km. In this area, erosion accounts for 21.1% of the total length of the coastline, with the common speed usually below 5 m/year, but showing an increasing trend. Recently, some new erosion cases have appeared (around 1999 to 2000) with rapid speeds, even reaching 40–60 m/year and sometimes 150–250 m/year. Erosion intensity here is categorized into four levels: weak (<5 m/year) account for 52.2%; moderate (5–15 m/year) account for 24.3%; strong (15–30 m/year) account for 10.0%, and very strong (>30 m/year) account for 13.5% of the total length of erosion. Each year, due to erosion, the coast loses up to 389.9 ha (Pham and Van Cu 2001). From the Len estuary to the Truong Giang creek, located in Nga Son district, Thanh Hoa, a characteristic erosion situation has emerged (Quan and Van Hung 2016). This area shows a trend of the coastline advancing inland, especially in the North of the Truong Giang creek. South of the Truong Giang creek, the coastline is experiencing accretion. From the Truong Giang creek to the Hoi estuary (Ma River), the coastline remains stable. The alluvial ground at the Hoi estuary is gradually increasing each year. From Sam Son to the Ghep estuary (Yen River), the coastline is relatively stable with slow accretion. At the Ghep estuary, there is strong accretion and concurrent erosion. In the Trạp area, Quang Xuong district (Thanh Hoa), the river widened, and the coastline from the Trạp estuary to Nghi Son island is gradually accreted, creating a connection between the island and the mainland. According to the research results by Nguyen and Mai 2021, the current situation at the Lach Ghep estuary area in Thanh Hoa indicates a strong scale and intensity of accretion and erosion. Particularly, the southern part of the Lach Ghep estuary is experiencing erosion with an accretion rate of up to 9.82 m/year, while the northern part of the Lach Ghep estuary is accreting at the rate of 23.51 m/year in the Quang Nam area (Nguyen and Mai 2021). The Can estuary area demonstrates an interplay between accretion and erosion. From the Can estuary to the Lo estuary, the coastline faces various erosion situations, especially near the river mouths. From the Lọ estuary to the Hoi estuary (Ca estuary flowing into the sea), the coastline is experiencing accretion, particularly at the Hoi estuary, and concurrently, the river is changing direction. The stretch from the Sot estuary to the Ron Cape has undergone minimal changes, with only minor alterations at the Gia Hoi estuary and the Khau estuary. At the Khau estuary, there is a relatively large accretion area. Based on statistics, the total length of eroding coastlines from Quang Binh to Thua Thien—Hue exceeds 100 km, accounting for about 24% of the total length of the coastline in this region. The entire area has 67 eroding coastline segments, with erosion lengths ranging from under 200 m to over 6000 m. The erosion phenomenon has spread and appeared at many locations along the coast of the provinces. Specifically, Quang Binh province has about 32 eroding coastline segments, Quang Tri province has 29 eroding segments, and Thua Thien—Hue has many eroding coastline segments, with the total erosion length accounting for about 30/127 km of the coastline. Notably, the changes in the two estuaries Thuan An and Tu Hien threaten both the lives and property of residents, as well as negatively impacting the natural stability of the Tam Giang—Cau Hai lagoon, affecting the infrastructure and socio-economic aspects of the coastal area (Cuong et al. 2018).

5.2.3 Coastal Erosion and Accretion in the Southern Region

The phenomenon of accretion and erosion along the coastline and estuaries in the Southern region of Vietnam is similar to that in the Northern delta region. It typically occurs strongly during the cold dry season with northeast monsoon winds or the southwest monsoon season. Specifically, along the southwestern coast, accretion and erosion occur strongly during the southwest monsoon season. The intensity of accretion and erosion develops slowly in coastal sections with numerous offshore islands, such as the Ha Tien coastline in the southwestern region. This phenomenon may be related to the dynamics of waves and marine currents or the small tidal amplitude in this area (Le Xuan et al. 2004). The coastal area from South Central to Southern Vietnam, starting from Son Hai (Ninh Thuan) and ending in Ha Tien (Kien Giang), has been recorded with a total of 71 erosion segments, totaling up to 478,120 m, equivalent to 52% of the total coastline length. The erosion rate in this area exceeds 2.5 m per year, impacting an area of up to 15,738 ha. The erosion intensity is classified into three levels: moderate level (2.5 to 5 m per year) account for 12.8%; strong level (5 to 10 m per year) account for 43.7%; and very strong level ($\geq$ 10 m per year) account for 43.5% of the total length of eroding areas. From Son Hai to Vung Tau, erosion occurs continuously, although not on a large scale, but with strong intensity. Areas like Phuoc The, Lien Huong, and the Phan Ri estuary are strongly affected by erosion and show an increasing trend recently. From Vung Tau to Ca Mau, erosion has a large scale, length, and high intensity, with prolonged and complex variations. In particular, the area from Vung Tau to Soai Rap estuary, although not having a large erosion length, has a very strong intensity and poses dangers due to its direct impact on port activities and maritime transportation. The area from Soai Rap estuary to Tranh De estuary has a large erosion length, strong intensity, and is hazardous due to the dense population concentrated in strongly eroding areas. Finally, from Tranh De estuary to Ca Mau Cape, there is a very large erosion length, lasting for an extended period, and with very strong erosion intensity. From Ca Mau to Ha Tien, although there is no large-scale erosion length, the average intensity is high and not dangerously alarming (Bui Hong 2001).

In the research results of "Characteristics of Erosion/deposition in the Coastal Strip of the Mekong Delta" by Mau et al. (2014), it is pointed out that the coastal stretch from Can Gio to Bac Lieu is facing severe erosion conditions in areas such as Can Gio (Ho Chi Minh City), Go Cong Dong (Tien Giang), and Ganh Hao (Bac Lieu). However, there are also areas identified as undergoing accretion processes. In Ca Mau province, the entire eastern coastal strip has been continuously impacted by erosion from 1965 to 1995, except for the Ca Mau cape, which advances into the sea by more than 100 m every year. Recently, the entire coastal strip in Ca Mau, including Ca Mau cape, is experiencing a serious erosion situation (Mau et al. 2014). The research results by Quang et al. show that the erosion process is taking place in the Can Gio coastal strip, from Dong Hoa to Can Gio capes, with an average speed ranging from 5–12 m per year and sometimes increasing to 22 m per year. Strongly affected areas include sand dunes from Dong Hoa cape to Can Gio cape, the Long Tau estuary, and the eastern coast of Thanh An island commune. Vung Tau city has undergone significant changes since 1989, including erosion and accretion phenomena. The area around the Lap estuary is facing the strongest erosion rate, with over 800 m of coastline lost. From Vung Tau to Ca Mau cape is an area with an accretion phenomenon. The material supply for this stretch is mainly alluvium from the Mekong and Dong Nai river systems. The mixed river-sea factors play an important role in the coastline formation process. The particularity of this stretch is the strong development of mangrove forests. Accretion and erosion on the estuary's coast are occurring strongly at varying degrees. The curvature coefficient of this coastal stretch is determined to be $K = 1.25$ (K is the ratio between the actual length of the coastal stretch and its length measured along the bird's flight path). The primary source of alluvium comes from the Dong Nai and Mekong River systems (Bui and Ly 2016). The coastal strip of Tien Giang province is approximately 30 km long and has a flat terrain. Here, coastal erosion typically occurs during the moonsoon season, when the impact is nearly perpendicular to the coastline, creating concentrated wave fields in the East and Northeast directions. The semi-diurnal tidal level also creates large wave pressure and high-speed wave currents carrying mud offshore. In recent years, the erosion situation on the Cong Dong forest coastal strip, especially in Tan Dien commune, has become very serious, with an annual erosion rate of about 31 m. Despite constructing protective works with a length of over 3 km and conducting annual maintenance for these sections, many sections of the coast continue to experience erosion, affecting protective dikes (Thanh and Nguyen 2016).

Regarding the coastline in the Ben Tre region, there is a tendency to expand towards the sea, but the expansion rate is not significant. Erosion and accretion phenomena are continuous and alternate over time. From November to March, during the headwind season, the wave field with a height of up to 2 m directly impacts the coastline, leading to strong erosion in this coastal stretch. Conversely, from June to September, the wind blows from the mainland to the sea, accompanied by coastal meteorological currents with an average speed of 0.4 to 0.5 m/s, carrying sediment from estuary, creating accretion along the coastal strip and forming large extensive mudflats (Vo 2021). The coastal area of Ca Mau is in a stage of strong and complex development. In the west, the accretion process is predominant, while in the east, the coastline undergoes significant changes, and erosion conditions prevail (Tinh and Phong 2017). In summary, in the coastal area of Southern Vietnam, the phenomena of erosion and accretion are occurring simultaneously and interacting with each other. The degree of erosion and accretion in each province varies. Currently, in the coastal areas of river mouths in the Southern region, accretion is still prevalent compared to erosion. Examples of erosion include Ho Chi Minh City, Tien Giang, and Tra Vinh, while Ca Mau province is a typical example of accretion. Ca Mau province has the largest accretion area, followed by Kien Giang, Bac Lieu, and Soc Trang provinces. Kien Giang province has a moderate accretion area, while Ben Tre province has the smallest accretion area in the current region (Pham 2005).

5.3 Mechanisms of Policies and Legal Frameworks for Disaster Prevention, Coastal Erosion, and Accretion in Vietnam

The mechanisms represent the concerted efforts of the government and local authorities to alleviate the impacts of natural phenomena, safeguarding both local communities and coastal ecosystems. Policies and laws related to disaster prevention, coastal erosion, and accretion in Vietnam are consistently evolving to address environmental changes and the coastal erosion scenario. These measures encompass: (a) Formulation of a national strategy for coastal erosion and accretion, outlining specific goals, measures, and action plans for managing and mitigating the impacts of erosion and accretion; (b) Establishment of environmental and coastal erosion laws and decrees within Vietnam's legal framework, covering aspects such as sand exploitation control, land use management, and identification of erosion-prone areas; (c) Implementation of land management and land use policies to control and minimize erosion, incorporating strict monitoring, prohibition of illegal sand exploitation, and rigorous coastal construction management; (d) Development of natural disaster warning and response systems, involving the construction of warning stations, transportation infrastructure, and swift-response plans for immediate attention; (e) Emphasis on international cooperation, wherein Vietnam collaborates with the global community to research and implement erosion and accretion prevention solutions. This collaboration involves sharing experiences, adopting advanced methods, and seeking technical and financial support from other countries and international organizations to enhance erosion prevention efforts.

In 2020, the Prime Minister of Vietnam endorsed the Project for Prevention and Control of Riverbank and Coastal Erosion until 2030. This initiative aims to actively manage, prevent, and control erosion along riverbanks and coasts, fostering stable and sustainable socio-economic development in these areas and contributing to security and national defense. The project delineated specific tasks and solutions, including: (a) Enhancing and revising policies and laws related to erosion prevention, coupled with awareness campaigns to reduce risks; (b) Establishing a comprehensive database on riverbank and coastal erosion through surveys, construction, and planning; (c) Implementing strict management of erosion-related activities, such as sand and gravel exploitation, construction, housing, and water exploitation; (d) Upgrading monitoring and forecasting systems to minimize erosion risk factors, providing timely warnings and coping strategies; (e) Researching and applying scientific and technological solutions for region-specific conditions, along with urgent and long-term structural and non-structural measures. International cooperation in research, data sharing, and experience exchange is actively pursued, utilizing technical and financial support. (f) Actively allocating state budget funds and mobilizing resources beyond the state budget to build infrastructure in key erosion-prone areas, safeguarding lives, property, and essential infrastructure, and enhancing international cooperation in research, data sharing, and experience exchange for riverbank and coastal erosion prevention, utilizing technical and financial support from countries and international organizations.

Additionally, Decision No. 01/2011/QD-TTg emphasizes the need for accurate, comprehensive, and timely categorization and handling of erosion levels. The decision aims to rectify, prevent, and limit erosion to maintain stability and highlights the risks of failing to address erosion promptly. The legal framework outlines responsibilities, classifications, and priority orders for addressing erosion levels systematically, presenting a strategic approach to the challenges of riverbank and coastal erosion in Vietnam. Combating accretion and coastal erosion is recognized as a challenging, urgent, and long-term task, requiring Party and State attention aligned with disaster prevention strategies, regional development planning, and adaptation to changing climate conditions. In summary, Vietnam has established a comprehensive policy and legal framework, reflecting national and international collaboration to protect its coastal areas from the challenges posed by erosion and accretion.

5.4 The Case Studies on Coastal Erosion and Accretion in Vietnam

5.4.1 Coastal Erosion and Accretion from Central Region to Thua Thien-Hue

Coastal erosion in the Central region, particularly in Thua Thien-Hue, is currently undergoing a complex and abnormal development with significant damage. The total length of eroded coastlines is over 328.16 km, accounting for 21.1% of the total coastline length. Although accretion is less severe than erosion, it predominantly occurs at the major estuaries in the Central region, such as the southern entrances of Lach Truong, Lach Hoi, and the northern entrances of Lach Ghep, Hoi, Sot, Khau, Tung, Nhat Le, Sa Huynh, Tam Quan, De Gi, and Ca Ty estuaries (Tien et al. 2002).

In the stretch from Ron cape (Ha Tinh) to Chan May cape (Thua Thien-Hue), large-scale erosion is taking place, with eroded coastlines accounting for about 24% of the total coastline length. Erosion intensity is notably strong, increasing in recent times, and the lengths of eroded sections range from under 200 m to over 6000 m. Notably, at Thuan An and Tu Hien estuaries (Thua Thien-Hue), which are typical areas for abnormal erosion-accretion phenomena, significant impacts on local communities, infrastructure, and socio-economic activities are observed (Cuong et al. 2018). The occurrence of coastal erosion is due to the high abnormal erosion tendency in the sandy sections along the estuary and lagoon entrance. Coastal erosion on the side of the estuary and lagoon, along with accretion and encroachment on the opposite side, results in the shifting of the estuary's position over distances ranging from several kilometers to tens of kilometers within a period of a few years to several decades, or even hundreds of years. Over the past half-century, the Tu Hien estuary has experienced accretion during the periods 1953–1959, 1979–1990, and 1994–1999. The rhythm of filling and opening the river mouth has occurred much faster than before. Due to accretion and displacement, the Thuan An estuary has existed at various positions and carried different names throughout its history. From the location of the old dam (1931) to the present, the estuary axis has moved northward at a rate of 15 m/year, while the accreted beach has shifted northward at a rate of up to 40 m/year. The accretion from the south to the north, coupled with the shifting of Thuan An estuary, has created instability in the 7 km-long coastal stretch (Duc et al. 2002). A study by Cuong et al. (2018) on the longitudinal changes in Thua Thien Hue's coastline from 1973 to 2016, using Landsat satellite imagery, identified severely eroded and sedimented sections. The affected areas include Phong Hai, Quang Cong, Quang Ngan, Hai Duong, Thuan An, Phu Thuan, Phu Vang (Thuan An estuary); Vinh Hai, Vinh Hien (Tu Hien estuary), posing a threat to the natural stability of the Tam Giang-Cau Hai lagoon system. The research is divided into three stages corresponding to human intervention activities influencing the coastal areas as follows:

During the period 1973–1995: when economic activities were limited due to Vietnam's closed economic system beginning to open. Coastal dynamics were primarily driven by natural factors. Over the 22-year period along the nearly 90 km stretch of coastline starting from the northern part of Dien Loc commune, Phong Dien district to Lang Co bay, the coastal area experienced significant fluctuations. The total eroded area reached 330.57 ha, while the accreted area was considerably smaller, only 34.39 ha. The areas most affected by coastal erosion were the Thuan An and Tu Hien estuaries. At Thuan An estuary, the northern coast in Hai Duong commune, Huong Tra district, experienced severe erosion, resulting in the loss of 33.7 ha conversely, the accreted area was only 5.1 ha. The erosion rate near the estuary was very high at 14.3 m/year, while the accretion rate at a location 1.5 km north of the estuary was only 5 m/year. On the southern coast of Thuan An, in the town of Thuan An, continuous erosion occurred with an average horizontal erosion rate of 9.2 m/year, causing the loss of 45 ha of land, significantly impacting the economic and living activities of the local residents. Tu Hien estuary, located in Vinh Hien commune, Phu Loc district, witnessed continuous accretion on the northern side with an accreted area of 8.83 ha and an average accretion rate of 3.38 m/year. Meanwhile, the entire southern coastline experienced severe erosion, resulting in the loss of 34.3 ha of land. The location with the highest erosion rate was approximately 200 m south of the estuary, with an average erosion rate of 15.48 m/year.

During the period 1996–2005: when economic development was underway, and protective structures, such as sand dikes and wave-reducing dikes, were constructed at Thuan An estuary. In these nine years, significant changes occurred in the coastal dynamics of Thua Thien Hue. The total accreted area was 96.0 ha, while the eroded area was 70.6 ha. At the northern coast of Thuan An estuary in Hai Duong commune, the accreted area of 23.8 ha was more than twice the eroded area of 10.93 ha. The largest accretion area was located 1.5 km north of the estuary, with an average accretion rate of 23.7 m/year. On the southern coast of Thuan An, the coastline continued to erode by 16.4 ha, with the highest erosion rate of 10.8 m/year occurring near the estuary. At Tu Hien estuary, the northern coast no longer experienced accretion as in the previous stage. During this period, both coasts of Tu Hien estuary experienced erosion, with a total eroded area reaching 22.1 ha, and the accreted area being very small, almost negligible at 0.2 ha.

The period 2006–2016 is the stage after the construction of rectification and bank protection works in the estuary and coastal areas of Thua Thien Hue. During this time, when certain structures at Thuan An estuary were completed and put into use, they altered the sediment transport process and transformed the coastal area. The total accreted area was 85.1 ha, while the eroded area was 55.1 ha. On the northern coast of Thuan An estuary, after the completion of wave-reducing dikes, the coastal area experienced severe erosion near the estuary at a rate of 11.25 m/year, resulting in the loss of 11.5 ha. On the southern coast, the construction of an additional sand dike in 2010 helped transform the coastal area from erosion to accretion, with a significant increase in the accreted area to 21.4 ha, while the eroded area decreased significantly to only 2.0 ha. The location right next to the sand dike experienced the most significant accretion, with an average accretion rate of 26.64 m/year. At Tu Hien estuary, the northern coast continued to erode by 5.9 ha, while the southern coast experienced erosion near the estuary but accreted at a location 1 km away with the highest accretion rate of 9.0 m/year (Fig. 5.1).

According to statistics, the entire Thua Thien-Hue province, there is currently about 10 km of coastline facing widespread erosion. Many residential areas and important economic zones along the sandy coast are highly sensitive, requiring relocation or heightened vigilance against unusual coastal erosion, especially during major disturbances such as storms and floods. Some precarious residential areas in Thua Thien-Hue due to coastal erosion include Hai Duong, Thuan An and Phu Thuan.

The filling up estuaries and lagoons along the coast not only eliminates sea access for hundreds of boats at each estuary but also leads to frequent flooding, prolonged freshwater inundation. This results in direct damages to communities, affecting the agricultural and aquaculture sectors. This practice disrupts the ecological balance, reduces biodiversity, and increases environmental pollution. When the Tu Hien estuary was filled in December 1994, it caused extensive harm, damaging 300 ha of shrimp farms, flooding around 1000 ha of rice fields, and rendering approximately 300 large fishing boats without sea access. The filling of Tu Hien estuary was also a significant factor contributing to a catastrophic flood in Thua Thien-Hue with a water level rise of 5.94 m in Hue, causing substantial loss of life and property in the first six days of November 1999. Flooding is a disaster that accompanies the accretion of estuaries and lagoon. In the coastal areas of Thua Thien-Hue, over the past half-century, there have been seven major floods, and six of them occurred during periods when Tu Hien estuary was being filled (Tien et al. 2002).

In recent years, erosion and accretion prevention efforts in the Central region in general and Thua Thien-Hue province in particular have received immediate and long-term attention through strategic and non-structural solutions, as well as infrastructure projects. Alongside effective coastal management, rational regional planning, and coastline protection, the province has implemented non-structural measures

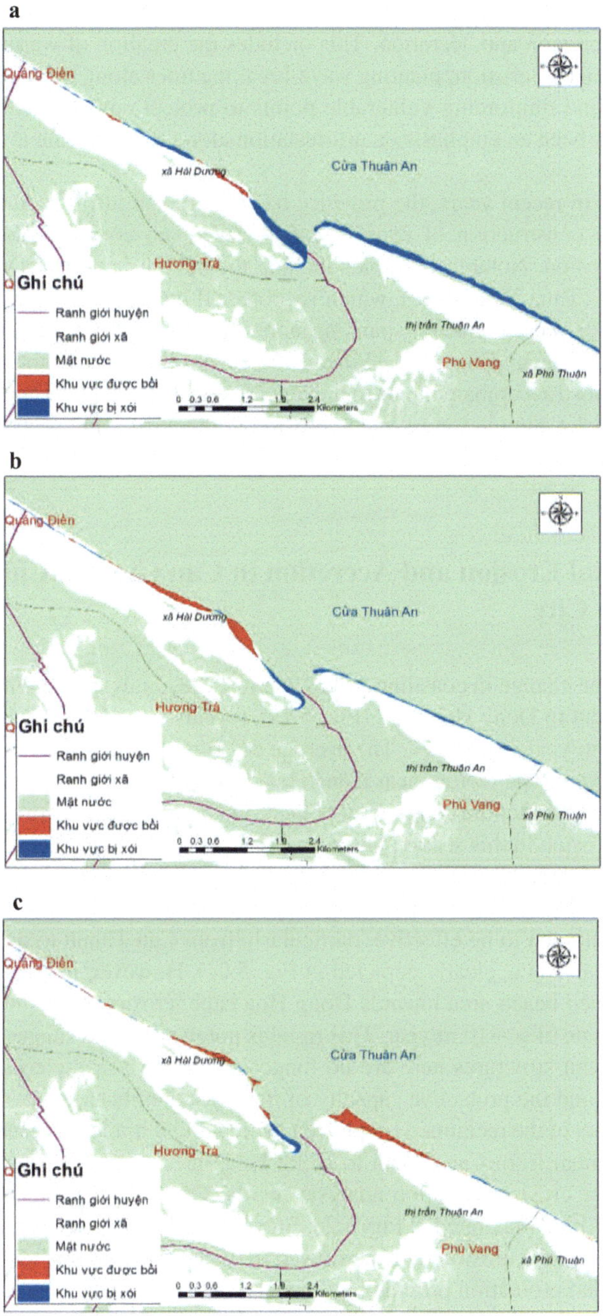

Fig. 5.1 **a** Map of coastal changes at Thuan An estuary in the period 1973–1995. **b** Map of coastal changes at Thuan An estuary in the period 1996–2005. **c** Map of coastal changes at Thuan An estuary in the period 2006–2016

such as awareness campaigns and education for communities on prevention and treatment of erosion and accretion. This includes the creation of warning maps for areas at high risk of erosion, planting wave-breaking trees along the Tam Giang-Cau Hai estuary, and reinforcing vulnerable points to protect embankments. Additionally, there has been an emphasis on afforestation along the coastline to limit coastal erosion.

Moreover, in recent years, the province has undertaken infrastructure measures, including the construction of over 71 km of river and coastal embankments to counteract erosion. Notable examples include the stretch from Hai Duong to Hoa Duan estuary (Thua Thien-Hue), which is protected by a T-shaped groynes system combined with wave breakers, spanning approximately 6.5 km (embankments in Thuan An, Phu Thuan, Giang Hai). Since 2022, the province has been implementing a project to construct embankments to prevent erosion through Phu Thuan commune, Phu Vang district, aiming to respond to erosion, protect lives and property, and help stabilize the livelihoods of residents.

5.5 Coastal Erosion and Accretion in Can Gio, Ho Chi Minh City

The data on the change in coastline from 1983 to 1988, analyzed specifically for the middle segment to Dong Hoa cape (Fig. 5.2a), reveals substantial erosion at a high rate of −100 m/year in this area. The average erosion rate along the entire stretch is up to −66 m/year. The section near Dong Hoa cape experiences the most significant erosion. However, after the completion of the Can Gio embankment projects (around the 1990s), erosion in this section has been effectively mitigated. The coastline data from 1988 to 2005 reflects this trend, with intermittent erosion and relatively low rates of < 10 m/year (Fig. 5.2b). In the subsequent period (2005–2014), these embankment structures continued to be effective, particularly from Can Thanh to the 30/4 beach (reclaimed beach area, clearly depicted in Fig. 5.2c). However, in the southern part of the reclaimed beach area towards Dong Hoa cape, erosion seems to reappear at an increased rate of > −10 m/year. This trend is noteworthy and suggests that either the embankment structures here are no longer as effective or the erosion rate has increased beyond the protective capacity of the coastal embankment. The potential negative impact of the reclaimed beach area, which might hinder the eastward movement of sediment from Can Thanh to Dong Hoa, needs further investigation. The comprehensive erosion-accretion analysis for the Can Thanh-Dong Hoa segment from 1983–2017 is illustrated in Fig. 5.2d. Erosion has been intense along this entire stretch. Although the erosion rate has slowed in the later period due to the relatively effective coastal protection provided by the Can Gio embankment structures, when not considering the artificially reclaimed areas, the entire region experiences erosion at a rate of up to −11 m/year (Ngo and Nguyen 2019). The results align with the study conducted by Anthony et al. on the coastline changes in the Southern region of

Vietnam during the period 1990–2015 (Fig. 5.3). Figure 5.3 illustrates that the area from Dong Hoa Point to Can Thanh experienced erosion trends ranging from 10 to 15 m during the 1990–2015 period.

The causes of erosion in this area can be divided into two main groups, including natural factors and human-induced factors. Natural factors include the influence

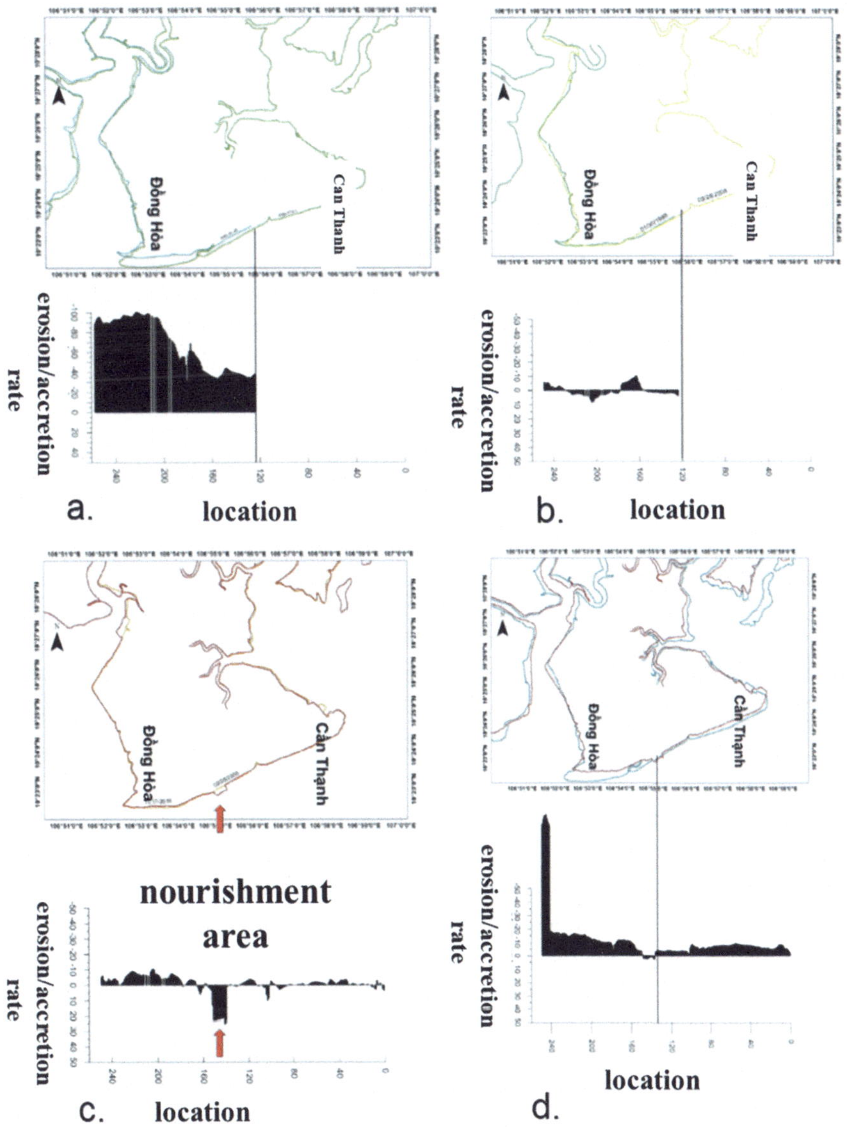

Fig. 5.2 Coastline changes in the Can Thanh-Dong Hoa section (**a** period from 1983–1988; **b** period from 1988–2005; **c** period from 2005–2015; **d** period from 1983–2015)

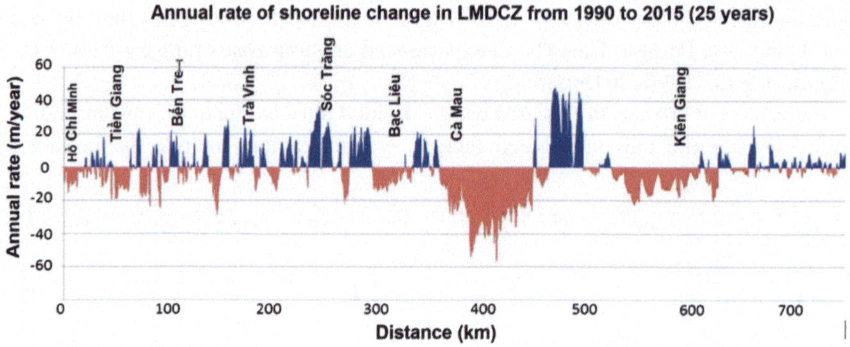

Fig. 5.3 Rate of coastline change (erosion/accretion) in the Southern region in the period 1990–2015 (Anthony et al., 2015)

of wave patterns, wind, currents, tides, and natural disasters such as storms. The Dong Hoa-Can Thanh beach area is strongly affected by the seasonal wind patterns, including the Northeast monsoon and the Southwest monsoon, making it susceptible to the impact of waves from both the Northeast and Southwest directions (Ngo and Nguyen 2019). Additionally, being the downstream area of the Dong Nai River system, with four main river mouths flowing into the sea, creates an interaction between the upstream flow from the Dong Nai River and the tidal regime from the East Sea (South China Sea). Therefore, the flow rate in this area is significant and complex. In the past, this area has also been directly impacted by tropical storms/typhoons, making it one of the natural causes contributing to intense changes in the erosion process in the region. Human-induced causes may include activities such as sand mining, mangrove deforestation, construction of water regulation reservoirs in the river basin, as well as the presence of port facilities and the impact of waves generated by vessels in the area.

5.6 Conclusion

Erosion and accretion at estuaries and coastal areas in the coastal provinces of Vietnam are currently undergoing extremely complex and unpredictable changes due to the influence of both natural and human factors. Based on statistics on the current situation of erosion and accretion throughout Vietnam, as well as two case studies in the central and southern regions of the country, several observations regarding this phenomenon can be made: coastal accretion and erosion occur concurrently, but coastal erosion takes precedence. This situation significantly impacts the daily lives of local communities and particularly affects coastal infrastructure. The current state of coastal erosion is intense, causing substantial disruptions to the lives of the population and, notably, to coastal structures. Despite proactive responses from the government and local authorities, the unpredictable nature of coastal erosion continues to impose severe consequences on coastal regions. In summary, the conditions of coastal erosion and accretion in Vietnam demand stringent management and effective response measures at both the national and local levels to minimize damages and ensure the stability of coastal areas.

References

Bui VT, Ly HM (2016) Detection of shoreline changes: a geospatial data analysis in Vung Tau City, Southern Vietnam (in Vietnamese). J Sci Technol Devel 19(1):131–137

Bui Hong L (2001) Research on forecasting and preventing landslides on the Southern coast (in Vietnamese)

Cambers G (2002) Coping with beach erosion: with case study from the Caribbean. UNESCO Publication, Paris, p 119

Cuong VD, Hung NT, Hung NV, Luan NT (2018) The application of remote sensing technology and GIS for studying the dynamics of erosion, sedimentation in coastal areas, and river mouth in Thua Thien Hue province (in Vietnamese). Hydraulic Sci Technol J (48):1–11

Cuong VD, Hung NT, Hung NV, Luan NT (2018) Applying remote sensing and GIS technology to study the evolution of erosion and sedimentation in coastal areas and estuaries of Thua Thien Hue province (in Vietnamese). J Irrig Sci Technol 48:1–11

Duc TT, Van Dien T, Chien DD (2002) Inlet change in Tam Giang-Cau Hai lagoon and coastal flood (in Vietnamese). Collect Mar Res Works XII:119–128

Duc TT, Cu NH, Huy DV, Lan DT (2002) Inlet change in Tam Giang-Cau Hai lagoon and coastal flood. Mar Sci Technol (3):33–43

Duc TT, Ngoc AN, Huu CN, Van Dinh H, Hoai ND, Kim ANT (2008) Basic geological features of Tien Yen-Ha Coi bay (in Vietnamese). Marine Resour Environ. Sci Technol

Dung HM, Nhan HA, Ly CD (2016) Developing indicators to assess climate change adaptable in Ba ria-Vung tau province and proposed improve solutions (in Vietnamese). VNUHCM J Sci Technol Devel 19(2):109–121

Hai Ha Quang, Tuyen Nguyen Ngoc. Coastal erosion of Can Gio, Ho Chi Minh City in the context of global climate change (in Vietnamese). Science & Technology. 2011;(14):17–27

Hoanh TP (2014) A study of natural environmental hazards in Dong Thap province and solutions to prevent and mitigate the damage (in Vietnamese). Sci Mag (61):185–197

Le Xuan H, Mai TA, Ho CH (2004) Characteristics of the current state of sedimentation and erosion of the coast and estuaries from Vung Tau to Ha Tien (in Vietnamese). VNU J Sci: Nat Sci Technol (4):73–81

Le Mau D, Trung PB, Le Trinh P (2014) Erosion/deposition features along Mekong river delta coast (in Vietnamese). Nat Sci Technol Publ House 154–163

Mai C, Stive MJ, Van Gelder PH (2009) Coastal protection strategies for the Red River Delta. J Coastal Res 25(1):105–116

Ngo TN, Nguyen TC (2019) Functional zoning of coastal areas of Ho Chi Minh City. Department of science and technology of Ho Chi Minh City

Nguyen TTH, Mai LN (2021) Applied the digital shoreline analysis system (Dsas) and remote sensing documents to study coastal changes in the Lach Ghep estuary area in Thanh Hoa (in Vietnamese). Research, application, and development of national geospatial data infrastructure: the role of modern mapping technology

Nicholls RJ, Wong PP, Burkett V, Codignotto J, Hay J, McLean R. Ragoonaden S, Woodroffe CD, Abuodha PAO, Arblaster J, Brown B, Forbes D, Hall J, Kovats S, Lowe J, McInnes K, Moser S, Rupp-Armstrong S, Saito Y (2007) Coastal systems and low-lying areas. http://ro.uow.edu.au/scipapers/164

Pham TH (2005) Forecasting the phenomenon of coastal and river mouth erosion and sedimentation and prevention solutions (in Vietnamese). Rep State-level Sci Technol

Pham TH, Van Cu N (2001) Research, forecast, and prevent landslides on the Central Coast (From Thanh Hoa to Binh Thuan) (in Vietnamese). The state-level scientific and technological project report: 5B

Quan NC, Van Hung P (2016) Characteristics of dynamic geomorphology of coastal-river mouth zones of Ma river of Thanh Hoa province (in Vietnamese). Sci Earth 38(1):59–65

Thanh LT, Nguyen HV (2016) Hydrodynamic developments in the Go Cong Dong-Tien Giang coastal area, under the effect of groynes and underground breakwaters using GEO-TUBE (in Vietnamese). Estuar, Coast Shelf Sci 363–365

Tien PH, Thanh TD, Long BH (2002) Research results on erosion and sedimentation in Vietnam's coastal estuaries (in Vietnamese). Marine Sci Technol 2(4):12–26

Tinh TV, Phong DH (2017) Using remote sensing images and GIS to study coastline changes in Ca Mau cape area (in Vietnamese). Vietnam J Hydrometeorol (12):35–42

Van Tho N (2020) Coastal erosion, river bank erosion and landslides in the mekong delta: causes, effects and solutions. Geotech Sustain Infrastruct Dev. Springer

Van Cu N, Shibayama T (2014) Assessment of Vietnam coastal erosion and relevant laws and policies. In: Coastal disasters and climate change in Vietnam. Elsevier, pp. 81–106

Van Cu N, Son NT, Thao DT, Son NH (2017) Real changes of coastline from Hai Phong to Ninh Binh and exploitation of its resources (in Vietnamese). Hue Univer J Sci 126(7A)

Vo TN (2021) Current status of use and management of coastal alluvial land in Ben Tre province (in Vietnamese) 97–108

Chapter 6
Impacts and Responses of Climate Change to the Coastal Areas of the Mekong River Delta in Vietnam

Le Anh Tuan

Abstract The Vietnamese Mekong River Delta (VMD), formed by the silt and other sediments deposits from the upstream Mekong River flow via its tributaries, is a very low-lying flat area combined with a high density network of rivers and canals. The Delta has more than 700 km shoreline, with special ecological and hydrodynamic characteristics, which are very important to the local livelihoods sources. The coast's morphological features are strongly influenced by changes within the river and the sea, with erosion and alluvial deposition due to the natural interaction of current flows. In last three decades, the VMD coastal areas are facing serious environmental challenges and climate change, particularly the increasing of saline intrusion, the discharging of polluted wastewater, the shrinking of mangrove forests and the degrading of natural resources due to anthropogenic impacts through agro-aquaculture productive activities. Presently, the coastal banks erosion and other abnormal weather phenomena seem more serious in their processes that threat the existing of many local ecosystems and residential lands. Concerning with the long-term sustainable development of the VMD's coastal areas, the relation among water—land—natural habitats—climate resilience should be of major interests for finding adaptive solutions in regional and local scales. It is estimated that by 2050, as many as one million people will be at risk of being displaced. It is therefore the overall purpose of this assessment and report is to strengthen the resilience of VMD's coastal provinces to the impacts of climate change through ecosystem-based adaptation. In this approach, concept "Ecosystem-based Adaptation" (EbA) is understood as flexible using and managing overall natural ecosystem services in order to support local communities' activities in increasing resilience and reducing the vulnerability of people and the environment to climate change. The combination between both maintaining ecosystem function and considering livelihoods in the context of natural and environmental change processes should be widely accepted. In facts, there are some pros and cons in EbA measures application. Currently, there are many existing creative climate change adaptations, both "hard" and "soft" measures, developed by local residents. Scientists and INGOs staff have supported them to optimize their

L. A. Tuan (✉)
Can Tho University, Can Tho, Vietnam
e-mail: latuan@ctu.edu.vn

V. Maccarrone and M. Fadzil Akhir (eds.), *Coastal Sustainability*, Coastal Research Library 39, https://doi.org/10.1007/978-3-031-75749-5_6

cultivation systems and provided more scientific processes and technological transfers. Central and provincial governments have tried to improve the policies and to integrate climate change adaptation solutions into provincial socio-economic development plans. However, the EbA application seems going rather slower than expected. Multiple mechanism and approaches should be integrated when developing adaptation strategies and actions. This should include: natural system adaptation; land-use planning; community and social adaptation; economic instruments; policies and plans; institutional mechanisms; and engineering solutions. There is a critical need for immediate implementation of no-regret and climate deficit reducing adaptation measures for coastal areas. This should aim to help people improve their key livelihoods in harmony with maintaining and restoring the key supporting ecosystems. It is now widely recognized that adaptation should:

- Promote long-term local solutions for people and nature—starting with no-regrets actions and reducing the adaptation deficit this means focusing on reducing non-climate stresses;
- Involve local communities and address governance and justice issues;
- Develop multi-partner, multi-sector, multi-scale strategies;
- Build upon existing Natural Resources Management (NRM) good practices; recognize climate is not static;
- Adopt adaptive management approaches and support local innovation;
- Integrate Ecosystem-based Adaptation with wider climate adaptation strategies; and
- Communication and education, sharing and exchange.

Keywords Climate change · Coastal areas · Ecosystem-based adaptation · Mekong River Delta · Impacts · Response

6.1 Background

Originating from the snowy mountains of Tibet, the Mekong River passes the narrow valleys and the jungle plateaus of Myanmar, coming to borders of Thailand, Laos, and downing to low-lands of Cambodia, then to the wetland delta of Vietnam before entering to the East Sea. The total drainage area of the Mekong River Basin is 810,000 km^2. The Mekong River is called the "Mother Stream" as a natural source for feeding at least 60 million people. Many scientists and residents living along the Mekong River countries, including China, Myanmar, Thailand, Laos, Cambodia and Vietnam, have noted as their increasing in the frequency of abnormal weather occurrences.

The Mekong Delta is the last downstream area of the Mekong River basin before flowing into the East Sea. This is a new, young geological land formed 6000–8000 years ago (Ta et al. 2005). The total area in the Mekong Delta in Vietnam is more than 40,000 km^2 of land area and a population of over 17 million people. The Delta is recognized as the largest agricultural and fishery production region in the country.

Almost all activities of local people depend greatly on water resources, especially the water resources of the Mekong River. The Mekong Delta has two sides bordering the Pacific Ocean to the East and the Gulf of Thailand to the West with a total coastline of more than 750 km, not including the coastal length of the islands. Thanks to the amount of alluvium from the Mekong River and the ocean currents in the East Sea, over the past 7000 years since its formation, the delta has expanded into the sea at an average of about 30 m/year (Liu et al. 2017), and in the last 3000 years, it has expanded at the river mouths at a rate of about 16 m/year and in the Ca Mau Peninsula at 26 m/year. At that time, the mangrove forest in Ca Mau Cape extended its roots into the sea about 40–70 m/year each year. Due to its location at the downstream end of the Mekong River system, the Mekong Delta has a very low and flat terrain, with an average elevation of 1.2–1.5 m above mean sea level. In general, the Delta with a rather weak foundation structure, so it is prone to displacement, subsidence and landslides when there are significant changes in water resources and activities that exceed the bearing capacity of the ground. The coastal area of the Mekong Delta is understood to be the area adjacent to the coast and is the area affected by salinity intrusion from the sea into the mainland all year round (Fig. 6.1).

Although worthy of being a prosperous agricultural region, over the past two decades, the delta has been facing three major challenges related to the water environment, which are the adverse effects of the global climate change, transboundary water impacts from water exploitation projects on the Mekong River and internal unsustainable production activities. Determining sustainable solutions to water environment problems has many challenges, requiring a lot of efforts from the people of the delta and investment policies of the central government. Macro managers need to be aware of these risks early and have appropriate responses to the problem.

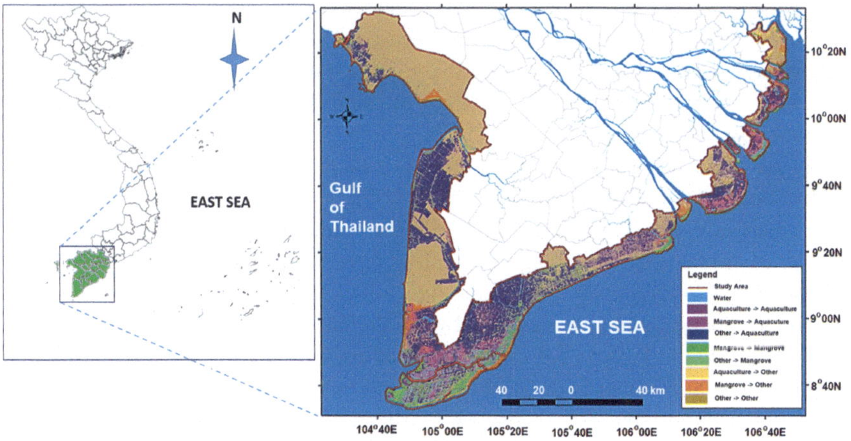

Fig. 6.1 Location maps of Vietnamese Mekong River Delta and the land use coastal areas of the Delta (cited from Phan and Stive 2022)

6.2 Literature Review

The phenomenon of global climate change is becoming more and more evident and there is much evidence that extreme weather factors such as heat, unusual rain, changes in flow and rising sea levels are increasing in intensity and frequency, greatly affecting life and production. The phenomenon of global warming continues to become more and more serious and shows no signs of abating, although there have been many warnings from scientists and civil society organizations. The consequences of global warming are the melting of ice and rising sea levels, causing many low-lying areas to be submerged and saline. The Ganges–Brahmaputra delta in Bangladesh, the Mekong delta in Vietnam and the Nile delta in Egypt are top vulnerability mega-deltas in the world at risk through coastal erosion and land loss, primarily as a result of decreased river sediment delivery and accentuated rates of sea-level rise (Ericson et al. 2006). The Intergovernmental Panel on Climate Change (IPCC 2007) has assessed the Mekong Delta in Vietnam that is facing the most adversely affected and vulnerable to climate change on people's lives and livelihoods. Climate change, rising sea levels and impacts on transboundary water resources are and will be huge challenges that threaten Vietnam's socio-economic development, causing the Mekong Delta's agriculture to become unsustainable. In parallel, the construction and operation of a series of hydroelectric dams in China and Laos is a factor, along with climate change, as a double risk that changes the hydrological characteristics of the Mekong River in an abnormal direction, causing many negative impacts related to sediment load, fish resources and ecosystems (Renaud and Kuenzer 2012). According to the estimation of the Ministry of Natural Resources and Environment of Vietnam (MONRE 2018), from now until the end of this century, if the average sea level rises by 50–70 cm, at least 25% of the agricultural land area in the coastal area of the Mekong Delta will be submerged and about 75% of the current cultivated area will be affected by salinity in the dry season and about 40–50% of the agricultural area will be affected by salinity even in the rainy season, making it difficult to grow rice. In addition, the abnormality of rainfall, air temperature and weather abnormalities will cause more major challenges for the region. Natural abnormalities will cause losses in productivity and output or increase input costs for agricultural production.

In order to cope with and minimize the potential risks, challenges and consequences of climate change, the United Nations has called on the whole world to unite and join hands to effectively implement the United Nations Framework Convention on Climate Change, which was ratified by the Vietnamese government on November 16th, 1994, and the Kyoto Protocol on responding to global climate change, which was ratified on September 25th, 2002. The Vietnamese government has ratified these international commitments and has promptly directed the research, development and promulgation of the National Target Program to respond to climate change. With the forecast results on the risk of being severely affected by climate change, the provinces and cities of the Mekong Delta have been selected as priority to implement the National Target Program. Since 2005, after the Kyoto Protocol came into effect, a number of policies and laws directly related to climate change have been issued, such as Directive No. 35 of the Prime Minister on the implementation of the Kyoto Protocol under the United Nations Framework Convention on Climate Change; Resolution No. 60 of the Government assigning the Ministry of Natural Resources and Environment to preside over and develop the National Target Program to Respond to Climate Change (2007), Official Dispatch No. 158/2008/QD-TTg of the Prime Minister issued on December 2nd, 2008 on approving the National Target Program to Respond to Climate Change is an important document for implementing measures to respond to climate change in Vietnam. There are 4 main contents of the National Target Program to Respond to Climate Change, specifically: (1) Assessing the impacts of climate change on sectors, industries and localities in each period; (2) Developing a feasible action plan to effectively respond to climate change for each period to ensure sustainable development of the country; (3) Taking advantage of opportunities to develop the economy towards low carbon; and (4) Joining the international community in efforts to mitigate climate change and protect the Earth's climate system.

This was followed by a series of more specific climate change response policies, with particular attention to coastal areas, including the National Strategy on Climate Change in (2011), and the National Strategy on Green Growth (2012). On November 17th, 2017, the Prime Minister issued Resolution 120/NQ-CP on sustainable development of the Mekong Delta in response to climate change as a policy strategy to help the Mekong Delta reposition its development process and activate activities in the spirit of "following nature", expressed in 4 guiding viewpoints:

(a) Creating sustainable and prosperous development, on the basis of proactive adaptation, promoting potentials and strengths, transforming challenges into opportunities for development, ensuring a stable and prosperous life for people as well as preserving the unique traditional cultural values of the Mekong Delta; focusing on protecting land, water and especially people.
(b) Changing the development mindset, shifting from purely agricultural production thinking, mainly rice production to the mindset of developing a diversified agricultural economy, meeting market demand, shifting from development based on quantity to quality; Building new rural areas associated with strong development of high-tech agriculture, organic agriculture and clean agriculture associated with value chains and brand building; focusing on developing processing industry and supporting industry associated with agricultural economic development. Agricultural products not only ensure food security but also have high nutritional value, serve disease prevention and treatment, creating famous brands.
(c) Respect the laws of nature, be suitable for actual conditions, avoid harsh interference with nature; choose models that adapt to nature, are environmentally friendly and develop sustainably with the motto of proactively living with floods, inundation, brackish water, and saltwater; research, develop scenarios and have effective response solutions to natural disasters such as storms, floods, droughts and saltwater intrusion, with the most unfavorable situations due to climate

change and development of the upper Mekong River. Economic development is closely linked to social development, poverty reduction, job creation, social security, environmental protection and new rural construction. Encourage the participation of all stakeholders, ensure organic cohesion within the region as well as close linkage with the Southern Key Economic Zone and the Mekong Sub-region.

(d) Sustainable development of the Mekong Delta for the common benefit of the country, the Mekong Sub-region and the world and is the cause of all people, encourage and mobilize all social classes, components, international partners and businesses to participate in the development process.

The Vietnamese Government has also had investment policies for coastal erosion prevention projects, specifically Decision 957/QD-TTg dated July 6th, 2020 on approving the "Project to prevent riverbank and coastal erosion until 2030" (Prime Minister 2020). For implementing the Project, the Decision No. 1162/QD-TTg has signed on additional capital from the central budget reserve in 2023 for provinces in the Mekong Delta to promote the implementation of projects to prevent and combat riverbank and coastal erosion (Prime Minister 2023).

6.3 Approaches and Data

This study approaches the issue in a comprehensive way by collecting evidence based on scientific reports from universities and research institutes, hydrometeorological monitoring stations, collecting stories from localities on damage caused by natural disasters, landslides and changes related to climate change and sea level rise. In addition, many field trips in the Mekong River Delta along the coastal areas from Tien Giang, Ben Tre, Tra Vinh, Soc Trang, Bac Lieu, Ca Mau and Kien Giang were conducted from 2011 to 2023 to collect data and information opinions from local people.

6.4 Findings and Discussion

According to the results of climate change predictions in the Mekong Delta from mathematical models of many scientists and the Vietnamese Government (Tuan and Chinvanno 2011; Nguyen et al. 2017; Ministry of Natural Resources and Environment 2021), analysis of meteorological data from coastal provinces from 1980 to 2017 shows an average temperature increase of about 1.9 °C over the past three decades (see Fig. 6.2 as a case study in Tien Giang province). According to the Ministry of Natural Resources and Environment (2018), both RCP 4.5 (low-medium) and RCP 8.5 (high) scenarios predicted for the period 2016–2035 show that the average annual temperature in almost coastal areas tend to increase by 0.35 °C and

0.68 °C if compared with the baseline data in in the periods 1986–2005, respectively. As for the climate change scenario of the Ministry of Natural Resources and Environment (2021), in the period 2046–2065, the temperature increase is 1.3 (0.9–2.0) °C for the RCP4.5 scenario and 1.9 (1.4–2.7) °C for the RCP8.0 scenario. Details of the average monthly air temperature according to the RCP 4.5 and RCP 8.5 scenarios for the period 2016–2035 in coastal provinces are shown in Fig. 6.3.

The rainy season tends to come later and the rainfall at the beginning and middle of the season tends to decrease, while the rainfall at the end of the season increases. Both RCP 4.5 (low-medium) and RCP 8.5 (high) scenarios analysis for the precipitation change in the period 2016–2035, 2046–2065 and 2080–2099 show rainfall percentage variability, compared with baseline data in 1986–2005, during the rainy seasons and dry season as in Table 6.1. In general, according to the RCP 8.5 scenario, the annual

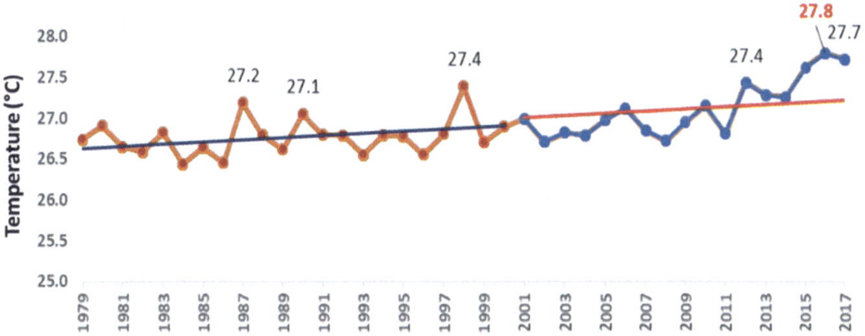

Fig. 6.2 Trend of increasing average annual air temperature in Tien Giang province during the period 1979–2017

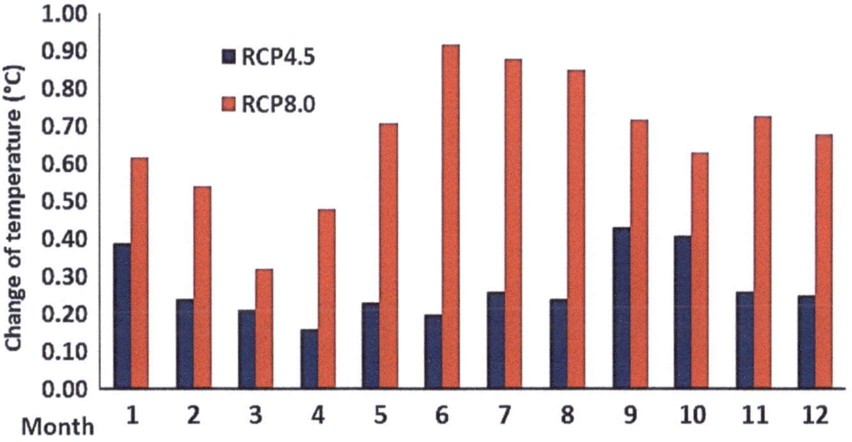

Fig. 6.3 Temperature change (°C) in the coastal areas of the Mekong River Delta in Vietnam under the RCP4.5 and RCP8.0 scenarios in the period 2016–2035

rainfall change in all three periods shows the general result that rainfall in coastal provinces tend to increase and is higher than the RCP 4.5 scenario.

The Ministry of Natural Resources and Environment (2021) has simulated the rising water level of the East Sea of the Mekong Delta of Vietnam for the years up to the end of the twenty-first century based on different RCP scenarios, the results compared with the baseline data of the period 1986–2005 gave the results in Table 6.2. It is estimated that from now until the end of this century, if the average sea level rises from 0.50 to 1 m, at least 25% of the agricultural land area in the coastal area of the Mekong Delta will be submerged and about 75% of the current cultivated area will be affected by salinity in the dry season and about 40–50% of the agricultural area will be affected by salinity even in the rainy season, making it difficult to grow rice. The flooded area of the provinces in the Mekong Delta corresponding to the sea level rise is shown in Table 6.3.

Rising sea levels, subsidence risks in the Delta and the decline in water resources from the Mekong River upstream are factors that increase the salinity intrusion into coastal areas, both in salinity concentration and distance from the sea to the mainland. The main cause of subsidence in the coastal areas of the Mekong Delta is the lack of surface fresh water and rainfall, forcing local people to overexploit underground water, causing the ground level to drop. Field observations combined digital elevation models show that the current subsidence in the Delta is more serious because it causes

Table 6.1 Projected precipitation change (%) under RCP 4.5 and RCP 8.5 for the different periods in the future

Seasons	Scenarios	2016–2035	2046–2065	2080–2099
Whole year	RCP 4.5	1.4–7.0	1.0–7.0	2.0–7.4
	RCP 8.5	0.3–4.5	0.1–4.0	3.2–9.3
Rainy (May–Oct)	RCP 4.5	0.4–1.8	0.2–1.2	0.9–2.3
	RCP 8.5	0.01–1.5	0.01–1.5	1.0–2.4
Dry (Nov–Apr)	RCP 4.5	1.2–6.2	0.3–8.1	3.2–5.4
	RCP 8.5	2.0–4.4	0.1–3.6	1.3–7.2

Source MONRE (2021)

Table 6.2 Sea level rise according to RCP scenarios for the East Sea Region (Unit: cm)

Scenario	Timelines of the 21st century							
	2030	2040	2050	2060	2070	2080	2090	2100
RCP2.6	14 (8–19)	20 (11–26)	25 (14–33)	30 (17–41)	34 (20–48)	38 (23–55)	42 (25–62)	46 (28–70)
RCP4.5	12 (8–18)	18 (11–25)	23 (14–33)	29 (18–42)	35 (22–51)	42 (26–61)	48 (30–71)	55 (34–81)
RCP8.5	14 (10–19)	20 (14–27)	28 (20–37)	34 (23–47)	43 (28–59)	52 (35–72)	64 (42–88)	77 (51–106)

Source MONRE (2021)

Table 6.3 Estimated percentage of flooded area in Mekong Delta provinces corresponding to sea level rise

Province	Areas (ha)	Flooding rate (% of area) according to sea level rise									
		10 cm	20 cm	30 cm	40 cm	50 cm	60 cm	70 cm	80 cm	90 cm	100 cm
Long An	449,100	0.00	0.00	0.31	0.49	0.61	1.36	2.85	7.12	12.89	27.21
Tien Giang	251,061	0.13	0.71	1.43	2.57	3.79	6.71	12.58	25.23	37.57	47.80
Ben Tre	239,481	0.55	1.43	2.52	4.08	6.74	10.19	15.11	21.46	27.83	35.11
Tra Vinh	235,826	0.50	0.61	0.89	1.28	2.29	4.95	11.51	22.22	32.79	43.88
Vinh Long	152,573	0.00	0.34	0.61	0.91	1.31	2.02	3.66	8.28	18.34	32.03
Dong Thap	337,860	0.00	0.00	0.17	0.21	0.36	0.69	0.96	1.28	1.94	4.64
An Giang	342,400	0.00	0.00	0.03	0.05	0.08	0.13	0.29	0.49	0.90	1.82
Long An	449,100	0.00	0.00	0.31	0.49	0.61	1.36	2.85	7.12	12.89	27.21
Can Tho	143,896	0.06	0.17	0.31	0.50	0.99	2.88	9.97	26.69	44.89	55.82
Hau Giang	162,170	0.00	0.75	3.42	10.31	18.83	29.37	38.50	45.88	53.21	60.85
Soc Trang	331,188	1.78	2.91	5.13	8.32	11.32	14.97	20.25	26.91	33.13	55.41
Bac Lieu	266,901	0.71	2.87	6.66	12.14	20.08	27.78	36.84	46.31	54.38	61.87
Ca Mau	522,119	7.21	14.06	20.17	28.73	40.31	48.05	56.81	64.42	73.58	79.62
Kien Giang	634,878	0.66	3.38	12.63	23.67	36.82	48.85	75.68	66.16	71.69	75.68
Whole Delta	4,069,453	1.29	2.97	5.92	9.86	14.86	19.69	27.94	31.94	38.80	47.29

Source MONRE (2021)

the ground level to drop faster than the rate of sea level rise (Minderhoud et al. 2019), at a rate of 1–4 cm a year or an average of about 3 cm/year, and even in some places like Ca Mau city, the subsidence rate is approximately 5 cm/year (INDRA-GISAT 2019). This rate is 10 times faster than the rate of sea level rise.

In the dry season, drought and saltwater intrusion from estuaries into rivers and canals increase, especially during El Nino years, when the water level from the upper Mekong River is reduced due to low rainfall, salinity follows the tide and penetrates deeper into the fields through the canal and river system, creating difficulties for water supply for daily life, livestock and agricultural cultivation. Assessing salinity intrusion in coastal areas is complex because it depends on the flow from the upper Mekong River, where much of the water is controlled by hydropower dams, water extraction for agricultural cultivation in the upper region, and the impact of internal water control works. Figure 6.4 compares the level of salinity intrusion with a concentration of 4 ppt into the river branches in the Mekong Delta in 4 years with the presence of El Nino phenomenon in the West Pacific Ocean and Fig. 6.5 shows the 4 ppt salinity isoline into the inland delta.

The coastal area of the Mekong Delta is home to mangrove forests that are very significant in terms of ecology and biodiversity. The area of mangrove forests in the Mekong Delta is 73,372.04 ha (2019), accounting for 50.5% of the country's mangrove area. The mangrove trees here are mainly *Rhizophora apiculata*, *Avicennia marina*, *Sonneratia caseolaris*, *Nypa fruticans* and *Avicennia officinalis*. Mangrove forests in the coastal areas of the Mekong Delta have been declining in both area and quality. The causes of mangrove forest decline are due to natural factors such as reduced sedimentation, landslides, waves, wind, storms, etc. and human factors from forest exploitation and aquaculture activities. The place with the largest mangrove forest area in the delta is Ca Mau Cape, which is more than 50,000 ha. At Ca Mau

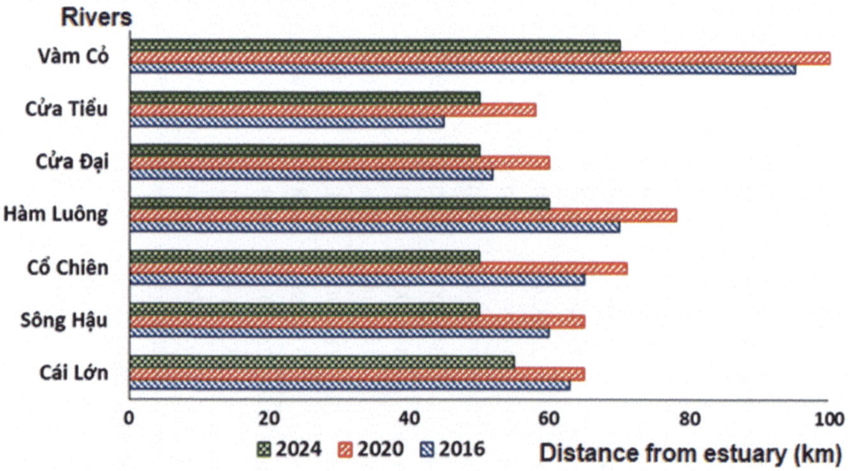

Fig. 6.4 Comparison distances due to saltwater intrusion from the estuary into the inland of the Mekong Delta in years with the El Nino phenomenon, i.e. 2016, 2020 and 2024

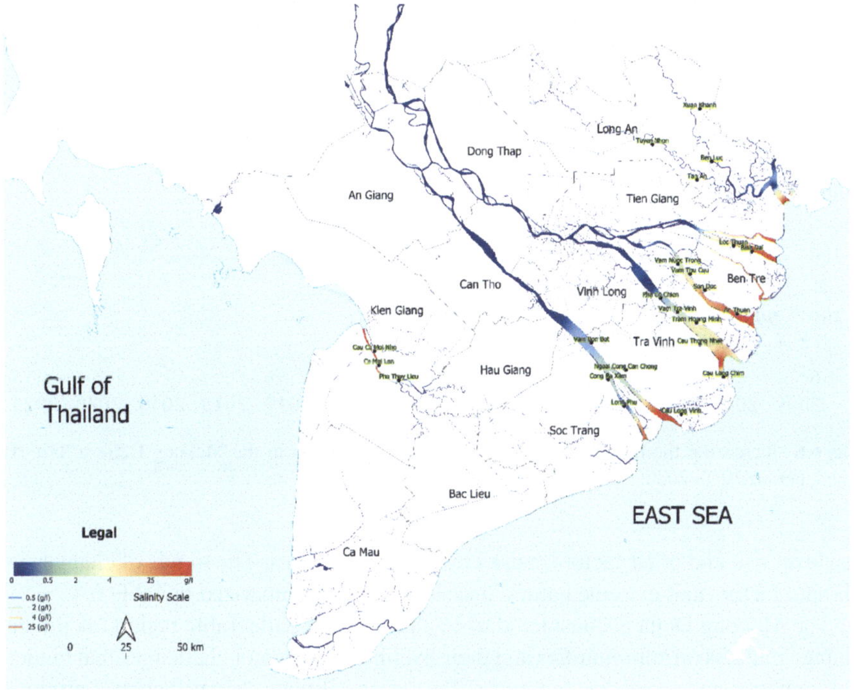

Fig. 6.5 Map of highest saline intrusion from the East Sea entering to the main river branches of the Mekong Delta from February 1, 2024 to April 28, 2024. *Source* DRAGON Institute, Can Tho University (2024)

Cape, the phenomenon of rising sea levels and coastal erosion has changed the coastline of the East Sea at a rate of 500–5000 ha/10 years and increased sedimentation in the western region of the Gulf of Thailand.

Most of the silt and sand are retained by hydroelectric dams in the upper Mekong River, causing a lack of sediment in the downstream area, causing the phenomenon of "sediment-starved water" which increases the pressure of the flow, causing erosion of rivers, canals and coastal areas (Allison et al. 2017). From 1992 to 2012, fine silt has decreased from approximately 160 million tons/year (Miliman and Syvistsky 1992; Jean-Paul et al. 2013) to 74.1% to the Delta, still in 43.1 Mt/yr occurred in 2012–2015 after China built 7 mega-dams in the upper Mekong. This is the most basic reason the number of erosion sites in the Delta increasing significantly (Fig. 6.6).

In addition to the simulations of climate change in the Mekong Delta (Tuan and Chinvanno 2011), the increase in wind speed from the sea is one of the factors that cause storm risks and can impact the coastal areas of southern Vietnam, which are very vulnerable. This climate analysis and simulation of the change in wind speed by month in the year can be seen in Fig. 6.7. In general, the coastal areas of the Mekong Delta are suffering severe damage due to the unpredictable changes of climate change,

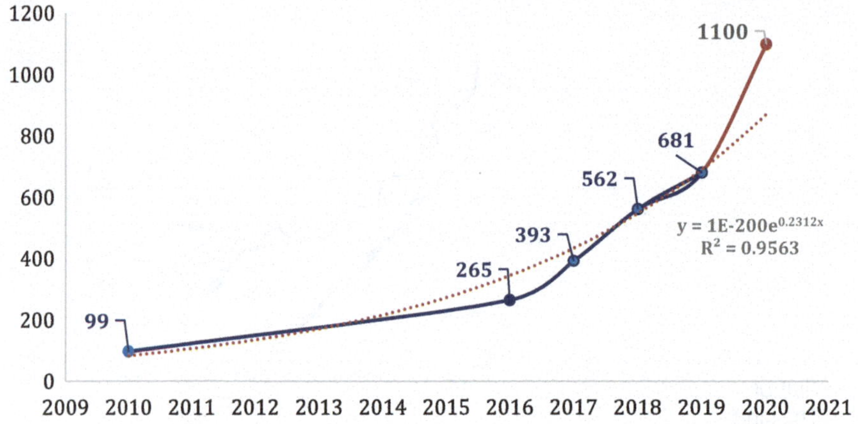

Fig. 6.6 Increasing the number of erosion and landslide points in the Mekong Delta in last 10 years, period 2010–2020

sea level rise and other factors caused by human activities. The trends of changes in climate factors and extreme natural disasters can be summarized in Table 6.4.

The Mekong Delta is considered to be an extremely vulnerable region to climate change impacts on communities and their livelihoods as major challenges that hinder the economic and social development of the region. Climate change increases production costs, increases risks in harvesting, processing and distributing agricultural products to markets, while extreme weather events increase health, migration and social structure issues in the region, not only in the inland areas but also in the coastal areas, with overlapping impacts that influence each other because livelihoods and social issues are interrelated. When the low air temperature increases by 1 °C, rice productivity will decrease by about 10% (Peng et al. 2004). Assuming this happens, the Mekong Delta could lose 2–4 million tons of rice per year due to regional warming alone. The model also predicts that in the next 30 years, the average total rainfall in the Mekong Delta will also decrease by 10–20%, making it more difficult to provide fresh water for rice cultivation. Abnormally high temperatures will greatly affect coastal aquaculture (shrimp, clams, mussels, etc.) and may affect the spawning grounds of many coastal marine species. The increase in drought and heat in the dry season will increase the demand for domestic water for people while the water source is mostly salty. This is a huge challenge for water supply plants, quick surveys show that in the dry season, about 50,000–100,000 families lack fresh water for daily use. Each year in the entire delta region, currently about 500 ha of land are lost due to landslides and erosion, causing damage of tens of thousands of billions of Vietnamese Dong.

The Mekong Delta is undergoing transformation under the adverse impacts of change, requiring economic structural reforms, investment policy reforms in parallel with the requirements of transforming the production system towards nature, valuing all freshwater, brackish water and saltwater resources. Any decisions for the Delta

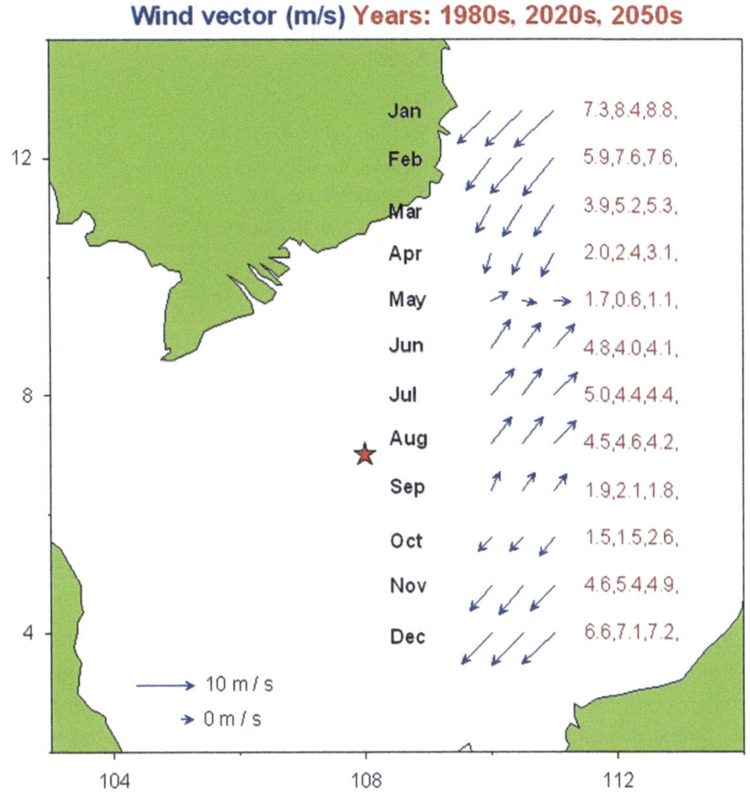

Fig. 6.7 Projected monthly average wind velocity and direction in the East Sea based on the recorded data in 1980s, projection to 2020s and 2050s

from now on and in the future must consider the selection factors according to the principle of "no regrets" or "little regrets", following the sprits of Integrate Ecosystem-based Adaptation with wider climate adaptation strategies. All solutions to respond to climate change and prevent natural disasters must consider both non-structural and structural factors. The Mekong Delta needs to have a multi-stakeholders combination of governments—farmers—scientists and businesses, along with the mutual linkage of donors (such as NGOs, Civil Society Organizations) and financial institutes as development banks (supporting lending activities, credit, incentive programs, etc.).

Using indigenous knowledge in coastal areas is a relatively easy, low-cost and effective solution to cope with climate change and natural disasters and to build upon existing Natural Resources Management (NRM) good practices. Folk solutions to store fresh water from the rainy season to the dry season such as using ponds, canals, water tanks, and jars are used by people in each family. Local solutions will reduce government support for the construction of additional water supply systems, bringing fresh water from the upper regions to the coastal areas. Agricultural production areas at risk of salinity are converted into brackish or saltwater aquaculture areas, while

Table 6.4 Trends of changing climate factors in coastal provinces of the Mekong Delta

No.	Items	Signals	Tendency
1	Starting time of rainy season	Rainy season is about 10–15 days late	↗
2	Heavy rain	Rainfall over 100 mm/day	↗
3	Temperature fluctuation	Increase the max–min temperature difference in the year	↗
4	High temperature damages production	Daily temperature > 35 °C	↗
5	Sea level rise–saline intrusion	Increase salinity level and saltwater intrusion	↗
6	Coastal and river bank erosion	Increase points and areas of erosion	↗
7	High tide	Rising tide levels in river and sea	↗
8	Unusual natural disaster	Record unpredictable storms and tornadoes	↗
9	Total yearly rainfall	Three-year moving average trend	↘
10	Water level in the started and middle rainy season	Rainfall in the started and middle rainy season	↘

areas with seasonal saltwater contact are cultivated according to the rice-shrimp agricultural model. Increasing forest planting and protecting mangroves are promoted in each locality, combined with some engineering measures such as embankments to minimize landslides and erosion. Plans to minimize the damage caused by natural disasters and prevent landslides related to the restoration of mangrove ranges and research to expand suitable native plant species are extremely necessary and urgent. Reproduction and population redistribution plans for coastal areas are carefully considered according to climate change and natural disaster factors. At the same time, raising awareness of the community and local authorities is regularly carried out. The natural disaster monitoring and warning system is always active, especially during times when extreme natural disasters are likely to occur. A new program that is being promoted and developed in coastal areas is the development of renewable energy, mainly wind and solar energy. Wave energy and tidal energy are potential energy sources that coastal areas can pay attention to develop in the future.

6.5 Conclusion

All countries are living in a world is facing the changes very rapidly, including natural conditions, market economy, social factors, epidemics as well as political policies are always changing and unstable. This makes community livelihoods strongly affected, both positively and negatively. The Mekong Delta is considered an area that is at high risk due to the negative impacts of climate change and rising sea levels, including the external impacts related to the development of hydropower dam chains in the upper Mekong River. In order to survive and develop, people in this delta have been creative in minimizing and adapting to adverse changes to their livelihoods and living conditions according to different conditions and circumstances. However, most of the forms of livelihood conversion to adapt to climate change that are popular today are still only short-term and medium-term coping activities. One difficulty that needs to be overcome and resolved is the shortage of experts with skills in policy analysis and implementation of integrated climate change response solutions with increased livelihoods in the provinces.

The general recommendation is to continue to link up, share information, exchange experiences to replicate or improve existing models and enhance leadership capacity and skills for people, which needs to be maintained continuously at present. Civil society organizations in localities play a leading role in applying diverse livelihood solutions in many fields in their localities. Retraining and advanced training for people in the community is one of the recommendations that need to be noted. Provinces and cities need to have long-term investment budgets for climate change response and disaster prevention activities, including solutions to support corresponding livelihood transformation. It is essential to integrate climate change into economic and social development programs in sectors and localities. In addition, it is necessary to mobilize financial, technical and human resources from government and non-government organizations to increase the chances of the fight against climate change being more effective and sustainable.

References

Allison MA, Nittrouer CA, Ogston AS, Mullarney JC, Nguyen TT (2017) Sedimentation and survival of the Mekong Delta: a case study of decreased sediment supply and accelerating rates of relative sea level rise. Oceanography 30(3):98–109

DRAGON Institute, Can Tho University (2024) Map of highest saline intrusion in the Mekong Delta from January 1, 2024 to April 28, 2024 (unpublished)

Ericson JP, Vorosmarty CJ, Dingman SL, Ward LG, Meybeck M (2006) Effective sea-level rise and deltas: causes of change and human dimension implications. Glob Planet Change 50(1–2):63–82

INDRA-GISAT (2019) Baseline report, climate change. JV HaskoningDHV Nederland B.V – GIZ, Feb 2020

IPPC (2007) IPCC fourth assessment report: climate change 2007. In: Working group II: impacts, adaptation and vulnerability, C3. Megadeltas: their vulnerabilities to climate change. https://archive.ipcc.ch/publications_and_data/ar4/wg2/en/xccsc3.html

Jean-Paul B, Goichot M, Gaillot S (2013) Geography of sand and gravel mining in the lower Mekong River. EcoGeo 26. https://doi.org/10.4000/echogeo.13659

Liu JP, DeMaster DJ, Nguyen TT, Saito Y, Nguyen VL, Ta TKO, Li X (2017) Stratigraphic formation of the Mekong River Delta and its recent shoreline changes. Oceanography 30(3):72–83. https://doi.org/10.5670/oceanog.2017.316

Miliman JD, Syvitsky JPM (1992) Geomorphic/techonic control of sediment discharge to the oceans: the importance of small mountain rivers. J Geol 100(5):525–544

Minderhoud PSJ, Coumou L, Erkens G, Middelkoop H, Stouthamer E (2019) Mekong delta much lower than previously assumed in sea-level rise impact assessments external link. Nat Commun 10:2019

MONRE (2018) Climate change and Sea level rise Scenarios for Viet Nam. Ha Noi, 187 p. https://www.imh.ac.vn/files/doc/2017/CCS%20final.compressed.pdf

MONRE (2021) Vietnam climate change scenarios. Ministry of Natural Resources and Environment, Publishing House of Natural Resources—Environment and Maps

Nguyen VT, Khiem MV, Vu VT, Nguyen DM (2017) Current and projected climate change in the Mekong Delta. Vietnam J Sci Technol Eng 59(2):92–96

Peng S, Huang J, Sheehy JE, Laza RC, Visperas RM, Zhong X, Centeno GS, Khush GS, Cassman KG (2004) Rice yields decline with higher night temperature from global warming. PNAS 101(27):9971–9975

Phan MH, Stive MJF (2022) Managing mangroves and coastal land cover in the Mekong Delta. Ocean and Coastal Management, 219, Article 106013. https://doi.org/10.1016/j.ocecoaman.2021.106013

Prime Minister (2020) Project to prevent riverbank and coastal erosion until 2030 (Decision No. 957/QĐ-TTg)

Prime Minister (2023) Additional capital from the 2023 central budget provision for provinces in the Mekong Delta to implement the project on river and coastal erosion prevention (Decision No. 1162/QĐ-TTg)

Renaud FG, Kuenzer C (2012) The Mekong Delta system: interdisciplinary analyses of a river delta. Springer, 480 p

Ta TKO, Nguyen VL, Masaaki T, Iwao K, Yoshiki S (2005) Holocene delta evolution and depositional models of the Mekong River Delta, Southern Vietnam. In: River deltas—concepts, models, and examples, SEPM Special Pub. No. 83, SBN 1-56576-113-8

Tuan LA, Chinvanno S (2011) Climate change in the Mekong River Delta and key concerns on future climate threats. In: Stewart MA, Coclanis PA (eds) Environmental change and agricultural sustainability in the Mekong Delta. Advances in global change research, vol 45, no 3, pp 207–217

Chapter 7
Emerging Pollutants in Marine Environment: Occurrence and Environmental Concerns

Meng Chuan Ong, Tuan Fauzan Tuan Omar, Roswati Md Amin, Lee Siang Hing, and Adiana Ghazali

Abstract Emerging pollutants is a chemical or substance that is not regulated under existing environmental regulation and law. These pollutants have been identified to give adverse effects on the environment and human itself. The sources of these pollutants may come from various sources, including agricultural practices, industrial processes, urban runoff, and consumer care products. In this review, several major emerging pollutants, namely endocrine disrupting chemicals, microplastics and heavy metals will be focussed to the occurrence and environmental concerns especially in our region waters. Therefore, comprehensive information on the pollutants threat to ecological and human health, as well as their fate and behaviour in the environment is crucial to solve the threats and fill in the gaps faced worldwide.

Keywords Paraben · Pharmaceutically active compounds · Microplastics · Heavy metals

7.1 Introduction

Emerging pollutants are a class of contaminants that are being increasingly detected in the marine environment. It includes heavy metals, pharmaceuticals waste, personal care products, microplastics, and other synthetic chemicals. While these pollutants are not new, they are emerging as a major environmental concern due to their increasing use and potential impacts on marine ecosystems (Häder et al. 2020; Thanigaivel et al. 2023).

M. C. Ong · T. F. T. Omar · R. M. Amin · L. S. Hing · A. Ghazali (✉)
Faculty of Science and Marine Environment, Universiti Malaysia Terengganu, 21030 Kuala Nerus, Terengganu, Malaysia
e-mail: adiana@umt.edu.my

M. C. Ong · T. F. T. Omar · R. M. Amin · A. Ghazali
Ocean Pollution and Ecotoxicology Research Group (OPEC-RG), Faculty of Science and Marine Environment, Universiti Malaysia Terengganu, 21030 Kuala Nerus, Terengganu, Malaysia

V. Maccarrone and M. Fadzil Akhir (eds.), *Coastal Sustainability*, Coastal Research Library 39, https://doi.org/10.1007/978-3-031-75749-5_7

One of the major concerns with emerging pollutants in the marine environment is their potential to harm marine wildlife (Avio et al. 2017). Many of these pollutants have been shown to have negative impacts on the reproductive and developmental systems of marine animals. Another concern with emerging pollutants in the marine environment is their potential to accumulate in the food chain (Baralla et al. 2021). Many of these pollutants are persistent and bioaccumulative, meaning that they can build up in the tissues of marine organisms over time. This can lead to biomagnification, where the concentration of the pollutant increases as it moves up the food chain. This is a particular concern for marine predators, such as dolphins and killer whales, which can accumulate high levels of pollutants in their tissues (Jepson and Law 2016).

The presence of heavy metals in the marine environment is a major concern, as these pollutants can have toxic effects on marine organisms (Ali et al. 2019). Heavy metals, such as lead, mercury, and cadmium, can be introduced into the marine environment through industrial and agricultural activities, as well as from natural sources such as volcanic activity (Wang et al. 2020a; Alengebawy et al. 2021; Yang et al. 2022). Once in the marine environment, these pollutants can accumulate in the tissues of marine organisms, leading to a range of negative impacts, including impaired growth and development, reproductive failure, and even death (Shahjahan et al. 2022). Additionally, heavy metals can enter the human food chain through the consumption of contaminated seafood, posing a risk to human health (Isangedighi and Davis 2019).

Pharmaceutical waste is another emerging pollutant that poses a significant threat to the marine environment (Ojemaye and Petrik 2018). Many pharmaceuticals are designed to be biologically active at low concentrations, and as such, they can have a range of negative impacts on marine organisms (Wang et al. 2021). For example, some pharmaceuticals have been shown to disrupt the endocrine systems of marine animals, leading to impaired development and reproductive function. Additionally, pharmaceutical waste can lead to the emergence of antibiotic-resistant bacteria in the marine environment, which can pose a risk to human health (Anand et al. 2021). Strategies for mitigating the impacts of pharmaceutical waste in the marine environment include better regulation of pharmaceutical manufacturing and disposal practices, as well as the development of more effective wastewater treatment technologies.

Microplastics are another emerging pollutant that is of increasing concern in the marine environment. These small plastic particles can be introduced into the marine environment through a range of sources, including the breakdown of larger plastic items, personal care products, and clothing fibers (Russell and Webster 2021; Atugoda et al. 2021; Azizi et al. 2021). Once in the marine environment, microplastics can be ingested by a range of marine organisms, leading to a range of negative impacts, including impaired growth and development, reproductive failure, and death (Gola et al. 2021). Additionally, microplastics can adsorb and concentrate other pollutants in the marine environment, leading to a range of negative impacts on marine organisms and the environment as a whole (Amelia et al. 2021). Strategies for mitigating the impacts of microplastics in the marine environment include better waste management practices, as well as the development of more effective filtration and treatment technologies for wastewater (Roy et al. 2022).

Overall, the presence of emerging pollutants in the marine environment is a complex issue that requires a multifaceted approach to address. Effective strategies for mitigating the impacts of emerging pollutants in the marine environment include better regulation of industrial and agricultural activities, the development of more effective wastewater treatment technologies, and increased public education and awareness on the impacts of emerging pollutants on the marine ecosystem. Additionally, further research is needed to better understand the long-term impacts of emerging pollutants on marine ecosystems and to develop more effective mitigation strategies.

7.2 Endocrine Disrupting Chemicals (EDCs)

The presence of EDCs in the environment has raised concerns over their potential impact on wildlife and human health. Studies showed that EDCs can interfere with the reproductive and developmental activities of wildlife, leading to reduced fertility (Marlatt et al. 2022), abnormal sexual development, and other health effects. EDCs have also been linked to health problems in humans, including reproductive and developmental disorders, reduce fertility (Marlatt et al. 2022), thyroid dysfunction, obesity, and cancer. EDCs can enter the environment through a variety of pathways. For example, industrial chemicals may be released into the air or water during manufacturing or disposal (Kumar et al. 2020), while pesticides were sprayed on crops or used to treat wood, which eventually will end up in the waterways. Once released into the environment, EDCs can persist for many years and can be transported over long distances by air and water movement (Godfray et al. 2019). They can also accumulate in the tissues of plants, animals, and humans through the food chain. Several common types of EDCs in the environment include parabens and pharmaceutically active compounds (PhACs).

7.2.1 Parabens

Parabens are a group of chemical preservatives commonly used in cosmetics, personal care products, and food products to prevent the growth of bacteria, mold, and yeast (Lincho et al. 2021). The most common types of parabens are methylparaben (MeP), ethyl paraben (EtP), propylparaben (PrP), butylparaben, or isobutyl paraben.

There has been some controversy surrounding the safety of parabens, as they have been showing to mimic estrogen in the body, as well as concern that exposure to parabens may be linked to certain health problems, such as breast cancer (Downs et al. 2023). However, many regulatory agencies, including the US Food and Drug Administration (FDA), have deemed parabens safe for use in cosmetics and food products at the levels typically used. The routes of entry of parabens into the environment can be from several sources, such as wastewater discharge, landfills, agricultural runoff, and atmospheric depositions. Parabens which are commonly used in personal care products, are often washed off in the shower or sink (Haman et al. 2015). These parabens can then enter the wastewater system and ultimately being discharged into rivers, lakes, and oceans.

Parabens might also enter the environment through landfill leachate (Evans et al. 2016). When personal care products containing parabens are disposed of in landfills, the chemicals can leach into the surrounding soil and groundwater. Another route of parabens into the environment is through agricultural runoff (Vale et al. 2022). Some personal care products containing parabens can end up in sewage sludge, which is often used as fertilizer on agricultural fields. The parabens then enter the soil and groundwater. In addition, parabens can also enter the environment through atmospheric deposition. When parabens evaporate from personal care products, they can be released into the air and ultimately deposited on soil and water surfaces.

Once parabens enter the environment, they can persist for long periods of time and can potentially accumulate in the food chain. Through biomagnification in food chains and tap water, parabens can enter human bodies (Vale et al. 2022). Studies have detected parabens in surface water, groundwater, sediments, and even in the tissues of fish and other aquatic organisms. However, the environmental impact of parabens is still not well understood and requires further research.

7.2.2 *Pharmaceutically Active Compounds (PhACs)*

The pharmaceutical industry has grown to become one of the largest and most recognised industries in the world. Pharmaceutically active compounds (PhACs) are a class of compounds which include hormones, antibiotics, and painkillers. All over the world, more than 4000 pharmaceuticals are used globally to treat clinical issues and improve livestock growth (Boxall et al. 2012). According to Alder et al. (2006), the global average per capita pharmaceutical consumption per year is estimated to be around 15 g. Hence, in developed countries, consumption might range from 50 to 150 g. It is also predicted that the global use of active chemicals totals 100,000 tonnes or more each year (Kümmerer 2004). Nonsteroidal anti-inflammatory medicines (NSAIDs), antibiotics, antidepressants, steroid hormones, anticonvulsants, lipid regulators, beta-blockers, and antineoplastics are among the therapeutic classes of these biologically active chemicals (Wang and Wang 2016). Though, pharmaceutical consumption is constantly growing over the world, owing primarily to population growth and ageing.

As the development increase, an increasing number of pharmaceuticals and their residues are being discharged and directly enters the water environment or municipal wastewater treatment plants (WWTPs). Due to the limitations in objective criteria such as technologies and budget, there are only a few types of wastewater treatment plants (WWTPs) in the world whereby PhACs can be eliminated, and some of them may not be effectively treated before being released into the aquatic environment (Peng et al. 2020). The pollutant PhACs can enter the marine environment by several pathways, including river input, atmospheric settling, agriculture and fisheries, coastal pollution discharge, and groundwater infiltration, among others, and has become a prominent kind of new marine organic pollutant. Sediment, soil, or sludge is the last stop for the PhACs to be collected. Pharmaceutical transportation in soil, and thus also their ability to pollute water sources, is controlled by parameters such as drug settling, the severity of "rain" occurrences, and type of soil, and is also connected to the adsorption trends (Białk-Bielińska et al. 2016).

Generally, pharmaceutically active compounds introduced into the environment and entire environment through a variety of point source pollution, including human waste and sewage disposal, surface runoff into water bodies, excess from WWTPs, and veterinary drugs used in agriculture and aquaculture or conveyed through livestock manure and sludge application. The untreated wastewater goes into the ocean, lake, river pond, etc., wastewater treatment plants (WWTPs). Pharmaceutical residues have also been found in soils, sediments, and sludges, and they are known to be persistent. Some compounds are totally biodegraded, whilst others strongly sorb and are persistent and physiologically active. Hence, this situation might be threatening as it can enter the human body through the food chain.

7.3 Microplastics

7.3.1 Distribution of Microplastic in Seawater

Pollution by unmanaged waste plastic is growing concern in aquatic environments. Whether intended or unintended, this contributes to the increasing plastic debris escape to the ocean called microplastic. Microplastics that are extremely small pieces of plastic debris (< 5 mm), are found in most commonly utilized products (primary microplastics) used in cosmetics and microfibers released via run-off from land or may originate from the fragmentation of larger plastic debris (secondary microplastics) via UV photo-degradation, wave action and physical abrasion (Boucher and Friot 2017). Lusher (Lusher et al. 2015) has documented the concentration of microplastics accumulation which almost reach to 100,000 particles per m^3. There were 4.8–12.7 million tons of microplastic in worldwide ocean environment and this number will keep multiply by the years (Jambeck et al. 2015). Out of this, 20% is originated from fishing debris and shipping container and 80% come from land base sources.

River recognized as one of the major pathways for microplastics into oceans (Lebreton et al. 2017). It is reported that Asia's river systems account for approximately 67% of the annual global total of plastics released into the oceans. Urban coastal areas have been found to be positively correlated with marine microplastic concentrations (Jang et al. 2020; Sugiura et al. 2021), However, this distribution is highly variable and region-specific which requires detail analysis to determine the fate and transport processes of microplastics, thus providing accurate holistic picture of microplastics in the ocean environment (Wang et al. 2020b).

7.3.2 Impact Towards Organisms

Due to the high concentration of microplastics at the ocean surface, organisms of higher trophic levels may inadvertently ingest microplastics directly or indirectly through trophic transfer from contaminated prey (Nelms et al. 2018). Hundreds of thousands of marine animals have been reported accidentally ingesting or becoming entangled in plastic, including zooplankton (Sarijan et al. 2018; Jaapar et al. 2021; Karbalaei et al. 2019).

Experimental studies have documented the potential of microplastics accumulation in the gut of zooplankton, which may directly interfere the digestive process and nutrient uptake, ultimately affecting their growth and reproduction (Jaapar et al. 2021; Cole et al. 2015). Size, shape and abundance are the factors that influence the uptake of microplastics (Wright et al. 2013). However, in nature, it is difficult to achieve high exposure concentrations of microplastics compared with that in the laboratory. Organisms may ingest the microplastics, but there will be no immediate toxic effects associated to it (Kim et al. 2019). In the field, microplastic ingestion has been recorded in fish, bivalves, invertebrates and zooplankton (Taha et al. 2021). The ingestion rate varies with organisms from as low as 0.003 particles per individual of zooplankton (Md Amin et al. 2020) in the southern South China Sea to up to 2.5 particles ingested per individual of benthic and pelagic fish in the South Pacific (Clere et al. 2022).

A potential of microplastics transfer to higher trophic level have been previously demonstrated both under laboratory condition and in the field (Lusher et al. 2015; Nelms et al. 2018; Wright et al. 2013; Setälä et al. 2014). Microscopy observations of mysid, *Neomysis integer* intestine showed the presence of zooplankton, *Eurytemora affinis* and microspheres after 3 h incubation. In the field, approximately half of wild grey seals (*Halichoerus grypus*) subsamples and a third of fish contained 1–4 microplastics, suggesting trophic transfer represents an indirect, yet potentially major pathway of microplastic ingestion for any species whose feeding ecology involves the consumption of whole prey, including humans.

7.4 Heavy Metals

Heavy metals in the environment refer to the presence of metallic elements in soil, water, sediment, air, and living organisms. They are considered as emerging pollutants due to their increasing occurrence and potential impact on human health and the environment. Naturally, their existence in the environment is originated from physical and chemical weathering such as root wedging, frost action, acidification, dissolution, and others. Aside from that, these elements are derived by a variety of anthropogenic sources, such as discharged of domestic wastewater, industrial processes, mining activities and agricultural practices. Such activities have the potential to be hazardous, which could lead to heavy metals pollution and have long-lasting effects on the environment and human health.

7.4.1 Heavy Metals in Sediment

Sediment is a contaminant reservoir and ecologically significant component of the aquatic ecosystem. It plays a key role in preserving the trophic status of any water body. Generally, sediment bed is commonly used to record the history of water pollution, in addition to having a significant impact on how polluted the water column becomes. In marine environment, sediment is known as main sink. It serves as a carrier of heavy metals as well as a possible non-point source, which might significantly impact the overlying waters (Chiaia-Hernández et al. 2022).

Heavy metals can occur in the sediment bed naturally, or as a result of human activities, such as urban runoff, industrial processes and mining activities. Since they are unable to decompose under natural conditions, they are persistent contaminants (Niu et al. 2009). Once deposited into sediment, heavy metals can strongly bind to the sediment particles and persist in the environment for long periods of time. The adsorbed heavy metals can be desorbed from sediment particles under certain environmental conditions, such as changes in pH, temperature, or salinity. Heavy metal's desorption can lead to the release of heavy metals into the surrounding environment, which can cause secondary pollution. Their occurrence in sediment can be an environmental concern as they are potentially causing harm to aquatic life and human health if they are released into the water column or air.

The degree of metals pollution in sediment bed can vary due to a range of factors, including the type and concentration of metals, the source and timing of the pollution. Additionally, variations in sediment's physical and chemical properties, as well as human activities and land use practices can contribute to changes in the degree of metals pollution over time. Furthermore, variation in the environmental conditions of the surrounding ecosystem which include changes in weather patterns may also affect the level of heavy metals pollution in an ecosystem.

Sediment pollution caused by heavy metals can be measured in terms of its severity. There are several methods are commonly practices in the worldwide including chemical analysis of the sediment, assessment of toxicology levels, and comparison to established environmental standards or guidelines. The sediment quality guidelines (SQGs), geoaccumulation index (Igeo) and pollution load index (PLI) are among of the common practices which have been used in the worldwide to define the degree of heavy metals pollution in sediment.

SQG is a key element of the framework for assessing sediment quality. It is typically being applied in the first tier of evaluation to foresee the potential threats of heavy metals contaminant in sediment bed. SQG values are essential for sediment quality assessment frameworks and are designed to enhance the safeguards of benthic ecosystems, fisheries and water quality of the ecosystem (Zhang et al. 2021). Even though SQG have been successfully assessed the sediment pollution in several area around the world, however, it is drawback by the contaminant's bioavailability in the sediment which may vary significantly in response to differences in sediment's physico-chemical properties. Besides SQGs, Igeo and PLI are commonly used to classify and map the ecosystem in accordance with the degree of pollution upon interpreting the concentration of heavy metals in sediment (Rezaei et al. 2019). The pollution level of heavy metals in sediment assessed by both methods is evaluated by comparing with the background value of an ecosystem. Such indices are useful to identify areas that are particularly vulnerable to heavy metals pollution and can help in developing targeted pollution control and remediation strategies.

7.4.2 Heavy Metals in Aquatic Organisms

Aquatic organisms are organisms that live in water, such as oceans, rivers, lakes, ponds, and other water bodies. They include a wide variety of species, ranging from microscopic plankton to massive whales. These organisms play important roles in the food web and ecosystem of aquatic environments, as well as important to human societies for food, recreation, and other purposes.

The presence of heavy metals an aquatic ecosystem can have a range of negative effects. This include toxicity to aquatic organisms, reduced the aquatic biodiversity, and negative impacts on human health when contaminated seafood is consumed. Heavy metals can accumulate in living organisms, such as animals and plants. They can be accumulated through various pathways such as absorption from contaminated sediment or water, ingestion of contaminated food, or uptake through respiration. Aquatic organisms are particularly vulnerable to the bioaccumulation of heavy metals as they are directly expose to the aquatic environment, which is often the site of heavy metals pollution.

Some common heavy metals that can bioaccumulate in aquatic organisms include mercury (Hg), lead (Pb) and cadmium (Cd). Mercury is a highly toxic metal that is commonly accumulate in fish. High concentration of Hg enters the aquatic environment from a variety of sources, including industrial processes, breakdown of Hg-containing products and others. Once in the water, Hg can be transformed into methyl-Hg (Me-Hg), a highly toxic form of Hg that is easily absorbed by organisms. Methyl-Hg can bioaccumulate in aquatic organisms and when consumed by human can cause a range of health effects, including neurological damage (Rodrigues et al. 2019).

Pb can enter the aquatic environment through sources such as mining, leaded gasoline and industrial processes. It can be toxic to aquatic organisms, and has been linked to decreased growth, reproduction, and survival in fish (Park et al. 2020). Other than that, Cd can enter the aquatic environment through mining, smelting and fertilizer use. It is toxic to aquatic organisms and can damage the gills and kidneys of fish (Zhang and Reynolds 2019).

In an aquatic ecosystem, heavy metals are taken up by aquatic plants and small organisms, such as plankton and algae at low concentration. As small organisms are consumed by larger organisms, the concentration of accumulated heavy metals in the larger organisms will increase. As a result, top predators can accumulate very high concentration of heavy metals in their bodies. This process continues up the food chain and known as biomagnification process.

The biomagnification of heavy metals in aquatic organisms can have serious consequences for organisms and ecosystem as a whole. High concentration of heavy metals can severely affect organisms, and even can cause death in affected organisms. Furthermore, as human consumed the top aquatic predators for their protein source, this high concentration of heavy metals can pose a risk to human health as well. Exposure to toxic heavy metals can cause a range of adverse health effects, such as neurological and developmental problems, cancer, and organ damage.

7.5 Conclusion

In the contemporary era, our oceans and marine ecosystems face unprecedented challenges due to the relentless release of pollutants into these delicate environments. Through comprehensive exploration and analysis, this book has illuminated the intricate web of interactions between emerging pollutants, the marine environment, and the human populace. The pollutants we release into the marine environment have a way of coming back to haunt us, as they find their way into our food chain and drinking water sources. Thus, the well-being of the oceans and the well-being of humanity are inexorably intertwined. Our actions today will shape the quality of life for future generations. The role of policy and regulation in addressing emerging pollutants cannot be overstated. Governments, international organizations, and the scientific community must work together to develop and implement stringent measures to monitor, control, and ultimately reduce the presence of these contaminants in our oceans. Public awareness and education also play a pivotal role in galvanizing support for change.

Acknowledgements The editors and authors of this book wish to thank the research grant co-founded by the Erasmus+ Programme of the European Union and MARE, Marine Coastal and Delta Sustainability for Southeast Asia to provide funding to complete this book. Authors also wish to acknowledge the reviewers for their time and effort and constructive recommendations that enhanced the value of this book.

References

Alder L, Greulich K, Kempe G, Vieth B (2006) Residue analysis of 500 high priority pesticides: better by GC–MS or LC–MS/MS? Mass Spectrom Rev 25(6):838–865

Alengebawy A, Abdelkhalek ST, Qureshi SR, Wang MQ (2021) Heavy metals and pesticides toxicity in agricultural soil and plants: ecological risks and human health implications. Toxics 9(3):42

Ali H, Khan E, Ilahi K (2019) Environmental chemistry and ecotoxicology of hazardous heavy metals: environmental persistence, toxicity, and bioaccumulation. J Chem 2019:6730305

Amelia TSM, Khalik WMAWM, Ong MC, Shao YT, Pan HJ, Bhubalan K (2021) Marine microplastics as vectors of major ocean pollutants and its hazards to the marine ecosystem and humans. Progr Earth Planet Sci 8:12

Anand U, Reddy B, Sing VK, Singh AK, Kesari KK, Tripathi P, Kumar P, Tripathi V, Simal-Gandara J (2021) Potential environmental and human health risks caused by antibiotic-resistant bacteria (ARB), antibiotic resistance genes (ARGs) and emerging contaminants (ECs) from municipal solid waste (MSW) landfill. Antibiotics (Basel) 10(4):374

Atugoda T, Vithanage M, Wijesekara H, Bolan N, Sarmah AK, Bank MS, You S, Yong SO (2021) Interactions between microplastics, pharmaceuticals and personal care products: implications for vector transport. Environ Int 149:106367

Avio CG, Gorbi S, Regoli F (2017) Plastics and microplastics in the oceans: from emerging pollutants to emerged threat. Mar Environ Res 128:2–11

Azizi N, Khoshnamvand N, Nasseri S (2021) The quantity and quality assessment of microplastics in the freshwater fishes: a systematic review and meta-analysis. Regional Stud Mar Sci 47:101955

Baralla E, Demontis MP, Dessi F, Varoni MV (2021) Another concern with emerging pollutants in the marine environment is their potential to accumulate in the food chain. Animals 11(11):3239

Białk-Bielińska A, Kumirska J, Borecka M, Caban M, Paszkiewicz M, Pazdro K, Stepnowski P (2016) Selected analytical challenges in the determination of pharmaceuticals in drinking/marine waters and soil/sediment samples. J Pharm Biomed Anal 121:271–296

Boucher J, Friot D (2017) Primary microplastics in the oceans: a global evaluation of sources (eds: Lundin CG, de Sousa JM). IUCN, Gland, Switzerland. https://portals.iucn.org/library/sites/library/files/documents/2017-002-En.pdf

Boxall AB, Rudd MA, Brooks BW, Caldwell DJ, Choi K, Hickmann S et al (2012) Pharmaceuticals and personal care products in the environment: what are the big questions? Environ Health Perspect 120(9):1221–1229

Chiaia-Hernández AC, Casado-Martinez C, Lara-Martin P, Bucheli TD (2022) Sediments: sink, archive, and source of contaminants. Environ Sci Pollut Res 29:85761–85765

Clere IK, Ahmmed F, Remoto PLJG, Fraser-Miller SJ, Gordon KC, Komyakova V, Allan BJM (2022) Quantification and characterization of microplastics in commercial fish from Southern New Zealand. Mar Pollut Bull 184:114121

Cole M, Lindeque P, Fileman E, Halsband C, Galloway TS (2015) The impact of polystyrene microplastics on feeding, function and fecundity in the marine copepod Calanus helgolandicus. Environ Sci Technol 49(2):1130–1137

Downs CA, Amin MM, Tabatabaeian M, Chavoshani A, Amjadi E, Afshari A, Kelishadi R (2023) Parabens preferentially accumulate in metastatic breast tumors compared to benign breast tumors and the association of breast cancer risk factors with paraben accumulation. Environ Adv 11:100325

Evans WA, Davies PJ, McRae C (2016) The occurrence of methyl, ethyl, propyl, and butyl parabens in the urban rivers and stormwaters of Sydney, Australia. Environ Sci: Water Res Technol 2(4):733–742

Godfray HCJ, Stephens AEA, Jepson PD, Jobling S, Johnson AC, Matthiessen P, Sumpter JP, Tyler CR, McLean AR (2019) A restatement of the natural science evidence base on the effects of endocrine disrupting chemicals on wildlife. Proc Biol Sci 286(1897):20182416. https://doi.org/10.1098/rspb.2018.2416

Gola D, Tyagi PK, Arya A, Chauhan N, Agarwal M, Singh SK, Gola S (2021) The impact of microplastics on marine environment: a review. Environ Nanotechnol Monit Manage 16:100552
Häder DP, Banaszak AT, Villafañe VE, Narvarte MA, González RA, Helbling EW (2020) Anthropogenic pollution of aquatic ecosystems: emerging problems with global implications. Sci Total Environ 713:136586
Haman C, Dauchy X, Rosin C, Munoz J-F (2015) Occurrence, fate and behavior of parabens in aquatic environments: a review. Water Res 68:1–11
Isangedighi IA, Davis GS (2019) Heavy metals contamination in fish: effects on human health. J Aquatic Sci Mar Biol 2(4):7–12
Jaapar AN, Md Amin R, Bhubalan K, Sohaimi ES (2021) Changes in the development and reproductive output of *Nitokra lacustris pacifica* (Crustacea: Copepoda) Yeatman, 1983 under short- and long-term exposure to synthetic and biodegradable microbeads. J Polym Environ 29:4060–4072
Jambeck RJ, Geyer R, Wolcox C, Siegel TR, Perryman M, Andrady A, Narayan R, Law KL (2015) Plastic waste inputs from land into the ocean. Science 347(6223):768–771
Jang M, Shim WJ, Cho Y, Han GM, Song YK, Hong SH (2020) A close relationship between microplastic contamination and coastal area use pattern. Water Res 15(171):115400
Jepson PD, Law RJ (2016) Marine environment. Persistent pollutants, persistent threats. Science 352(352):1388–1389
Karbalaei S, Golieskardi A, Hamzah HB, Abdulwahid S, Hanachi P, Walker TR, Karami A (2019) Abundance and characteristics of microplastics in commercial marine fish from Malaysia. Mar Pollut Bull 148:5–15
Kim SW, Chae Y, Kim D, An Y-J (2019) Zebrafish can recognize microplastics as inedible materials: quantitative evidence of ingestion behavior. Sci Total Environ 649:156–162
Kumar M, Sarma DK, Shubham S, Kumawat M, Verma V, Prakash A, Tiwari R (2020) Environmental endocrine-disrupting chemical exposure: role in non-communicable diseases. Front Public Health 8:553850
Kümmerer K (2004) Resistance in the environment. J Antimicrobial Chemotheraphy 54(2):311–320
Lebreton LCM, van der Zwet J, Damsteeg JW, Slat B, Andrady A, Reisser J (2017) River plastic emissions due to the world's oceans. Nat Commun 8:15611
Lincho J, Martins RC, Gomes J (2021) Paraben compounds—part I: an overview of their characteristics, detection, and impacts. Appl Sci 11(5):2307
Lusher A (2015) Microplastics in the marine environment: distribution, interactions and effects. In: Bergmann M, Gutow L, Klages M (eds) Marine anthropogenic litter. Springer, Cham, pp 245–307
Marlatt VL, Bayen S, Castaneda-Cortès D, Delbès G, Grigorova P, Langlois VS, Martyniuk CJ, Metcalfe CD, Parent L, Rwigemera A, Thomson P, Van Der Kraak G (2022) Impacts of endocrine disrupting chemicals on reproduction in wildlife and humans. Environ Res 208:112584
Md Amin R, Sohaimi ES, Anuar ST, Bachok Z (2020) Microplastic ingestion by zooplankton in Terengganu coastal waters, Southern South China Sea. Mar Pollut Bull 150:110616
Nelms SE, Galloway TS, Godley BJ, Jarvis DS, Lindeque PK (2018) Investigating microplastic trophic transfer in marine top predators. Environ Pollut 238:999–1007
Niu H, Deng W, Qunhe W, Chen X (2009) Potential toxic risk of heavy metals from sediment of the Pearl River in South China. J Environ Sci 21:1053–1058
Ojemaye CY, Petrik L (2018) Pharmaceuticals in the marine environment: a review. Environ Rev 27(2). https://doi.org/10.1139/er-2018-005
Park K, Han EJ, Ahn G, Kwak IS (2020) Effects of thermal stress-induced lead (Pb) toxicity on apoptotic cell death, inflammatory response, oxidative defense, and DNA methylation in zebrafish (*Danio rerio*) embryos. Aquat Toxicol 224:105479
Peng Q, Song J, Li X, Yuan H, Liu M, Duan L, Zuo J (2020) Pharmaceutically active compounds (PhACs) in surface sediments of the Jiaozhou Bay, North China. Environ Pollut 266(Pt 3):115245
Rezaei A, Hassani H, Mousavi SBF, Hassani S, Jabbari N (2019) Assessment of heavy metals contamination in surface soils in Meiduk copper mine area, SE Iran. J Earth Sci Malaysia 3(2):23–31

Rodrigues PdeA, Ferrari RG, dos Santos LN, Conte Junior CA (2019) Mercury in aquatic fauna contamination: a systematic review on its dynamics and potential health risks. J Environ Sci 84:205–218

Roy P, Mohanty AK, Misra M (2022) Microplastics in ecosystems: their implications and mitigation pathways. Environ Sci Adv 1:9–29

Russell M, Webster L (2021) Microplastics in sea surface waters around Scotland. Mar Pollut Bull 166:112210

Sarijan S, Azman S, Said MIM, Andu Y, Zon NF (2018) Microplastics in sediment from Skudai and Tebrau river, Malaysia: a preliminary study. MATEC Web Conf. EDP Sci 250:06012

Setälä O, Fleming-Lehtinen V, Lehtiniemi M (2014) Ingestion and transfer of microplastics in the planktonic food web. Environ Pollut 185:77–83

Shahjahan M, Taslima K, Rahman MS, Al-Emran M, Alam SI, Faggio C (2022) Effects of heavy metals on fish physiology—a review. Chemosphere 300:134519

Sugiura M, Takada H, Takada N, Mizukawa K, Tsuyuki S (2021) Microplastics in urban wastewater and estuarine water: importance of street runoff. Environ Monit Contam Res 1:54–65

Taha ZD, Amin RM, Anuar ST, Nasser AAA, Sohaimi ES (2021) Microplastics in seawater and zooplankton: a case study from Terengganu estuary and offshore waters, Malaysia. Sci Total Environ 786:147466

Thanigaivel S, Vickram S, Dey N, Jeyanthi P, Subaiya R, Kim W, Govarthanan M, Karmegam N (2023) Ecological disturbances and abundance of anthropogenic pollutants in the aquatic ecosystem: critical review of impact assessment on the aquatic animals. Chemosphere 313:137475

Vale F, Sousa CA, Sousa H, Santos L, Simões M (2022) Parabens as emerging contaminants: environmental persistence, current practices and treatment processes. J Clean Prod 347:131244

Wang J, Wang S (2016) Removal of pharmaceuticals and personal care products (PPCPs) from wastewater: a review. J Environ Manage 182:620–640

Wang Y, Duan X, Wang L (2020a) Spatial distribution and source analysis of heavy metals in soils influenced by industrial enterprise distribution: case study in Jiangsu Province. Sci Total Environ 710:134953

Wang T, Hu M, Song L, Yu J, Liu R, Wang S, Wang Z, Sokolova IM, Huang W, Wang Y (2020b) Coastal zone use influences the spatial distribution of microplastics in Hangzhou Bay, China. Environ Pollut 2660:115137

Wang H, Xi H, Xu L, Jin M, Zhao W, Liu H (2021) Ecotoxicological effects, environmental fate and risks of pharmaceutical and personal care products in the water environment: a review. Sci Total Environ 788:147819

Wright SL, Thompson RC, Galloway TS (2013) The physical impacts of microplastics on marine organisms: a review. Environ Pollut 178:483–492

Yang J, Sun Y, Wang Z, Gong J, Gao J, Tang S, Ma S, Duan Z (2022) Heavy metal pollution in agricultural soils of a typical volcanic area: risk assessment and source appointment. Chemosphere 304:135340

Zhang H, Reynolds M (2019) Cadmium exposure in living organisms: a short review. Sci Total Environ 678:761–767

Zhang Y, Li H, Yin J, Zhu L (2021) Risk assessment for sediment associated heavy metals using sediment quality guidelines modified by sediment properties. Environ Pollut 2021(275):115844

Chapter 8
Measurement Tools for Determination of Coastal Vulnerability in Malaysia

Vasukey Palany Kumar, Hee-Min Teh, Daniele La Rosa, and Pietro Scandura

Abstract Coastal Vulnerability Index (CVI) is a one of the commonly used tools to evaluate the level of exposure or vulnerability in a certain coastal area, as well as the sustainability of the ecosystem diversity. The scoring of the CVI reflects the level of risks that exist in a coastal area. Very often, CVI of a coast serves as a reference or basis to the relevant authorities in decision-making and to promote a holistic approach to coastal management. Originally developed by Gornitz, the CVI considers multiple factors that cause physical vulnerability to a coast. Subsequently CVI is further improvised to address other pertinent factors that are deemed to be significant. In Malaysia, National Coastal Zone Physical Plan adopted a CVI that comprehensively addresses physical, socio-economic, and biodiversity characteristics of a coast, in evaluating the vulnerability of all the coasts in the country. The outcomes of this assessment can be presented in the form of a CVI global map for the ease of interpretation. This chapter outlines the Malaysian's CVI measurement tools, and the assessment rubrics.

Keywords Climate change · Coastal Vulnerability Index (CVI) · Physical · Socio-economic · Biodiversity · Mapping · Malaysia

V. P. Kumar (✉)
Department of Irrigation and Drainage Malaysia, Kuala Lumpur, Malaysia
e-mail: vasukey@water.gov.my

H.-M. Teh
Civil and Environmental Engineering Department, Universiti Teknologi PETRONAS, Perak, Malaysia
e-mail: heemin.teh@utp.edu.my

D. La Rosa · P. Scandura
Department Civil Engineering and Architecture, University of Catania, Catania, CT, Italy
e-mail: dlarosa@darc.unict.it

P. Scandura
e-mail: pscandu@dica.unict.it

V. Maccarrone and M. Fadzil Akhir (eds.), *Coastal Sustainability*, Coastal Research Library 39, https://doi.org/10.1007/978-3-031-75749-5_8

8.1 Introduction

8.1.1 Perspective and Challenges on Coastal Vulnerability

The coastal zone, where land and sea converge, is a unique and dynamic region demanding special attention due to its distinctive environmental conditions and diverse array of natural resources. The continuous modifications in the shape and position of the shoreline, influenced by dynamic environmental factors and human interventions, contribute to the ongoing evolution of coastal landscapes (Amri Mohd et al. 2018). Coastal vulnerability assessment, crucial for understanding the impact of climate change on coastal communities, involves a wide range of natural and anthropogenic processes. Innovative index systems, tailored to evaluate specific hazards or combining multiple factors, serve as essential tools in grasping the true nature of vulnerability and guiding effective mitigation efforts (Oloyede et al. 2022).

The escalating threat of global sea level rise raises significant concerns for coastal vulnerability, projecting substantial increases by mid-century alongside rapid urbanization and intensified human activities (Zhang et al. 2021). This amalgamation heightens the vulnerability of coastal regions, placing communities and properties at greater risk from natural hazards. Vulnerability, denoting a system's susceptibility to hazard impacts and ensuing social changes, remains subject to ongoing debate due to intricate relationships among its components. Research communities respond to this urgency by conducting vulnerability assessments, producing valuable outcomes like vulnerability maps as tools for stakeholders (Bukvic et al. 2020). Southeast Asia, grappling with coastal vulnerability, employs diverse approaches, from mathematical formulas to weighted variables, influencing decision-making processes (Noor and Abdul Maulud 2022). Navigating these complexities is crucial for a comprehensive understanding that guides effective decision-making, enhancing coastal resilience against impending challenges.

Sea level rise stands as a pivotal factor in assessing coastal vulnerability, with profound implications for the physical, biodiversity, and socioeconomic aspects of coastal regions. Projections depict a global mean sea level increase from 0.09 to 0.88 m between 1990 and 2100, posing significant challenges worldwide (Bukvic et al. 2020). The consequences of rising sea levels are diverse and impactful. Coastal erosion accelerates as waves penetrate farther inland, leading to the loss of valuable land and infrastructure, particularly affecting communities reliant on tourism, fisheries, and other coastal industries. This erosion can trigger population displacement, property loss, and necessitate costly coastal defense measures. Sea level rise also heightens the risk of storm flooding, intensifying the impact of storm surges during tropical cyclones. Vulnerable coastal communities face increased flooding, causing property damage, infrastructure disruption, and potential loss of life. Low-lying areas are particularly susceptible, enduring socioeconomic challenges post-flooding. In summary, sea level rise is a complex phenomenon with far-reaching consequences, necessitating comprehensive strategies to mitigate impacts and enhance the resilience of coastal communities and ecosystems. Understanding these interdependencies is crucial for developing effective responses (Bukvic et al. 2020).

To attain a comprehensive understanding of vulnerability in coastal areas, an integrated approach considering physical, biodiversity, and socio-economic aspects is crucial for assessing how hazards impact communities (Oloyede et al. 2022). Effective coastal zone management involves identifying areas susceptible to sea-level rise, and the widely used Coastal Vulnerability Index (CVI) by Gornitz (1991) plays a pivotal role. The CVI considers factors such as geomorphology, shoreline changes, coastal slope, sea-level rise rate, wave height, and tidal range. As coastal vulnerability assessments evolve, advancements include modifications to CVI parameters, resulting in composite and multiple-scale vulnerability indices. The multiple-scale CVI approach incorporates sub-indices based on specific parameter features, combining socioeconomic vulnerability index (SVI) and physical vulnerability index (PVI) to obtain an overall CVI (Noor and Abdul Maulud 2022). Changes in physical, socioeconomic, and biodiversity parameters as Fig. 8.1 offering a visual insight into the dynamic nature of coastal vulnerability assessments. These evolving indices and visualizations aid in categorizing coastlines, formulating appropriate management strategies, and ensuring a holistic approach to coastal vulnerability analysis. Effective coastal zone management necessitates the integration of hydrology and coastal engineering to address natural phenomena like tides, storm surges, rising sea levels, and tsunamis, which pose risks such as erosion and flooding. Understanding factors like tidal gradients, geology, and geomorphology is essential for assessing the fragility of these areas and their adaptability to environmental changes. Human activities, including coastal development, land reclamation, and sand mining, can also exacerbate these hazards, heightening the vulnerability of coastal regions.

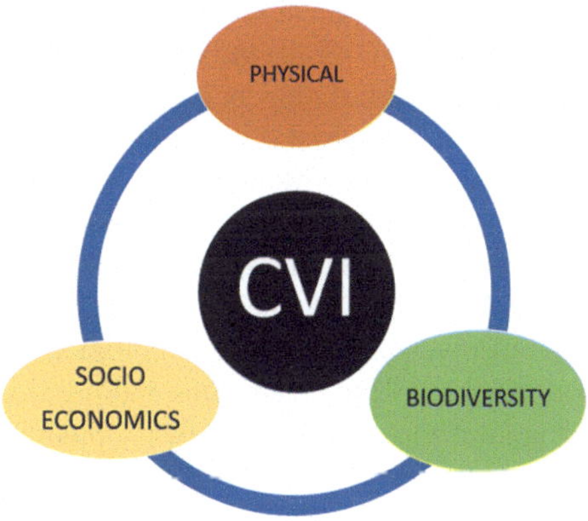

Fig. 8.1 Coastal vulnerability based on physical, socioeconomics, and biodiversity parameters

8.2 Coastal Vulnerability in Malaysia

Sea-level rise (SLR) induced by climate change has emerged as a critical concern, posing substantial threats to coastal regions and human activities. The issue is particularly acute in densely populated coastal areas, such as those in Peninsular Malaysia. Despite the urgency of the matter, comprehensive studies on the Coastal Vulnerability Index (CVI) in the east coast of Peninsular Malaysia are lacking, as emphasized in existing literature (Amri Mohd et al. 2018).

The global sea level has exhibited a notable increase, as highlighted in the 2019 Intergovernmental Panel on Climate Change (IPCC) report, which reported a mean global sea level rise of approximately 15 cm (6 in.) in the twentieth century and a current rate of about 3.6 mm per year (0.14 in. per year) since the early 1990s. It is imperative to acknowledge regional disparities in sea level rise across Malaysia, as projections indicate a range of 0.67–0.71 m by 2100 (RCP8.5), in contrast to the global projection of 0.52–0.98 m (Nahrim, 2017). Consequently, certain areas may experience sea level rise rates that deviate from the global average, emphasizing the necessity for tailored adaptation strategies in addressing this environmental challenge.

In the Malaysian context, the National Coastal Zone Physical Plan (NPP-CZ), 2021 plays a pivotal role in providing a comprehensive planning framework for coastal zones. Despite its present limitation, primarily focusing on development activities along the coast, the NPP-CZ aims to encompass diverse values and interactions within dynamic environments. For consistency and alignment, it is essential that the goals and strategies outlined in the physical plan, both at the state and national levels, are substantiated by justifications for conservation or development in coastal zones. This approach ensures a holistic understanding of policy formulation and development requirements, contributing to a comprehensive and sustainable approach. Overall, the NPP-CZ serves as a vital tool for sustainable coastal development, promoting an integrated approach that balances development with the preservation of coastal ecosystems.

Moreover, within Malaysia's coastal zone, million people, constituting 23.9% of the total population, reside. The population is expected to grow alongside economic and urban activities, necessitating careful management to ensure coastal resources can sustain increasing demands for food and shelter. Malaysia's coastal areas are hubs of diverse and intricate ecological networks, featuring dynamic ecosystems and services. Varied landforms and profiles contribute to the biological richness of the coastal zone, exemplified by the muddy terrains and mangrove swamp forests on the west coast of Peninsular Malaysia, serving as vital habitats and nurseries.

In pursuit of sustainable development, the control of reclamation activities is pivotal. The revitalization of abandoned or outdated development sites can significantly reduce the necessity for extensive reclamation, thereby preserving the ecological balance of these areas. An essential aspect of this endeavour involves understanding coastal profiles, their sources, and potential risks, allowing for well-planned interventions. It is imperative to continuously review, update, and accurately represent data, underscoring the importance of adopting comprehensive analysis tools such as the Coastal Vulnerability Index (CVI). The CVI plays a crucial role in assessing the strength, resistance, and risks within coastal realms, thereby facilitating informed decision-making processes.

Moreover, the Coastal Vulnerability Index (CVI) serves two primary functions. Firstly, it establishes a systematic framework for evaluating disaster risk in coastal areas, homing in on pertinent aspects of vulnerability. This involves a scoring system that classifies different coastal regions based on risk levels, supported by an interactive mapping and data analysis platform. Secondly, the CVI provides invaluable assistance to policymakers and project implementers in planning and executing development plans by offering profound insights into vulnerabilities and risks. Comprising three components—the Physical Coastal Vulnerability Index (CVIP), Biodiversity Coastal Vulnerability Index (CVIB), and Socio-Economic Coastal Vulnerability Index (CVIS)—the CVI evaluates physical vulnerability, focuses on biodiversity preservation, and addresses socio-economic aspects.

Furthermore, the formula expressions for computing the Coastal Vulnerability Index (CVI) in Eqs. 8.1, 8.2, 8.3, and 8.4 play a pivotal role in quantifying and assessing vulnerability to physical, biodiversity, and socioeconomic factors. These formulae serve as a critical instrument, empowering researchers, policymakers, and coastal managers to methodically gauge and compare vulnerability levels among diverse coastal regions. By inputting specific data and variables into these equations, a numerical CVI score can be derived, acting as a quantitative indicator of a coastal area's susceptibility to potential risks and hazards. This comprehensive understanding is imperative for informed decision-making and the development of strategies aimed at enhancing coastal resilience and adaptation.

Physical Coastal Vulnerability Index,

$$\mathrm{PCVI} = \sqrt{\left(\frac{p1 \times p2 \times p3 \times p4 \ldots \times pn}{n}\right)} \tag{8.1}$$

where the pn stands for physical parameter value and the n represents the number of parameters

Biodiversity Coastal Vulnerability Index,

$$\mathrm{BCVI} = \sqrt{\left(\frac{b1 \times b2 \times b3 \times b4 \ldots \times bn}{n}\right)} \tag{8.2}$$

where the bn stands for biodiversity parameter value and the n represents the number of parameters

Socioeconomic Coastal Vulnerability Index,

$$\text{SCVI} = \sqrt{\left(\frac{s1 \times s2 \times s3 \times s4 \ldots \times sn}{n}\right)} \tag{8.3}$$

where the *sn* stands for socioeconomic parameter value and the *n* represents the number of parameters

National Coastal Vulnerability Index,

$$\text{NCVI} = \frac{\left(\frac{1}{2} * \sqrt{((PCVI \times BCVI))}\right) + \left(\frac{1}{2} * \sqrt{((PCVI \times BCVI))}\right)}{2} \tag{8.4}$$

The evaluation of physical vulnerability in coastal areas encompasses critical factors such as geological characteristics, erosion patterns, flood susceptibility, and the preservation of sea swamp regions. A thorough examination reveals a significant finding: approximately 41% of the entire coastline of Peninsular of Malaysia falls within the high and very high vulnerability categories. This underscores the utmost importance of comprehending and addressing coastal vulnerability to effectively manage and mitigate risks in these regions.

Moving on to the Biodiversity Coastal Vulnerability Index (CVIB), it utilizes parameters to assess the vulnerability of marine biodiversity areas. These parameters include the existence and location of marine biodiversity areas, the official gazette status of these areas, the evaluation of conservation programs, and the assessment of endangered areas within the marine environment. The CVIB analysis aims to comprehensively understand the presence, conservation status, and potential threats faced by marine biodiversity in the NPP-CZ region. It plays a crucial role in identifying and preserving areas of biodiversity interest, ensuring the protection and restoration of biodiversity-rich regions. By basing decisions on factual studies, the CVIB analysis helps in selecting appropriate intervention measures and becomes an integral part of coastal vulnerability. According to the CVIB evaluation, the analysis reveals that 46% of the overall coastal area falls into the high and very high categories. The significance of biodiversity vulnerabilities on coasts cannot be overstated, considering coastal ecosystems are among the most biologically diverse and productive habitats on Earth. However, human activities, climate change, and other factors have led to significant threats to coastal biodiversity, making them highly vulnerable.

Thirdly, the Socio-Economic Coastal Vulnerability Index (CVIS) focuses on balancing development activities and addressing socio-economic conflicts while considering the impact of coastal disasters on municipal and economic aspects. It helps resolve conflicts and ensures the sustainability of socio-economic elements in coastal areas. The CVIS utilizes indicators that assess socio-economic risk at individual, societal, regional, and institutional levels. Socio-economic vulnerability mapping for coastal areas involves the consideration of various socioeconomic variables to gain insights into the susceptibility of coastal communities to different risks and challenges. Combining multiple socioeconomic variables into a Social Vulnerability Index (SVI) provides a comprehensive measure of a community's overall vulnerability to coastal hazards. According to the evaluation 40% of the overall coast falls into the high and very high categories.

The National Coastal Vulnerability Index (NCVI), amalgamating three key components: Physical, Biodiversity, and Socio-Economic, provides a comprehensive assessment of coastal vulnerability, considering the interplay between these factors. An analysis of overall vulnerability based on these three dimensions reveals 39% of the total fall within the high and very high vulnerability category. Figure 8.2 visually represents the map detailing the overall Coastal Vulnerability Index Ranking for Malaysia.

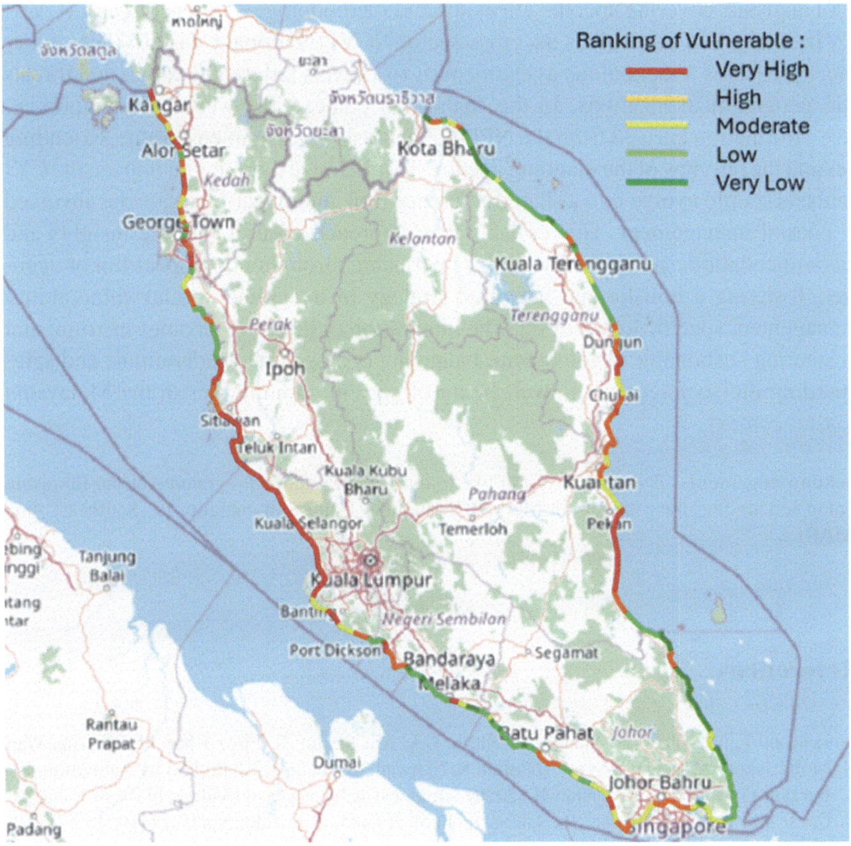

Fig. 8.2 Ranking of National Coastal Vulnerability Index (NCVI) in Malaysia

8.3 Conclusion

In conclusion, the application of the Coastal Vulnerability Index (CVI) stands as a cornerstone in the effective management of coastal areas, not only within Malaysia but also on a global scale (Pantusa, 2022). This tool's significance lies in its ability to seamlessly integrate risk assessments into early-stage development planning, facilitating the timely implementation of adaptation and intervention strategies. The CVI, when utilized in conjunction with proposed development components, emerges as an invaluable resource, swiftly identifying and rectifying disparities to minimize potential global risks (Plan Malaysia, 2021). The comprehensive breakdown of the CVI analysis, covering specific attributes and methodologies for each sector (CVIP, CVIB, and CVIS), as well as the cross-sector NCVI (National Coastal Vulnerability Index), ensures a meticulous assessment of risk levels tailored to contextual threats and geographical locations. In the Malaysian context, the wealth of comprehensive information derived from the NPP-CZ analysis holds immense value, extending beyond the purview of the planning agency. To maximize its potential impact, the CVI analysis should evolve into a shared reference tool among various agencies involved in coastal management. This collaborative approach ensures that the insights and recommendations generated by the CVI permeate throughout the spectrum of agencies, fostering a holistic and informed strategy in integrated coastal vulnerability management specific to Malaysia. The integration of this tool becomes instrumental in steering sustainable development, balancing ecological considerations, and safeguarding the socio-economic well-being of coastal communities in the Malaysian coastal context.

Acknowledgements The authors would like to thank the Eramus+ Programme of the European Union for funding the research on Marine Coastal and Delta Sustainability for Southeast Asia (MARE).

References

Amri Mohd F, Nizam Abdul Maulud K, Karim OA, Ara Begum R, Firoz Khan M, Shafrina Wan Mohd Jaafar W, Mastura Syed Abdullah S, Ekhwan Toriman M, Khairul Amri Kamarudin M, Barzani Gasim M, Abd Wahab N (2018) An assessment of coastal vulnerability of Pahang's Coast due to sea level rise. Int J Eng Technol 7(3.14). https://doi.org/10.14419/ijet.v7i3.14.16880

Bukvic A, Rohat G, Apotsos A, de Sherbinin A (2020) A systematic review of coastal vulnerability mapping. Sustainability 12(7). https://doi.org/10.3390/su12072822

Gornitz (1991) Vulnerability of U.S. to future SLR

NAHRIM (2017) Impact of climate change: sea level rise projection in Malaysia. National Hydraulic Research Institute of Malaysia, Selangor, Malaysia; Ministry Water, Land and Natural Resources, Putrajaya, Malaysia

Noor NM, Abdul Maulud KN (2022) Coastal vulnerability: a brief review on integrated assessment in Southeast Asia. J Mar Sci Eng 10(5). https://doi.org/10.3390/jmse10050595

Oloyede MO, Williams AB, Ode GO, Benson NU (2022) Coastal vulnerability assessment: a case study of the Nigerian coastline. Sustainability 14(4). https://doi.org/10.3390/su14042097
Pantusa D, D'Alessandro F, Frega F, Francone A, Tomasicchio GR (2022) Improvement of a coastal vulnerability index and its application along the Calabria Coastline, Italy. Sci Rep 12(1):21959. https://doi.org/10.1038/s41598-022-26374-w
Plan Malaysia (2021) National physical plan—coastal zone
Zhang Y, Wu T, Arkema KK, Han B, Lu F, Ruckelshaus M, Ouyang Z (2021) Coastal vulnerability to climate change in China's Bohai Economic Rim. Environ Int 147:106359. https://doi.org/10.1016/j.envint.2020.106359

Chapter 9
Environmental Buoys for Water Quality Assessment

Muhammad Fathurizqullah, Adi Maimun Abdul Malik, Shamila Azman, Arifah Ali, Jauhari Khairuddin, Khairul Anuar Mat Saad, and Amayrol Zakaria

Abstract This chapter explores the design and analysis of environmental buoys for water quality assessment, focusing on the essential role of monitoring water quality in preserving aquatic ecosystems and sustainable water resource management. The use of environmental buoys equipped with various sensors and instrumentation for continuous data collection and monitoring is examined along with the challenges and opportunities associated with buoy stability, motion characteristics and sensor integration. It covers topics on estimation of stability and motion values, material estimation, and sensor integration of inertial measurement units, providing insights into the critical factors influencing buoy design and performance. It delves into the comparative method framework for water quality monitoring networks and the utilization of computational fluid dynamics in buoy design analysis. By integrating engineering principles and practical considerations, it emphasizes the importance of well-designed buoys in water quality assessment and data collection.

M. Fathurizqullah · A. M. A. Malik · A. Ali
Faculty of Mechanical Engineering, Universiti Teknologi Malaysia, 81310 Johor Bahru, Johor, Malaysia

S. Azman (✉)
Faculty of Civil Engineering, Universiti Teknologi Malaysia, 81310 Johor Bahru, Johor, Malaysia
e-mail: shamila@utm.my

J. Khairuddin
Petronas, Kuala Lumpur, Malaysia

K. A. M. Saad
Malaysia Institute of Marine Engineering Technology, Universiti Kuala Lumpur, Perak, Malaysia

A. Zakaria
Department of Business and Management, Universiti Teknologi Mara, Cawangan Pulau Pinang, Malaysia

V. Maccarrone and M. Fadzil Akhir (eds.), *Coastal Sustainability*, Coastal Research Library 39, https://doi.org/10.1007/978-3-031-75749-5_9

9.1 Introduction

Monitoring water quality is crucial for preserving the wellbeing of aquatic ecosystems and ensuring sustainable management of water resources. Environmental buoys have become useful tools with a variety of sensors and instrumentation to continuously monitor the environment and the quality of the water to achieve this. With real-time data on physical parameters and environmental behavior in water bodies, these floating sensors act as data gathering and monitoring systems.

The size and shape of the buoy hull was evaluated throughout the investigation to determine how they may affect stability and the precision of the data collected. The knowledge gathered will advance the area of naval architecture more broadly and direct future buoy designs for better performance under various environmental circumstances. It has the potential to progress the field of river water assessment, resulting in enhanced monitoring procedures and the creation of cutting-edge technology. The results can improve the precision and efficacy of water quality evaluation, which will aid in the protection and conservation of essential water resources for both the benefit of the present and future generations. Water pollution poses a critical challenge in Malaysia, negatively impacting the sustainability of water resources. The adverse effects extend to plants, organisms, human health, and the country's economy. The considerable reduction in total water availability arises from the high cost of treating polluted waters and, in some cases, the inability to make them suitable for consumption. Despite the abundance of water resources in the catchment, the issue of river pollution hinders equitable supply to all users (Ling 2010).

The use of buoys is one method and efficient as monitoring platforms in marine and freshwater environments, including the types of sensors and equipment that are commonly used and the methods used to deploy and maintain these platforms. The potential benefits and drawbacks of using buoys for water quality monitoring in rivers, including considerations such as cost, reliability, and scalability.

Water quality monitoring is crucial for the assessment and preservation of river ecosystems and their surrounding environments. To ensure accurate and continuous data collection, environmental buoys equipped with various sensors and instrumentation have been widely used. These buoys serve as data collection and monitoring systems, providing valuable information on physical parameters and environmental conditions. (Dalla via, Mattia 2021) and also wave measurement (Wei 2005). Environmental buoys have proven to be an effective tool in river water assessment, offering real-time data on water quality and environmental behaviour. The design of the buoy is very wide because it depends on specification however, their performance is subject to various challenges, including the influence of environmental factors such as wave conditions, wind patterns, and sedimentation. These factors can impact the buoy stability, motion characteristics and sensor accuracy, potentially leading to drift and inaccurate data collection. Additionally, the incorporation of a remote operation system is vital to streamline data collection and transmission from the buoy to a centralized location. Efficient data transfer enables timely access to water quality information, empowering stakeholders in water resource management and environmental protection decision-making.

9.2 Environmental Buoys

Continuously monitoring water quality parameters in river systems, researchers and environmental managers gain insights into the impacts of natural processes, human activities, and environmental stressors on water quality. It helps in identifying pollution sources, evaluating the effectiveness of remediation measures, and understanding the ecological health of the river ecosystem. Additionally, continuous monitoring facilitates the early detection of water quality degradation, allowing for timely mitigation of actions and protection of human and aquatic life. There are several factors that could make it necessary for Malaysia to monitor the water quality of its rivers using buoys. For example, if the rivers in Malaysia are used for drinking water, irrigation, or other important purposes, it is important to ensure that the water is of high quality and safe for use. Buoys can help to identify any contaminants or other issues that may be present in the water and allow authorities to take steps to address them. Additionally, if the rivers in Malaysia are home to important aquatic ecosystems or are used for recreational activities, it is important to monitor the water quality to ensure that these ecosystems and activities are not adversely affected (Camara et al. 2019). Continuous monitoring of water quality parameters in river systems is essential for understanding the dynamic nature of water bodies and assessing their overall health. Various studies have emphasized the significance of continuous monitoring as it provides a comprehensive and real-time understanding of water quality trends, enabling prompt identification of potential issues and effective management strategies (Herrera et al. 2020).

Table 9.1 includes information on previous studies related to buoys. In all the buoys listed below, Mooring Weight is used to stand statically, ensuring accurate data collection while maintaining the buoy position.

However, the WSN buoy stands out in that it does not use mooring weights. Instead, it focuses on measuring in still water conditions, which may indicate that it's designed for a specific environmental setting. On the other hand, the CB-150 float, which was produced by Nexsens Technology in 2019, stands out as a comprehensive model. Despite its advanced features, installation with a dedicated team is required, which may indicate a more complex setup process compared to other floats. Finally, the Smart Buoy and Common Wave Buoy demonstrated problems with the telemetry system and buoy stability. This shows that these buoys face challenges in transmitting data efficiently and maintaining stability in dynamic water conditions.

In summary, while previous research has demonstrated the effectiveness of the Mooring Weight for static positioning and data collection, the WSN buoy stands out for its focus on measuring still water. The CB-150 buoy provides a complete solution but may require special installation. Meanwhile, Smart Buoy and Common Wave Buoy experienced problems related to telemetry and buoy stability. Further research and improvement in this area is essential to improve the overall performance and reliability of environmental buoys for river water assessment.

Table 9.1 Studies related to buoys

Buoy	Type	Published	Findings
	CB-150 and CB-250	Nexsens Technology 2019	Measurement on physical, chemical and meteorology parameters in sea and river water
	Wave Buoy	Liu et al. 2015	Measurement of wave height and wave energy
	Smart Buoy	T3ch flicks, 2019	Measurement on physical and meteorology parameters in Grenada Island

9.3 River Water Assessment Estimation Method

River water assessment involves the evaluation of various physical and hydrological parameters to gain insights into the dynamics of rivers and their surrounding environments. Physical parameters such as water temperature, air pressure, total dissolved solids (TDS), and air temperature provide valuable information about the current state of the river ecosystem. Water temperature, for example, affects the overall health and biodiversity of aquatic life, and variations in temperature can indicate changes in seasonal patterns or the presence of pollution sources. Air pressure measurements help in understanding atmospheric conditions that can influence water movement and weather patterns (Fig. 9.1).

Total dissolved solids (TDS) represent the concentration of dissolved minerals and salts in the water. Monitoring TDS levels aids in assessing water quality, salinity, and the presence of pollutants or contaminants. Air temperature, on the other hand, influences the overall climate and can impact river water temperature, evaporation rates and biological processes. Hydrological parameters such as current and wave characteristics and water depth provide essential information about the flow dynamics of the river. Current measurements help in understanding the speed and direction of water movement, which can affect sediment transport, navigation, and habitat

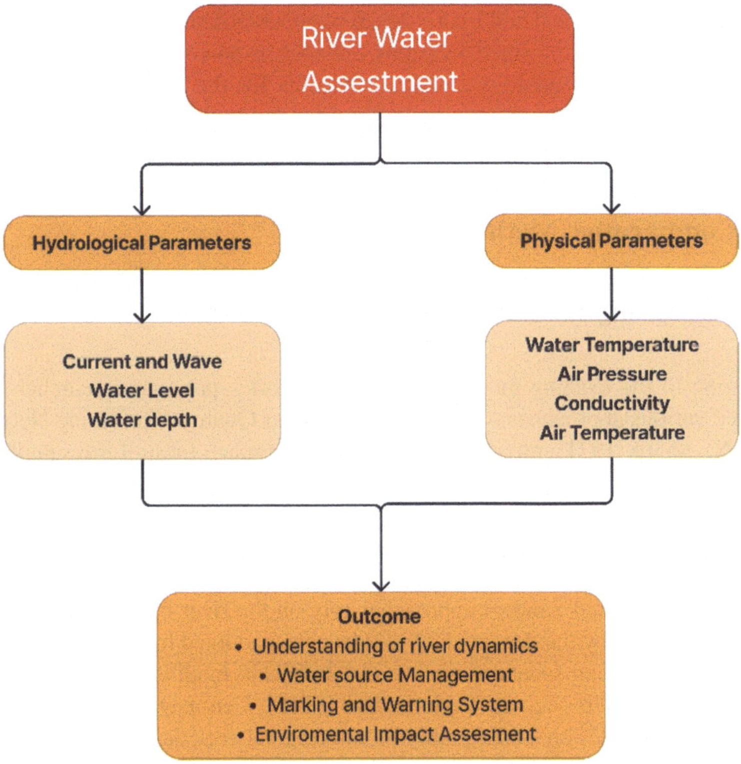

Fig. 9.1 Parameters of river water assessment

suitability. Wave characteristics indicate the presence and intensity of wave action, which can impact erosion, sedimentation, and the stability of riverbanks. Water depth measurements are crucial for assessing the overall water volume and determining variations in river stages. It aids in understanding flood potential, water resource management, and ecological impacts. These hydrological parameters contribute to the identification of critical areas prone to erosion, flooding, or sediment deposition, enabling the implementation of effective management strategies.

The benefits of river water assessment encompass a wide range of applications. Understanding river dynamics through the monitoring of physical and hydrological parameters helps in formulating sustainable water source management strategies. It assists in identifying potential risks, mitigating hazards, and optimizing water allocation for various uses. Additionally, the collected data supports the development of marking and warning systems for navigation, ensuring safe passage for vessels and preventing accidents. Furthermore, river water assessment plays a vital role in environmental impact assessment, enabling the evaluation of human activities' effects on the ecosystem and facilitating the implementation of measures to protect and preserve river environments (Poff et al. 1997).

By integrating physical and hydrological parameters in river water assessments, stakeholders can make informed decisions, enhance resource management practices, and promote the sustainable use of river systems for the benefit of both human populations and the environment.

9.4 Comparative Method Framework

9.4.1 Sampling Point Method

In addition to the existing guidelines and frameworks, previous researchers have proposed various approaches for the design of Water Quality Monitoring Networks (WQMN). Sharp (1971) introduced a systematic approach for uniform sampling of rivers and streams, which involves identifying optimal sampling locations by dividing the stream network into halves and identifying the centroids. Sanders et al. (1983) modified Sharp's method by incorporating pollution loadings and the number of outfalls instead of tributaries.

Both the Sharp and Sanders approaches rely on the river basin's topology. The modified approach by Sanders et al. (1983) has been adopted by several researchers for selecting sampling locations and sub-basins. This modified method has been utilized by researchers such as (Do et al. 2011, 2013; Cetinkaya and Harmancioglu 2011) for their studies on water quality monitoring network design.

The systematic approach proposed by Sharp (1971) for uniform sampling of rivers and streams involves a step-by-step process. The first step is to divide the stream network into successive halves. This is done by identifying specific locations or centroids that represent equal portions of the stream network. These centroids serve as potential sampling locations. The second step is to select the optimal sampling locations based on various factors such as accessibility, representativeness, and significance. The goal is to ensure that the selected sampling locations provide a comprehensive representation of the water quality characteristics within the river or stream.

Sanders et al. (1983) further modified Sharp's approach by incorporating pollution loadings and the number of outfalls. This modification considers the impact of pollution sources on water quality. Instead of considering the tributaries as in the original Sharp's method, the modified approach considers the distribution of pollution sources and the potential influence of outfalls on water quality. This modification enhances the relevance of the sampling locations for assessing water pollution and enables the identification of critical areas for monitoring and management.

9.5 Design Analysis Estimation Method

There are several methods to estimate the Stability and Motion value, starting from the Computational Fluid Dynamics and experimental method.

9.5.1 Computational Fluid Dynamics (CFD)

CFD (computational fluid dynamics) simulations. Reliable results can be obtained for partial problems. Wave resistance computations and related shape optimization of the forebody. CFD results are useful for comparing and ranking design alternatives. With the steady increase in computational speed and improvements in flow modelling, CFD will play an increasingly important role in ship resistance and propulsion predictions. Examples of CFD software are SolidWorks Flow Simulation, ANSYS CFX, FLOW 3D, and many others.

9.5.2 Numerical Analysis with Ansys AQWA

In the context of ANSYS AQWA, CFD (Computational Fluid Dynamics) simulations are used to analyze hydrodynamic diffraction and buoy response. CFD-based techniques, such as the panel method, are used to solve the equations governing fluid flow and the interaction of wave structures.

Hydrodynamic Diffraction to get the frequency response results (spectrum). This is in frequency domain. Hydrodynamic Response to get the response in time domain. Hydrodynamic diffraction refers to the wave phenomenon that interacts with a buoy and causes a change in the wave field around it. CFD simulations can accurately predict diffraction effects, providing insight into the wave forces and stresses exerted on the buoy structure. This information is essential for assessing structural response and designing buoy systems that can withstand wave loads (Fig. 9.2).

CFD simulation in ANSYS AQWA also allows analysis of the dynamic response of buoys in waves. By considering the fluid-structure interaction, the simulation can capture the motion and behavior of the buoy when it reacts to the incoming waves. This includes evaluating parameters such as the amplitude of motion, acceleration and hydrodynamic forces acting on the float.

Regular waves

$$\text{Airy Waves (linear wave)}\, a = A\cos\left(-\omega t + kx\right) \tag{9.1}$$

ω frequency in rad/s
k wave number

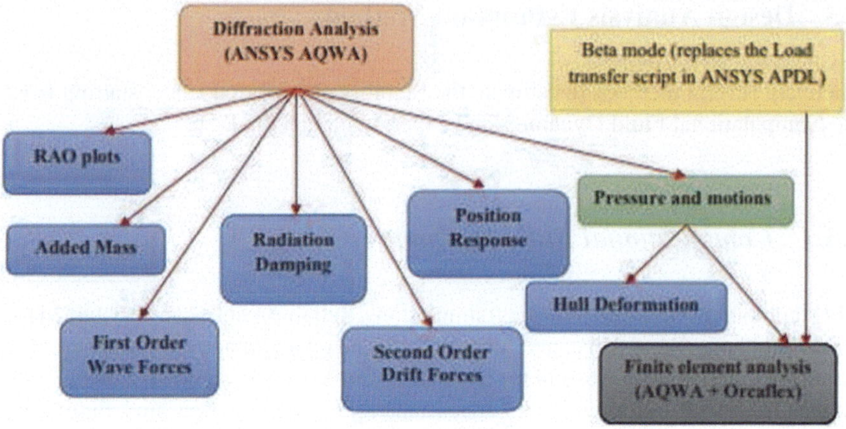

Fig. 9.2 Schematic diagram of numerical analysis and coupling model

Equation 9.1 **Regular Waves**
Stokes 2nd Order Waves

$$a = A\cos(-\omega t + kx) + 0.5\,k\,A^2\cos 2(-\omega t + kx) \tag{9.2}$$

Equation 9.2 **2nd Stokes Equation**
By leveraging the CFD techniques at ANSYS AQWA, engineers can gain a detailed understanding of the hydrodynamic behavior of buoys, assess their performance in various wave conditions, and optimize their design to improve stability and response characteristics.

9.6 Material Estimation

Weight is one of the basic parameters for a marine product, it should be determine in the early of the design. Although the calculation of total weight of a buoy design need a lot of complicated formula but estimation of total weight is needed and simple formula is used in the preliminary design of buoy as show below,

$$\Sigma Wt = Wh + Wb + We \tag{9.3}$$

Wt Total weight
Wh Hull or shell weight
Wb Ballast Weight
We Auxiliary equipments weight

Equation 9.3 **Material Estimation**

9.7 Characteristic Motion of Buoy

The characteristic motion of a buoy refers to its response and typical behaviour when subjected to wave-induced forces and environmental conditions. Buoy motion can vary depending on their design, buoyancy, mass distribution and hydrodynamic properties. Some of the characteristic movements observed in buoys include (Figs. 9.3 and 9.4) (Amaechi et al. 2022).

In designing a buoy, engineers must first estimate its weight, which encompasses various components. The buoy's structure, including its hull and support systems, contributes to its weight, and the materials used and dimensions of the buoy play a role in this estimation. Additionally, any equipment and instruments attached to the buoy, such as sensors, communi power systems, are considered for their weight. Furthermore, if the buoy is intended to carry a payload, such as scientific instruments or environmental monitoring devices, the weight of the payload is factored in as well. Accurate weight estimation is crucial because it affects the buoyancy and overall stability.

The stability analysis of the buoy is of paramount importance to ensure its safe and effective operation in varying environmental conditions. One key parameter assessed is the metacentric height (GM), which signifies the vertical distance between the center of gravity (G) and the metacenter (M). A larger GM indicates better initial stability and greater resistance to tilting or rolling. Another critical metric is the metacentric radius (B0M), which considers the inertia of the water plane and the buoy underwater volume (V). This factor determines the buoy stability characteristics, and a proper calculation of B0M helps engineers optimize the buoy design to withstand various forces while maintaining its upright position. Based on the flowchart (Wei 2005)

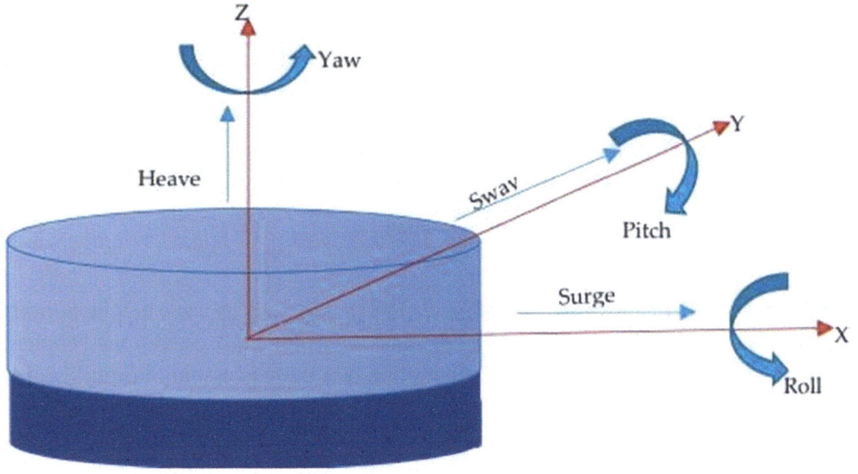

Fig. 9.3 The six degrees of freedom of a floating buoy

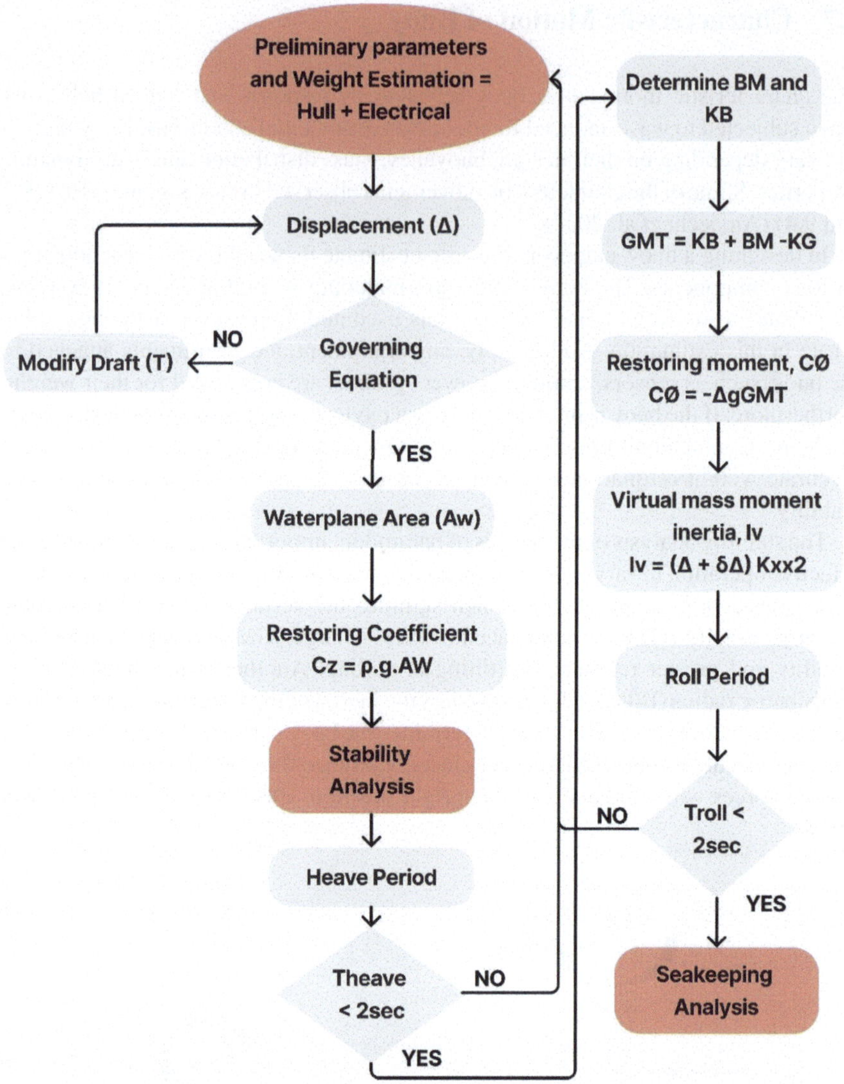

Fig. 9.4 Flowchart of calculation stability and motion

This characteristic movement is influenced by factors such as wave height, period, direction, buoy design, mooring system, and environmental conditions. Understanding and analysing the characteristic motion of a buoy is essential for assessing its stability, structural integrity and performance in various sea conditions. This helps in designing buoys that can withstand the forces and maintain their functionality and data collection capabilities even under challenging wave conditions.

9.8 Stability at Small Angles of Heel

When a buoy experiences small angles of heel (tilting) due to external forces like waves or wind, it follows a circular path centered at a point M. This circular motion is referred to as the "metacentric circle." In this context:

In summary, when the buoy tilts at small angles, it rotates around a virtual point called the metacenter (M), which is above the center of buoyancy (B). The metacentric height (GM) and the cation devices, and metacentric radius (KM) are essential parameters in evaluating the buoy's stability and motion in response to external forces.

$$\begin{aligned} GM &= KM - KG \\ KM &= KB_0 + B_0M \end{aligned} \tag{9.4}$$

Equation 9.4
B represents the buoy's initial position (upright) before any tilting occurs.

M is the metacenter, which is a virtual point above the center of buoyancy, representing the buoy effective center of rotation when it heals.

GM (metacentric height) is the vertical distance between the center of gravity (G) and the metacenter (M). It is a measure of the buoy's initial stability. A larger GM provides better stability against tilting.

KM is the vertical distance between the center of buoyancy (B) and the metacenter (M). It is called the "metacentric radius" and represents the distance from the buoy center to the metacenter in its initial upright position.

$$B_0M = \frac{I_f}{V_B}$$

$$I^f = \frac{1}{64}\pi D^4$$

$$\mathrm{GM} = (\mathrm{KB}_0 + \mathrm{B}_0\mathrm{M}) - \mathrm{KG} \tag{9.5}$$

Equation9.5
The inertia of the water plane, represented by the moment of inertia (Iw) about the axis of the center of the water plane O, plays a crucial role in characterizing the energy required to rotate the water surface around an axis as the buoy is displaced from the vertical position.

Motion Characteristics

Heave Motion Equations

$$\text{Natural heave period, } T_l = \frac{2\pi}{\omega_t}$$

$$\text{Natural heave period,} \quad \omega_t = \left[\frac{c}{m}\right]^{\frac{1}{2}} \tag{9.6}$$

Equation 9.6
Heave motion refers to the vertical displacement of the buoy due to the up-and-down wave motion. The float moves in response to alternating pressure changes caused by passing waves.

Roll Motion Equations

$$\text{Natural roll period,} \quad T_x = \frac{2\pi}{\omega_x}$$

$$\text{Natural roll frequency,} \quad \omega_x = \left[\left(\frac{c}{m}\right)\right]^{\frac{1}{2}} \tag{9.7}$$

Equation 9.7
Roll motion involves the rotation of the float around its longitudinal axis. It occurs when waves exert uneven pressure on the sides of a float, causing it to tilt and roll from side to side.

Heave motion refers to the vertical movement experienced by the buoy due to wave action. Engineers analyze the buoy's response to different wave heights, periods, and directions to assess its ability to withstand these forces without excessive vertical displacement. Effective buoyancy, well-distributed weight, and appropriate structural design are all important factors in minimizing heave motion. Ensuring that the buoy remains stable in its vertical position is crucial for its intended functions, such as accurate data collection or navigation assistance.

Roll motion, on the other hand, deals with the rotational movement of the buoy around its longitudinal axis. To address roll motion, engineers consider several factors. The metacentric height (GM) is particularly relevant, as a larger GM leads to a shorter roll period and quicker stabilization, ultimately reducing the extent of roll motion. The metacentric radius (B0M) also comes into play, influencing the buoy's resistance to rolling. By properly calculating B0M, engineers can optimize the buoy's design for enhanced stability and reduced roll motion. Moreover, the moment of inertia of the buoy's water plane plays a vital role in determining its roll response. A lower moment of inertia helps minimize the roll motion, contributing to the buoy's overall stability.

9.9 Sensor Integration of Inertial Measurement Unit

The Inertial Measurement Unit (IMU) is a module that consists of an accelerometer and gyroscope or a 6 Degree of Freedom (6DoF) IMU. It is used to obtain the sensor's orientation relative to the Earth's orientation (Pramana 2013; Munarto et al.

2022). Another type of IMU includes an additional magnetometer sensor, known as a Magnetic Angular Rate and Gravity (MARG) or a 9 Degree of Freedom (9DoF) IMU. The 9DoF IMU measures orientation with reference to gravity and the Earth's magnetic field's north pole.

9.9.1 Acceleration

An accelerometer is a sensor used to measure acceleration. Micro Electro Mechanical System (MEMS) is a technology that enables the fabrication of extremely small physical structures. The MEMS accelerometer used is of the capacitive accelerometer type. The sensor converts the mass displacement caused by sensor movement into a change in capacitance value, which is then converted into a voltage signal (Mohammed et al. 2018) (Fig. 9.5).

The advantage of the 9 DOF IMU over the 6DoF is that it provides a more stable output when operated with the Mad wick filter. The use of a 9DoF IMU aims to overcome the limitations of gyroscopes when calculating sensor orientation (roll, pitch, and yaw), specifically addressing the drift issue caused by the lack of a reference point in gyroscopes' measurements. In this case, the accelerometer and magnetometer

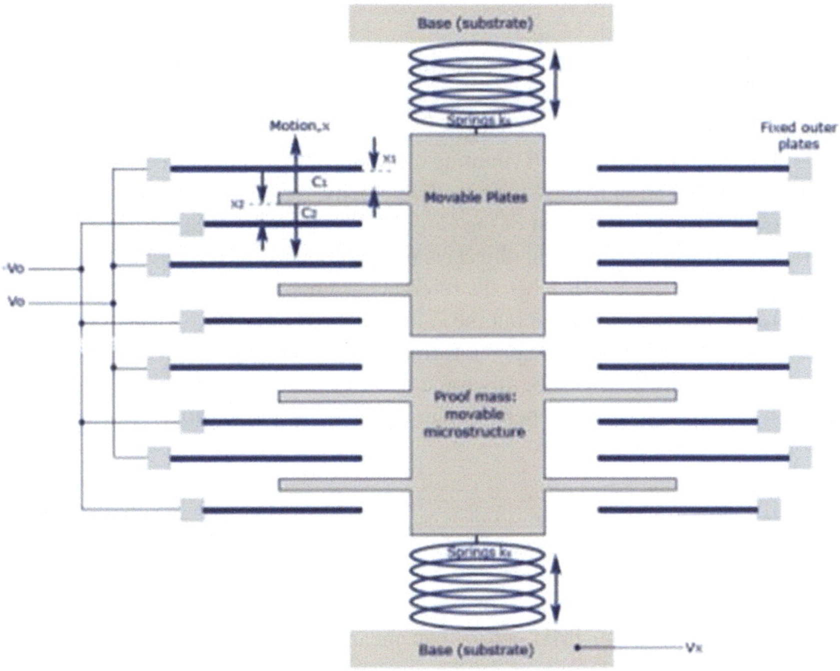

Fig. 9.5 MEMS of acceleration

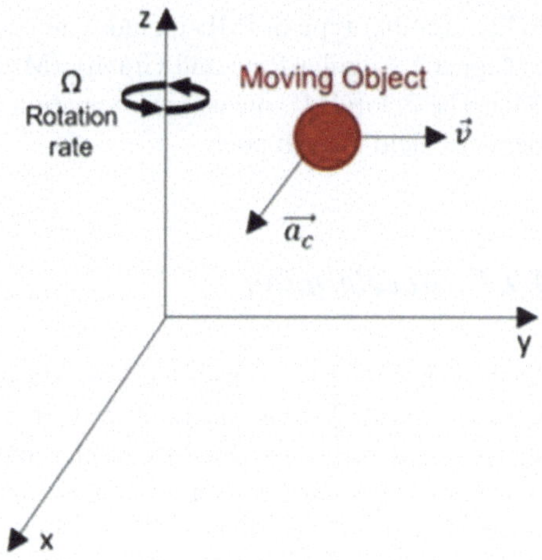

Fig. 9.6 Coriolis effect

measure gravity and the Earth's magnetic field as references to compensate for the gyroscope drift (Pramana 2013).

9.9.2 Gyroscope

A gyroscope is used to measure the angular velocity of an inertial reference frame. The simple working principle of a MEMS gyroscope involves using vibrating mechanical structures to detect changes in the sensor's angular position (Northrop 2014). A MEMS gyroscope with vibrating mechanical structures utilizes the principle of Coriolis acceleration to measure angular velocity.

A more detailed explanation of the Coriolis effect can be seen in Fig. 9.6. In the diagram, the red object (moving object) is moving linearly in the X, Y, Z plane with a velocity v along the Y-axis when the reference frame is stationary. However, when the reference frame is rotated about the Z-axis, the red object experiences a force in a direction perpendicular to the plane formed by the moving object's direction and the axis of rotation. In Fig. 9.6, this force has an acceleration ac and moves in the direction of the X-axis. This phenomenon is known as the Coriolis effect (Passaro et al. 2017).

9.9.3 Magnetometer

The magnetometer in an IMU sensor can measure the magnitude of the Earth's magnetic field around the sensor. There are various technologies used in magnetometers for detecting magnetic fields, with hall effect sensors being the most

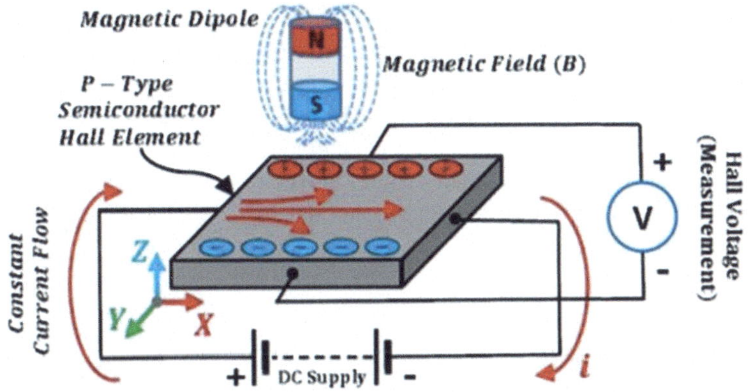

Fig. 9.7 Hall-effect magnetic sensor

common in the market. Other technologies include Anisotropic Magnetoresistance (AMR), Tunnel Magnetoresistance (TMR), and Giant Magnetoresistance (GMR). The MPU9250 utilizes the AK8963 as its magnetometer sensor, which leverages the hall effect phenomenon (Fig. 9.7).

The hall effect phenomenon can be visualized as follows: when a flat-shaped semiconductor object is subjected to a continuous electric current near a magnetic field, it causes electric charges (electrons and holes) to separate on both sides of the semiconductor plate (Northrop 2014) and (Karimi et al. 2020). The separation of electrons and holes to different sides results in a difference in voltage. This voltage difference is then converted to the magnitude of the magnetic field in the magnetometer sensor.

9.9.4 Orientation and Reference Frame

The aim of this research is to obtain the vertical displacement of the device in meters, calculated from the acceleration data along the z-axis relative to the Earth. However, during device testing, the device is not in an upright position relative to the Earth, resulting in varying values of the z-axis on the MPU 9250. Therefore, the sensor orientation needs to be adjusted based on a reference to the Earth. The acceleration data based on this Earth reference will be converted into vertical displacement data. The inertial reference frame is a reference frame that is originally located at the center of the Earth. This frame does not rotate, accelerates, and remains fixed relative to celestial bodies in the universe. The orientation of this frame, such as the z-axis, is parallel to the Earth's axis of rotation, the x-axis extends toward the mean vernal equinox, and the y-axis completes the right- hand orthogonal frame. The Earth frame is a reference frame originally fixed at the center of the Earth, with its axis parallel to the mean Greenwich meridian, and the y-axis completes the right-hand orthogonal frame. The Earth frame is also known as the Earth Centered Earth Fixed (ECEF) frame (Yang et al. 2021).

The navigation reference frame is a local reference frame with its origin coinciding with the sensor reference frame. The z-axis points upward normal to the ellipsoid, the x-axis points to the geodetic north, and the y-axis points to the geodetic east. This frame is referred to as the East North Up (ENU) frame. The body frame (b-frame) is a local reference frame with its origin coinciding with the center of the body (in this research, the device). The x-axis points to the front side of the device, the y-axis points to the right side of the device, and the z-axis completes the right-hand orthogonal frame (Yang et al. 2021).

9.10 Madgwick Filter

The Madgwick Filter is a filter developed from the Mahony Filter. The Madgwick Filter offers higher accuracy compared to the Mahony Filter, as it incorporates the use of a magnetometer sensor to reduce gyroscope biases (Munarto et al. 2021). The Madgwick Filter algorithm utilizes quaternion representations to estimate the orientation of the MARG sensor by implementing a gradient descent algorithm. The algorithm computes the gyroscope direction and measurement errors from the quaternion derivatives, while incorporating accelerometer and magnetometer data in the derived and optimized gradient descent algorithm (Hasan and Rahman 2020). The fundamental components of the Madgwick Filter algorithm include the orientation of angular velocity, the fusion process, compensation for magnetic distortion, and adjustable parameters (Madgwick et al. 2011).

9.11 Conclusion

The design is presented in various parts, system integration and Motion Characteristics.to provide a critical analysis of their strengths, limitations, and practical considerations. It examines progress, challenges, and opportunities in the field, identifying key trends, best practices and areas for further improvement.

Designing effective buoys requires a thorough understanding of various factors affecting their performance. Weight estimation accounts for the buoy's overall mass, encompassing its structure, equipment and potential payload. Stability analysis, relying on parameters like GM and B0M, ensures the buoy's ability to remain stable in different environmental conditions. Engineers also address heavy motion, striving to minimize the vertical movement caused by waves, and roll motion, optimizing design features to reduce rotational movement around the longitudinal axis.

A well-designed buoy, with accurate weight estimation, favorable stability characteristics, and careful consideration of heave and roll motion, can serve essential roles in marine navigation, data collection, environmental monitoring, and various scientific applications. By applying comprehensive engineering principles to buoy design, these floating structures can effectively perform their intended tasks and contribute to safer and more efficient marine operations.

References

Amaechi CV, Wang F, Ye J (2022) Numerical studies on CALM buoy motion responses and the effect of buoy geometry cum skirt dimensions with its hydrodynamic waves-current interactions. Ocean Eng 244:110378

Camara M, Jamil NR, Abdullah AFB (2019) Impact of land uses on water quality in Malaysia: a review. Ecol Process 8:10

Cetinkaya CP, Harmancioglu NB (2011) Assessment of water quality sampling sites by a dynamic programming approach. J Hydrol Eng 17(2)

Do HT, Lo SL Chiueh PT, Thi LAP, Shang WT (2011) Optimal design of river nutrient monitoring points based on an export coefficient model. J Hydro 406(1–2)

Do HT, Lo SL, Phan Thi LA (2013) Calculating of river water quality sampling frequency by the analytic hierarchy process (AHP). Environ Monit Assess 185:909–916

Hasan MA, Rahman MH (2020) Smart phone based sensor fusion by using Madgwick filter for 3D indoor navigation. Wireless Pers Commun 113(4):2499–2517

Herrera JR, Pradhanan SM, Guzman CD, Karki R (2020) Continuous water quality monitoring in river systems: importance, methods, and applications. Water 12(11):3217. https://doi.org/10.3390/w12113217

Karimi M, Babaians E, Oelsch M, Aykut T, Steinbach E (2020) Skewedredundant Hall-effect magnetic sensor fusion for perturbation-free indoor heading estimation. In: Proceedings 2020 fourth IEEE international conference on robotic computing (IRC), Taichung, Taiwan, pp 367–374

Ling JKB (2010) Water quality study and its relationship with high tide and low tide at Kuantan river. Universiti Malaysia Pahang

Madgwick SOH, Harrison AJL, Vaidyanathan R (2011) Estimation of IMU and MARG orientation using a gradient descent algorithm. In: Proceedings 2011 IEEE international conference on rehabilitation robotics, Zurich, Switzerland, pp 1–7

Mohammed Z, Elfadel IM, Rasras M (2018) Monolithic multi degree of freedom (MDoF) capacitive MEMS accelerometers. Micromachines 9(11):602 (53 Universitas Sultan Ageng Tirtayasa (Basel))

Munarto R, Wiryadinata R, Utama DP (2021) Implementation of AHRS (Attitude Heading and Reference Systems) with Madgwick filter as hexapods robot navigation. In: 2022 international conference on informatics electrical and electronics (ICIEE), Yogyakarta, pp 1–9

Northrop RB (2014) Introduction to instrumentation and measurements, 3rd edn. CRC Press, Boca Raton

Passaro VMN, Cuccovillo A, Vaiani L, De Carlo M, Campanella CE (2017) Gyroscope technology and applications: a review in the industrial perspective. Sensors (Basel) 171(10)

Pramana YA (2013) Implementasi sensor accelerometer, gyroscope dan magnetometer Berbasis Mikrokontroler Untuk Menampilkan Posisi Benda Menggunakan Inertial Navigation System. Bachelor Thesis, Dept. Electrical Eng., Universitas Komputer Indonesia, Bandung

Poff NL, Allan JD, Bain MB, Karr JR, Prestegaard KL, Richter BD, Sparks RE, Stromberg JC (1997) The natural flow resssgime. Bioscience 47(11):769–784

Sanders TG, Wang W, Chesters G (1983) Pendekatan sistematis untuk desain jaringan pemantauan kualitas air. Buletin Sumber Daya Air 19(2):179–186

Sharp W (1971) A topologically optimum water-sampling plan for rivers and streams. Water Resour Res 7(6):1641–1646

Wei NC (2005) Wave buoy for east Malaysian sea water. Universiti Teknologi Malaysia

Yang Z, Jing H, Kuang B, Zhong S, Qin J, Tong Y (2021) GPS and lidar fusion positioning based on unscented kalman filtering and long short-term memory cosidering gps failure. In: Proceedings 2021 IEEE 4th international conference on automation, electronics and electrical. Universitas Sultan Ageng Tirtayasa, Engineering (AUTEEE), Shenyang, China, pp 18–23

Chapter 10
Control Dangerous Goods Transported by Ship

Nguyen Xuan Long, Nguyen Cao Hien, Tran The Nam, Nguyen Thanh Son, Nguyen Thanh Dieu, and Phan Van Hung

Abstract Dangerous goods (DG) or Hazardous and Noxious Substances (HNS) transported by ship play an important role in global economic development. Along with that, the transportation of dangerous goods is at an increasing risk of accidents. The consequences of dangerous goods incidents are difficult to estimate. Therefore, the International Maritime Organization has made efforts to introduce measures to improve safety, security, prevention, and control of the transport of dangerous goods through Conventions, Codes, and Guidelines related regulations. In particular, the IMDG Code is mandatory from January 1, 2022, so controlling dangerous goods transported by ship is an urgent issue today. The book chapter focuses on an overview of Dangerous goods transported by ship; Risk assessment of DG pollution from ships; Contingency planning of DG incidents from ships; Response to DG incidents from ships; and Management after DG incidents from ships. The content of the book chapter is both theoretical and practical; it is the basis for state managers at seaports, officers, crew members, and related individuals and organizations to understand and apply them to prevent accidents as well as be ready to respond when DG pollution incidents occur comply with the Conventions by the law. The book chapter is systematically compiled, based on international legal regulations in the maritime domain, and consists of 6 sections: Introduction; Dangerous goods transported by ship; Risk assessment DG pollution from the ship; Contingency planning of DG incidents from the ship; Response DG incidents from the ship; Management after DG incident from the ship; and Conclusion. This chapter is useful to students, graduate students, scientists, lecturers, and experts in the maritime field.

Keywords Controlling · Dangerous goods · Hazardous and noxious substances · Transported by ship

N. X. Long · T. T. Nam · N. T. Son · N. T. Dieu · P. Van Hung (✉)
Vietnam Maritime University, 484 Lach Tray, Le Chan, Hai Phong, Vietnam
e-mail: phanvanhung@vimaru.edu.vn

N. C. Hien
Ministry of Transport, 80 Tran Hung Dao, Hoan Kiem, Ha Noi, Vietnam

V. Maccarrone and M. Fadzil Akhir (eds.), *Coastal Sustainability*, Coastal Research Library 39, https://doi.org/10.1007/978-3-031-75749-5_10

10.1 Introduction

Nowadays, over 50% of goods shipped by sea can be classified as dangerous, hazardous, or harmful to the environment. Several factors have contributed to the increase in dangerous goods shipments—replacement of natural materials with synthetics that require hazardous substances for production, development of new chemical products, and specialized vessel types dedicated to transporting specific cargo types in larger quantities (IPIECA-IOGP 2015). Despite relatively few major accidents recently, dangerous goods have been involved in some of shipping's worst disasters. Smaller incidents are also frequent, highlighting the risks that make regulation of dangerous goods vital (Shaverdi et al. 2014).

As dangerous goods often move through multiple transport modes, basic rules need to be aligned internationally. Concern over the risks has led to the progressive development of technical standards by organizations like the International Maritime Organization (IMO) to improve safety. Key instruments include the International Maritime Dangerous Goods (IMDG) Code which lays out detailed procedures for transport and the International Maritime Solid Bulk Cargoes (IMSBC) Code for bulk commodities. The IMDG Code covers classification, packaging, handling, storage, and documentation while the IMSBC Code focuses on assessing, testing, and classifying solid bulk cargoes.

A core regulating principle is the proper classification of dangerous goods based on the type of hazard into 9 classes: explosives, gases, flammable liquids, flammable solids, oxidizing substances, toxic substances, radioactive material, corrosives, and miscellaneous hazards. This enables the identification of risks and appropriate safety precautions. Other key provisions include specifications for packaging, labeling, documentation procedures, stowage and segregation requirements, emergency response plans, and training needs.

All parties in the transport chain have responsibilities—shippers must classify, label, and document dangerous goods properly and provide safety information. Carriers need to stow cargo safely, inspect equipment, and train crews. Ports must maintain facilities and implement emergency plans. Reporting incidents or defects is essential. Proper packaging, labeling, and placarding of transport units are also crucial.

Though major disasters are less common today, the diverse array of hazardous materials shipped, and complex global supply chains mean effective regulation, compliance, and vigilance in enforcing standards remain vital to prevent loss of life, property damage, and environmental harm. Implementing a uniform international framework while also addressing the special handling needs of different hazardous commodities is an ongoing process. However, commitment from all stakeholders is key to ensuring dangerous goods reach destinations safely.

To prevent and mitigate environmental impacts, a crucial process is assessing the risks through systematic analysis. Risk assessment involves identifying hazards, evaluating the likelihood and consequences of adverse events, and supporting decisions on appropriate safeguards. For ships carrying dangerous goods, risk assessment examines factors like type of substances, accident scenarios, vulnerability of affected areas, and effects on marine ecology. Quantitative models can estimate damage probabilities and impacts. Risk assessment provides a proactive way to understand pollution risks from dangerous goods transport and tailor preventive measures to local conditions. Undertaking robust assessments is key to enhancing maritime safety and sustainability.

To effectively respond to accidents involving hazardous materials, advanced contingency planning is essential. Contingency plans outline procedures to control, contain, and mitigate dangerous goods incidents on ships or in ports. They provide systematic guidance on actions like notifying authorities, isolating danger zones, deploying equipment, coordinating responders, and minimizing environmental harm. Well-developed contingency plans enable rapid, coordinated responses, minimizing risks to human life, property, and marine ecosystems.

The control activities after a dangerous goods incident, comprehensive management is vital. Key response priorities include attending to causalities, containing spills and leaks, securing unstable cargo, and isolating danger zones. Investigating causes, assessing damages, and reporting per statutory requirements are crucial. Implementing remedies like environmental decontamination and restoration is needed. Incident reviews identify lessons learned and promote continual improvement in dangerous goods safety management. Proper management drives optimal recovery and prevents recurrence.

10.2 Dangerous Goods Regulations

Dangerous goods refer to materials and substances that pose risks to health, safety, or property due to their chemical or physical properties. The main instruments include:

The SOLAS Convention: The International Convention for the Safety of Life at Sea, 1974 (SOLAS), as amended, deals with various aspects of maritime safety and contains in Chapter VII the mandatory provisions governing the carriage of dangerous goods in packaged form. The carriage of dangerous goods in packaged form shall comply with the relevant provisions of the International Maritime Dangerous Goods (IMDG) Code which is considered an extension to the provisions of SOLAS chapter VII.

The MARPOL Convention: the International Convention for the Prevention of Pollution from Ships, 1973, as modified by the Protocol of 1978 relating thereto (MARPOL), which contains in its Annex III mandatory provisions for the prevention of pollution by harmful substances carried by sea in packaged form, prohibits the carriage of harmful substances in ships except by the provisions of MARPOL Annex III, which are also extended in the IMDG Code.

The UN Recommendations on the Transport of Dangerous Goods

International Maritime Dangerous Goods (IMDG) Code:

The IMDG Code was developed as an international code for the maritime transport of dangerous goods in packaged form, to enhance and harmonize the safe carriage of dangerous goods and to prevent pollution to the environment. The Code sets out in detail the requirements applicable to each substance, material, or article, covering matters such as packing, container traffic, and stowage, concerning the segregation of incompatible substances.

The Code was initially adopted in 1965 as a recommendatory instrument. It was in 2002 that the General Assembly at its seventeenth session adopted resolution A.716(17) the IMDG Code and decided to give it a mandatory status under the umbrella of the SOLAS Convention, from 1 January 2004. However, some parts of the Code remain recommendatory.

paragraph 1.1.1.8 (Notification of infringements)

paragraphs 1.3.1.4 to 1.3.1.7 (Training)

Chapter 1.4 (Security provisions) except 1.4.1.1, which is mandatory.

Section 2.1.0 of chapter 2.1 (Class 1—Explosives, Introductory notes)

Section 2.3.3 of chapter 2.3 (Determination of flashpoint)

columns 15 and 17 of the Dangerous Goods List in chapter 3.2

the segregation flow chart and example in the annex to chapter 7.2

Section 5.4.5 of chapter 5.4 (Multimodal Dangerous Goods Form), insofar as the layout of the form is concerned.

Chapter 7.8 (Special requirements in the event of an incident and fire precautions involving dangerous goods)

Section 7.9.3 (Contact information for the main designated national competent authorities) Appendix B.

Since its introduction in 1965, the IMDG Code has undergone many changes, both in appearance and content, to keep pace with the ever-changing needs of the industry. Amendments that do not affect the principles upon which the Code is based may be adopted by the Maritime Safety Committee alone. Thus, IMO can respond to transport developments in a reasonable time.

Amendments to the IMDG Code are made (E&T Group) on a two-year cycle, these mainly originated from two sources; proposals submitted directly to IMO by Member Governments and amendments required to take account of changes to the United Nations Recommendations on the Transport of Dangerous Goods which sets the basic requirements for all the transport modes.

The IBC Code

The IGC Code

The INF Code

The BC Code

The International Grain Code

The Radioactive materials regulations

Key regulations include the International Maritime Dangerous Goods (IMDG) Code and the International Maritime Solid Bulk Cargoes (IMSBC) Code. The objective of these regulations is to establish procedures for proper handling, packing, labeling, documentation, and stowage of dangerous goods to minimize risks. Some core requirements include:

Proper classification of dangerous goods based on hazard classes and divisions. This helps identify the risks and determine appropriate precautions.

Packaging requirements like type of packages, labeling, and marking to facilitate identification and indicate risks.

Documentation procedures for transport documents, stowage plans, and shipper's declarations ensuring relevant parties have information.

Stowage and segregation provisions to prevent incompatible goods near each other and minimize hazards.

Reporting incidents involving dangerous goods and taking emergency response measures as per the EmS Guide.

The regulations apply to all parties involved in the transport of dangerous goods like shippers, carriers, and ports. They must take safety measures per the codes for packing, container loading, cargo handling, storage, and carriage.

With the extensive movement of dangerous goods by sea, following standardized regulations is crucial for the safety of life, cargo, and property and the protection of the marine environment. Implementation requires coordinated efforts between regulatory bodies, industry stakeholders, and the training of personnel.

Shippers must properly classify, document, pack, mark, and label dangerous goods as required by regulations like the IMDG Code. They must provide accurate shipping names, UN numbers, labels, markings, and other details so carriers can handle the cargo safely. For bulk cargoes like chemicals, shippers must ensure tanks are cleaned and provide instructions like maximum loading rates.

Shippers must also inform carriers about the hazards, precautions, and emergency response needed for the dangerous goods. Carriers rely on shippers, who know the cargo best, to provide complete safety information. Non-disclosure violates shippers' common law duty to warn and their statutory duties under laws like the Hague-Visby Rules. Providing incorrect or incomplete cargo details also breaches the shipper's duty not to mis-describe goods under Article III Rule 5 of Hague-Visby.

For packaged goods, shippers must ensure UN-approved packaging is used and the cargo is secured inside containers to withstand handling and the voyage. Improper packaging commonly leads to claims against shippers. Though stowage is normally the carriers' responsibility, for containerized cargo the shipper must load and secure the goods inside. Wrong stowage inside containers by shippers has caused many incidents.

The 24-h advance cargo declaration rule for U.S.-bound cargo also creates a shipper duty to provide detailed descriptions, packaging, weight, and shipper/consignee information to carriers well in advance of loading. This is critical for carriers to comply with U.S. customs requirements. Non-compliance can bring penalties against the carrier for incomplete manifest data.

Carriers have a paramount duty under Article III of Hague-Visby Rules to exercise due diligence to provide a seaworthy ship before and during the voyage. This includes properly equipping vessels to carry specific dangerous goods safely. The crew must also be trained and qualified for such cargo. Further, carriers must take due care of dangerous goods under Rule III-2, including careful handling and stowage to avoid damage. As stowage is in the master's control, the carrier remains responsible for proper stowage even if shippers load containers.

Carriers must verify shippers' dangerous goods documentation is complete before loading. They can reject cargoes lacking proper papers, packing, markings, or labels as per the IMDG Code. Checking for compliance provides an essential safeguard if shippers fail to fulfill duties. Similarly, under SOLAS carriers must not load undeclared or mis-declared dangerous goods. They must also issue the required dangerous goods manifest for each voyage.

However, carriers have no general common law duty to inspect cargo they do not know is dangerous. Their duties regarding unknown hazards are limited. Thus, shippers retain primary responsibility for the extensive IMDG Code compliance burden. Nonetheless, carriers should be aware of classes of goods with hidden hazards like self-heating properties and avoid complacency.

Overall, the distribution of responsibilities aims to ensure dangerous goods are transported safely. Both shippers and carriers have crucial roles at different points in the handling process. Current maritime liability conventions define these interlocking duties, which are supplemented by regulations and industry standards. Progressively tighter rules reflect heightened risk awareness, though cost and practicality require duties to not become unrealistically onerous. Achieving an equitable and viable sharing of the safety burden remains an ongoing challenge.

10.3 Dangerous Goods Transported by Ship

The term "Dangerous goods" or "Hazardous and Noxious Substances" (HNS) has a broad meaning, but also specific definitions in maritime law. The 2000 Protocol on Preparedness, Response, and Cooperation for Pollution Incidents by Hazardous and Noxious Substances defines HNS as any substance other than oil that can harm human health, marine life, amenities, or other legitimate uses of the sea if released. Therefore, HNS encompasses hazardous chemicals but also many substances not necessarily dangerous elsewhere. HNS cargoes are transported by sea in various ways—in bulk on gas, chemical, or bulk carrier ships, or packaged in containers, drums, and tanks on container vessels, ferries, and general cargo ships. While HNS has a wide-ranging meaning, the 2000 Protocol provides a clear definition focusing on the potential harm these substances can inflict on the marine environment.

The maritime transportation of dangerous and hazardous substances is governed by International Maritime Organization (IMO) regulations and codes. Transporting these cargoes requires understanding how they are moved in packaged, gas, liquid, or solid form.

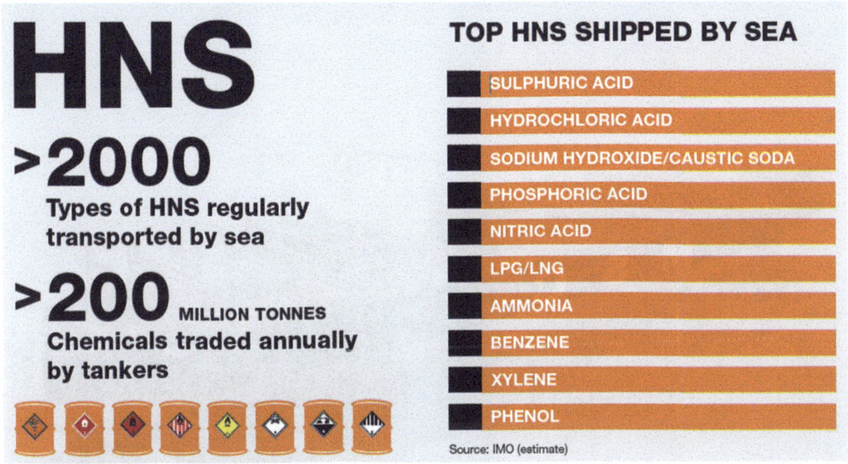

Fig. 10.1 Top DG/HNS shipped by sea

(i) ***In packaged form***, there are several ship types:

- **Containerships**: Containerships transport goods in containers. While mainly for general cargo, some can hold chemicals in portable ISO tanks. The largest can carry over 24,000 TEUs (Twenty-foot Equivalent Units). Tank containers carry liquids in a frame to stack with regular boxes (Fig. 10.1).

Smaller packages like drums and intermediate bulk containers (IBCs) may hold limited HNS quantities. Drums range from 20 to 200 L, made of plastic or steel. IBCs also contain solids and liquids.

Packages are often inside containers mixed with harmless goods. Other transport packages include sacks and boxes.

Proper packaging allows the transporting of diverse hazardous and noxious substances by sea while minimizing risks (Fig. 10.2).

- **General cargo ships**: One type of ship usually carrying cargo in relatively small consignments (i.e. crates, boxes, drums, or sacks), all placed in proximity in several large holds (Fig. 10.3).

- **Ro-Ro**: These are designed with facilities for quick loading and discharging with cargoes being moved on and off using road trailers or rail tank cars (Fig. 10.4).

(ii) **In liquid or gas form**, there are the following types of ships:

- **Chemical Tankers**—A specialized class of tankers designed to carry the most dangerous of liquid chemicals (Fig. 10.5).

Product Tankers—Ships generally built for carriage of a broader range of chemicals. Also referred to as parcel tankers because of the large number of individual tanks and pumping systems (Fig. 10.6).

Fig. 10.2 MSC Irina (24,346 TEU)

Fig. 10.3 General cargo ship

Gases

Gases are carried in liquefied form in gas carriers (gas tankers) where the gas is refrigerated or pressurized. The IGC Code (International Code for the Construction and Equipment of Ships Carrying Liquefied Gases in Bulk) identifies 32 specific substances. These include:

- *Liquefied Natural Gas (LNG)*

Fig. 10.4 Ro-Ro

Fig. 10.5 Chemical Tanker

Fig. 10.6 Product tankers

LNG is methane that is converted temporarily to liquid form for ease of storage or transport. International regulations require that LNG is transported in specially designed and constructed LNG carriers which have an excellent safety record (Fig. 10.7).

- Liquefied Petroleum Gas (LPG)

LPG is a variable mixture of propane and butane; normally propane (60%) and butane (40%), but the ratio can vary (Fig. 10.8).

– Other liquefied gases

These include ethylene, ammonia, vinyl chloride monomer, butadiene, and approximately 34 other substances.

Fig. 10.7 Liquefied Natural Gas

Fig. 10.8 Liquefied Petroleum Gas

The potential hazard from these substances if spilled at sea varies. Some (for example LNG, LPG, butadiene, and vinyl chloride monomer) are flammable, posing the risk of explosions or fires, while others (for example ammonia and chlorine) are toxic.

(iii) Solids

A bulk carrier, or "bulker", is a ship specially designed to transport unpackaged bulk cargo, such as grains, coal, ore, and cement in its cargo holds (Fig. 10.9).

Fig. 10.9 Bulk carrier

The regulations for the transport of solids in bulk are contained in the BC Code (Code of Safe Practice for Solid Bulk Cargoes), which is to be superseded by the International Maritime Solid Bulk Cargoes (IMSBC) Code, probably in 2011. The BC Code contains a list of 175 substances (with approximately 225 names) based on bulk materials that may liquefy during transport and possess specific chemical hazards. The major dry bulk cargoes, based on total tonnage transported at sea are:

Coal
Iron ore
Grain
Bauxite & alumina
Phosphate rock

Other dry bulk cargoes transported in lesser volumes include steel products, forest products, fertilizers, cement, scrap metals, sugar, soya meal, rice, pig iron, coke, and manganese. Although many of these materials have not been classified as dangerous goods, the release of these cargoes into the sea could, under some circumstances, cause negative effects on the marine environment or interfere with legitimate uses of the sea (Fig. 10.10).

ASEAN supplies around 50 million tons of LNG annually. The solid line in Fig. 10.11 shows that the LNG demand as a percentage of supply tends to increase. In addition, the chemical market in the ASEAN region (Fig. 10.12) is very bustling. Besides that, the ASEAN region is located in important seaborne trades, where maritime shipping cargoes and energy resources to the largest economies of Asia

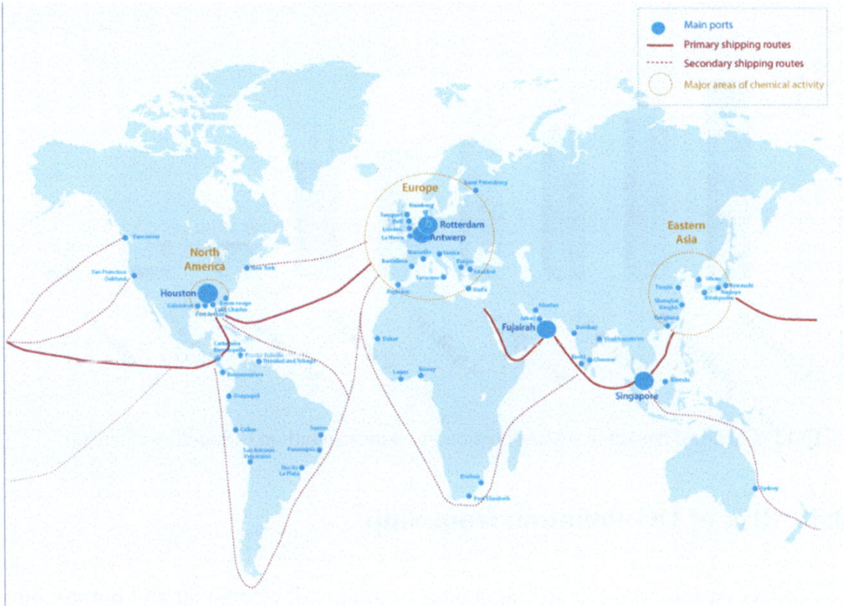

Fig. 10.10 HNS shipping routes

such as China, Japan, and Korea, resulting in a greater risk of incidental oil and HNS spills than any region in the global.

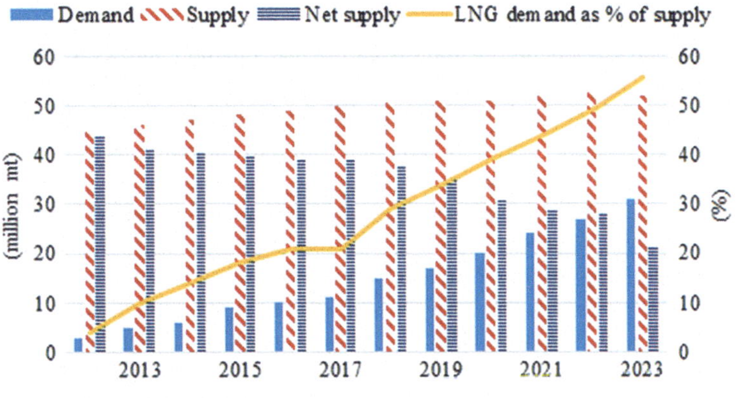

Fig. 10.11 LNG demand and supply in ASEAN region (S&P)

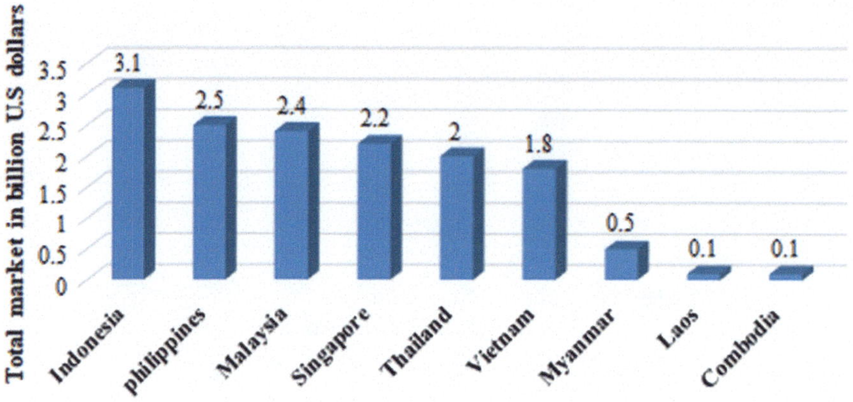

Fig. 10.12 Chemical market in ASEAN region by country in billion U.S. dollars (Statista)

10.4 Risk of DG Pollution from Ship

HNS or DG an important factor impacting economic development and human life, the region has been going through a significant increase in seaborne trade of oil and HNS (EIA U.S. 2016; Reynolds 2014; Behmiri and Manso 2013), contributing to great benefits to various localities, countries (Alidi and Al-Faraj 1994) in ASEAN regions. The risks of HNS or DG spill influences on marine ecosystems and coastal resources in the waters of other countries (Elliff and Kikuchi 2015; Chang et al. 2018), gave rise to losses of the fisheries, aquaculture, and tourism industries and consequences on social and environmental values. A well-known example of this is the mass fish deaths along the coast from Ha Tinh Province to Thua Thien-Hue Province in central Vietnam resulting from the phenomenon and Xyanua. HNS or DG pollution has become a general issue concern to the region (ITOPF 2017; Alvarez et al. 2014; Garza-Gil et al. 2006), and closer cooperation to HNS or DG pollution prevention and response within the region to minimize damage from HNS or DG spill incidents. Over two decades of working, the Regional Oil Spill Contingency Plan is undergoing internal approval by the ASEAN countries, which would oblige member states of ASEAN to cooperate. However, they have also been beset by one shortcoming a lack of HNS pollution response cooperation in the region.

This section examines the issue concerning historical pollution incidents in the ASEAN region caused by HNS or DG spill incidents, documents milestones in cooperation response to HNS or DG spills in the region, and proposes some recommendations for how to strengthen cooperation in the region (Fig. 10.13).

The rapid economic growth of ASEAN member states, making greater energy demands in Southeast Asia. Table 10.1 shows the status of the ASEAN countries over ten years from 2006 to 2016. The oil consumption of ASEAN member states has undergone a significant increase, the highest increase in Vietnam, peaking at

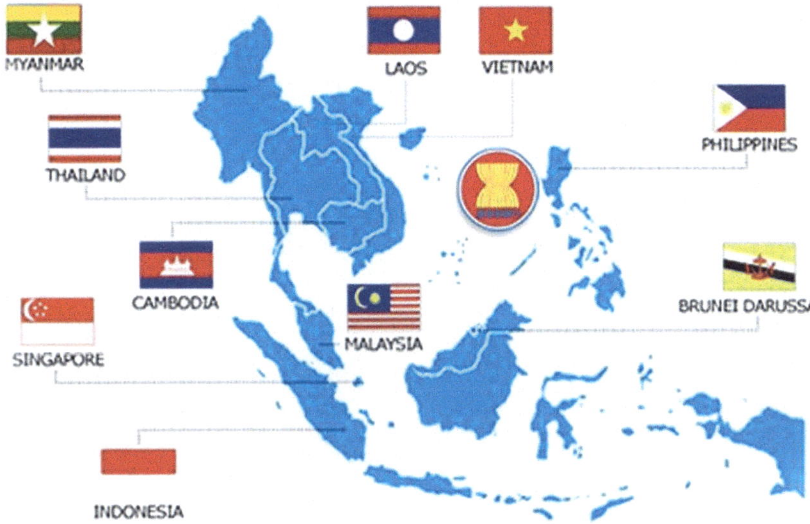

Fig. 10.13 Geographical ASEAN region in Oil and HNS spill response (infonet.vn)

about 69.7%. There have been no remarkable increases in oil refinery throughput of Indonesia Singapore, Thailand, and Vietnam in this period, about 18.4% and 15.4% respectively.

The increase in oil consumption shows the gross domestic product (GDP) growth rates of ASEAN countries from 2006 to 2015. Vietnam has experienced a 6.7% average GDP growth over five years from 2010 to 2015, the economies of the ASEAN region have a significant increase with an average GDP growth rate of 4.8% (Su 2015).

Table 10.1 indicates that there are many marine oil and HNS spills in the ASEAN region. Interestingly, in 1997 the tanker *Evoikos* collided with Crude oil tanker *Orapin Global* about 5 km south of the Singapore Port Limit, resulting in 28,463 tons of oil spills from tanker *Evoikos*. In 2008, 1000 tons of Sulphuric acid were spilled from M/V *Princess of the Stars* in the Philippines. On 7 July 2010, due to the fire on board Charlotte Maersk in the Strait of Malacca, 17.78 tons of Trichloroisocyanuric acid, 13.44 tons of Methyl ethyl ketone (MEKP), and 40 tons of Calcium hypochlorite were discharged into the sea. On Dec 29, 2013, the LNG carrier Al Gharrafa collided with containership *Hanjin Italy*. Figure 10.12 shows that the quantity of HNS spills in the ASEAN region is higher when compared to the Northwest Pacific Region (NOWPAP). Therefore, the enhancement of cooperation to control marine oil and HNS spill incidents around the ASEAN region appears to be necessary, taking into consideration NOWPAP (Korea, Japan, China, and Russia) experiences and case studies in the field of marine oil and HNS spill prevention and control (Fig. 10.14).

Table 10.1 Major HNS or DG spills in the ASEAN region

Name of spill	Type of spill	Spillage (tonnes)	Year	Country
JP4	DO, FO	1900	1973	Thailand
Showa Maru	Crude oil	3300	1975	Singapore/ Indonesia/ Malaysia
Aquarius	LNG gas	Unknown	1977	Indonesia
Elhami	Crude oil	2300	1987	Singapore
Nagasaki Spirit	-	13,000	1992	Malaysia
Neptune Aries	DO, FO	1800	1994	Vietnam
Tanker ONE	Crude Oil	140	1996	Thailand
Evoikos	HFO	28,500	1997	Singapore/ Malaysia
Multiara	Sangatla crude	150	1997	Indonesia
Princess of the Orient	FO, HFO	600	1998	Philippines
Mary Anne	INFO	711	1999	Philippines
Natuna Sea	Crude oil	7000	2000	Indonesia/ Singapore
CMA CGM Normandie	Fuel Oil	-	2001	Singapore
MT Singapura Timur	Bitumen	1500	2001	Malaysia
Indah Lestari	Phenol	630	2001	Johor Straits/ Malaysia
Neptank VII	FO	450	2002	Singapore
Dragon 1	FO	150	2004	Thailand
Solar 1	IFO180	8000	2006	Philippines
Unknown Spill	Crude oil	3000	2007	Vietnam
Princess of the Stars	Sulphuric acid	1000	2008	Philippines
Charlotte Maersk	Trichloroisocyanuric acid, Calcium hypochlorite solid, and methyl ethyl ketone peroxide liquid	71.22	2010	Malaysia
Bunga Alpinia	Methanol	6	2012	Malaysia
Dong Bac channel	Linear Ankyl Benzen (LAB)	300	2014	Vietnam
Vung Ang Factory	Phenol, Xyanua	-	2016	Vietnam

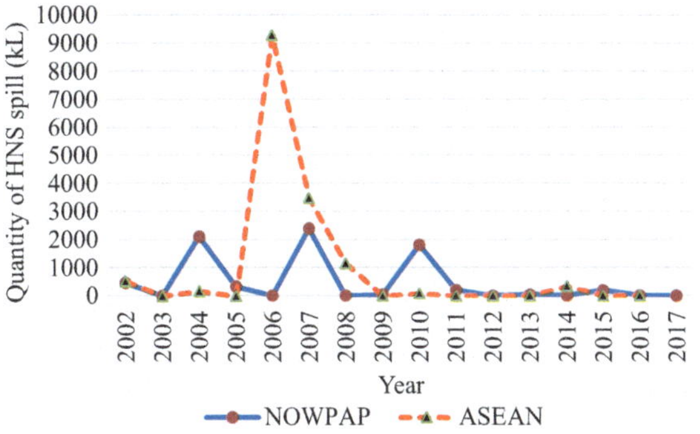

Fig. 10.14 Comparison of HNS spill in NOWPAP and ASEAN region

10.5 Contingency Planning of DG Incidents from Ship

International Convention and Legislation Framework.

The United Nations Convention on the Law of the Sea—UNCLOS was adopted on 10 December 1982. It entered into force on 16 November 1994. UNCLOS establishes extensive guidelines for the environment and management of marine natural resources. However, only a few of the provisions in UNCLOS refer to pollution by discharges of oil and hazardous and noxious substances as follows: To establish national regulations to prevent, reduce, and control pollution of the marine environment from vessels (Article 211); Regulate enforcement by Flag States (Article 217); Inspections and enforcement by Port States (Article 218 and 219); Enforcement by Coastal States (Article 220 and 221); and Pursue offenders (Article 111).

The mandate of IMO is to promote the safety and security of maritime shipping and the prevention of marine pollution by ships. The ratification status of ASEAN member states to various IMO conventions is very important because it pertains to HNS and oil spill preparedness and response. Table 10.2 shows the state of ASEAN countries' parties to international conventions on marine HNS and Oil Spill. Key conventions of the OPRC-90 and OPRC-HNS Protocol that adopted to facilitate effective preparation for combating oil and HNS pollution. Less than half of the ASEAN member states are parties to the OPRC-90 while Brunei, Cambodia, Indonesia, Lao, Timor-Leste, and Vietnam are not a party to the OPRC-90. An important part of the OPRC-90 and OPRC-HNS Protocol is the obligation of States Parties to ensure that oil and HNS contingency plans are devised and implemented for ships, offshore units, seaports, and oil/HNS handling facilities. Additionally, States that are parties to OPRC-90 and OPRC-HNS Protocol are required to build up their national capacity to prepare for and respond to Oil/HNS spill incidents either individually or through bilateral or multilateral cooperation, as appropriate, incorporating or encompassing shipping industries, industries dealing with oil/HNS, port authorities, national oil/HNS spills response centers, oil/HNS spills response private companies and other relevant entities such as the investment and development of equipment and chemicals to combating oil/HNS spills, being commensurate with the risk of oil/HNS spill incidents around its jurisdiction. Further, national training and exercise programs for experts in response to oil/HNS incidents at sea should be initiated with every regard to both quantity and quality of response manpower in every state party.

National Authorities

As shown in Table 10.3, the Marine Department of Brunei, Myanmar, Malaysia, and Thailand, the RGC of Cambodia, the Directorate General of Sea Transportation of Indonesia, the Coast Guard of the Philippines, the Maritime and Port Authority of Singapore and National Committee for Search and Rescue of Vietnam, are the primary government agencies responsible for oil spill response. Most of the countries, on-scene response activities are cooperation with other agencies that are a part of the public sector (see Table 10.3). ASEAN region is an oil exploration area, and large oil, and HNS shipping routes are passing through, leading to the high risk of large oil or HNS spill. A single country may not be able to meet the emergency response requirement, so comprehensive regional cooperation is needed.

Timeline of Cooperative Framework in the ASEAN Region

The Regional Contingency Plan needs to take into consideration preparedness in the sub-region as a platform of the regional framework. Sub-regional agreements between the ASEAN countries in Table 10.3 provide a comprehensive oil spill preparedness and response regime but are rarely updated and their status remains unclear. According to Table 10.3, the ASEAN Region had developed its own Contingency Plan for marine pollution due to oil spills in the early 1970s in the Straits of Malacca and Singapore. In addition, the Association of Southeast Asian Nations Oil Spill Response Action Plan (ASEAN OSRAP) was initiated in 1991 by the Japanese Government. This Action Plan was reinforced through many milestones: The six ASEAN member countries signed an MoU establishing the ASEAN OSRAP in 1993, Reinforce national stockpiles supported by Japan in 1994, four new member countries of ASEAN did not sign the ASEAN OSRAP MoU in 1995–1999, Global Initiative Program for Southeast Asia (GISEA) with four-times Regional workshop funded by IMO and IPIECA in 2009, 2011 and 2013 to enhance global preparedness and response capacity to respond to oil spill. As a result, on 28 November 2014, the ASEAN Transport Ministers (ATM) signed the MoU on ASEAN Cooperation Mechanism on Joint Oil Spill Preparedness and Response during the 20th ATM Meeting in Mandalay, Myanmar. The updated Plan is designed to ensure Control and Mitigation of Marine Pollution. However, this Plan did not concern HNS pollution response in the region while the action plans in other regions such as Northeast Asia (NOWPAP) and South Asia (SAS) are concerned with marine pollution caused by both oil and HNS spill incidents in the context of the region regime. Therefore, marine HNS spill response cooperation in the ASEAN region seems to be necessary (Tables 10.4 and 10.5).

The ASEAN region member states have undertaken cooperative efforts to protect the marine environment since the 1970s on both a bilateral and multilateral basis. There has been considerable focus on the development of oil spill preparedness and response in the ASEAN region. The efforts of the governments, regional bodies, the oil industry, and a range of other stakeholders and international organizations are in addition to the primary focus on the prevention of marine spills. However,

Table 10.2 State of ASEAN countries parties to international conventions on marine HNS and Oil Spill

	Conventions related to HNS and oil spill response	Conventions related to compensation								
	UNCLOS 1982	MARPOL 73/78	OPRC Convention	OPRC-HNS Protocol	London Convention	CLC 92	Fund Protocol 92	Fund Protocol 03	Bunker Convention	HNS Convention
Brunei	x	x	o	o	o	x	x	o	o	o
Cambodia	o	x	o	o	o	x	x	o	o	o
Indonesia	x	x	o	o	o	x	o	o	x	o
Malaysia	x	x	x	x	x	x	x	o	x	o
Myanmar	x	x	x	o	o	x	x	o	x	o
Philippines	x	x	x	o	x	x	x	o	o	o
Lao	x	o	o	o	o	o	o	o	o	o
Singapore	x	x	x	x	o	x	x	o	x	o
Thailand	x	x	x	o	o	x	x	o	o	o
Timor-Leste	x	o	o	o	o	o	o	o	o	o
Vietnam	x	x	o	o	o	x	o	o	x	o

Note x ratified; o: not ratified
Source http://www.imo.org/en/About/Conventions/StatusOfConventions/Pages/Default.aspx (As at 12/06/2023)

Table 10.3 Lead agencies, on-scene response organization, and national act for HNS and oil response

Country	Lead government agency	On-scene response	National/federal act
Brunei	Details as for spill notification point	Marine Department	National Oil Spill Contingency Plan (NOSCOP)
Cambodia	The Royal Government of Cambodia (RGC)	Ministry of Environment (MoE), (Department of Pollution Control (DPC))	Constitution of the Kingdom of Cambodia Law on Environmental Protection and Natural Resource Management
Indonesia	Directorate General of Sea Transportation (DGST)	DGST HQ	National Operations Centre for Oil Pollution (NOCOP)
Laos	-	-	-
Malaysia	Department of Environment (DOE) Malaysia Marine Department (MMD)	Marine Department of the Ministry of Transport Petroleum Industry of Malaysia Mutual Aid Group (PIMMAG)	Environment Quality Act (1974)
Myanmar	Department of Marine Administration (DMA)	DMA with the assistant of the Norwegian Coastal Administration (NCA)	Natural Disaster Management Law
Philippines	The Philippine Coast Guard (PCG)	Marine Environmental Protection Command (MEPCOM)	Oil Pollution Management Fund (OPMF)
Singapore	Maritime and Port Authority of Singapore (MPA)	EARL, SOSRC, and government agencies (Singapore Armed Forces, the Police Coast Guard, SCDF, etc.)	The Prevention of Pollution of the Sea Act (SPPA). Marine Emergency Action Procedure (MEAP)
Thailand	Marine Department (as part of the Ministry of Transport)	Department of Disaster Prevention and Mitigation, Oil Industry Environmental Safety Group Association (IESG), Royal Thai Air Force, The Royal Thai Navy, The Royal Thai Army, etc	National Oil Spill Contingency Plan
Timor	-	-	-
Vietnam	National Committee for Search and Rescue (VINASARCOM)	NOSRCEN, SOSRCEM, NASOS, etc	VPM No. 02/2013/ QĐ-TTg

Table 10.4 Multilateral sub-regional oil and HNS response cooperation

Geographic coverage	Sub-regional agreements	Member states	Date and location
Gulf of Thailand	Joint Statement of Cambodia, Thailand, and Vietnam on Partnership in Oil Spill Preparedness and Response in the Gulf of Thailand (GOT)	3: Cambodia, Thailand, Vietnam	Jan 12, 2006 Ha Noi, Vietnam
East Vietnam Sea (South China Sea) Waters within the National boundaries and/or jurisdiction of the Contracting Parties	Memorandum of Agreement between the Government of the Socialist Republic of Vietnam and the Government of the Republic of the Philippines on Cooperation in Oil Spill Preparedness and Response	2: Philippines, Vietnam	Oct 26, 2010 Ha Noi, Vietnam
Waters within the National boundaries and/or jurisdiction of the Contracting Parties	Memorandum of Understanding between the Republic of the Philippines and the Kingdom of Thailand on Joint Oil Spill Response	2: Philippines, Thailand	Nov 27, 1999
Malacca and Singapore Straits	Standard Operating Procedure (SOP) for Joint Oil Spill Combat in the Straits of Malacca and Singapore	3: Indonesia, Malaysia, Singapore	1981
Straits of Lombok, Makassar and Sulawesi Sea	Sulu-Sulawesi Sea Oil Spill Response Network	3: Indonesia, Malaysia, Philippines	
Waters within the national boundaries and/or jurisdiction of the Contracting Parties	Standard Operating Procedure for Joint Oil Spill Combat in the South China Sea including Brunei Bay	2: Brunei, Malaysia	1994

many issues to fulfill cooperation have been put on hold because of political, and social-economic aspects such as the political inertia of member states, overlapping maritime claims, the wide economic gap, and territorial sovereignty disputes.

10.6 Response DG Incidents from Ship

When DG/HNS is spilled at sea, measures must be taken to minimize physical damage to and pollution of marine resources and the environment. The measures of HNS spill responses are relatively limited and are often dependent on various factors such as physical characteristics of the HNS, meteorological conditions (sea state, sunlight, air temperatures), characteristics of the seawater, distance from shore, and considering safety (IMO 1988). The measure selected must also consider the protection of environmental sensitivity to marine and coastal resources and balance against

Table 10.5 Timeline in the development of regional preparedness in ASEAN

Date	Location	Milestone	Description
1991		Association of Southeast Asian Nations Oil Spill Response Action Plan (ASEAN OSRAP)	The ASEAN OSRAP was initiated in 1991 by the Japanese Government through the Japanese Association of Medical Sciences and Nippon Foundation under the ASEAN-Oil Spill Preparedness and Response (OSPAR) Project
May 20, 1993		The six ASEAN member countries signed an MoU establishing the ASEAN OSRAP	The main objective was to provide a cooperative plan for mutual assistance from Member States and organizations in the event of a major oil spill incident that exceeds the response capability of the national government. The geographic scope of the ASEAN OSRAP includes all the waters within the EEZ (Exclusive Economic Zone) of the ASEAN member countries (Brunei Darussalam, Indonesia, Malaysia, Philippines, Thailand, and Singapore)
1994		Japan sponsored the OSPAR (Oil Spill Preparedness and Response) to reinforce national stockpiles	1 billion yen was donated to the ASEAN countries for the purchase of oil spill equipment to reinforce national stockpiles in 11 key locations. The OSPAR project also enabled the development of the ASEAN Oil Spill Information Network System
1995–1999		Burma (now Myanmar), Cambodia, Lao PDR, and Vietnam joined ASEAN	Four new member countries of ASEAN did not sign the ASEAN OSRAP MoU
2009		The IMO set out to revitalize the MOU	The implementation of the original ASEAN OSRAP had met with challenges, as only six countries were party to the MoU. By holding a National Focal Points Meeting and a Workshop in 2009 hosted by Indonesia, the MoU was reviewed and a draft Strategy and Action Plan (SAP) was agreed to strengthen it

(continued)

Table 10.5 (continued)

Date	Location	Milestone	Description
Oct, 2011		Regional Workshop funded by IMO	A follow-up Regional Workshop funded by IMO made policy recommendations on the establishment of a Regional Cooperation Mechanism on oil spill preparedness and response including the development of a regional action plan. It was also agreed that the revised MOU would include all the 10 ASEAN member states within the Brunei Action Plan. This was followed up by another IMO-funded Regional Workshop a year later where the draft MOU for the ASEAN Cooperative Mechanism was drafted for submission to the 24th meeting of the ASEAN MTWG
Mar 19, 2013	Jakarta, Indonesia	The 4th IMO-funded Regional Workshop for the establishment of a Regional Cooperative Mechanism for Oil Spill Preparedness and Response in the ASEAN region	The outcome of the workshop included the final draft MoU on Joint Oil Spill Preparedness and Response in the ASEAN region for endorsement by the 35th session of the Senior Transport Officials Meeting (STOM) and an action plan to implement the provisions of the draft MoU
Aug 22–24, 2013	Yangon, Myanmar	The draft MoU was subsequently approved	The final text of the draft MoU was subsequently approved during the 26th Meeting of the ASEAN MTWG held in Myanmar August 22–24, 2013
Nov 28, 2014	Mandalay, Myanmar	Signing ASEAN OSPAR MOU	The ASEAN Transport Ministers (ATM) signed the MoU on ASEAN Cooperation Mechanism on Joint Oil Spill Preparedness and Response (ASEAN OSPAR MOU) during the 20th ATM Meeting in Mandalay, Myanmar
Mar 1, 2018	Singapore	ASEAN Regional Oil Spill Contingency Plan (ROSCP)	GISEA (at the 35th MTWG meeting) provided an update to the ASEAN Member States on the progress of the development of the draft ASEAN Regional Oil Spill Contingency Plan (ROSCP), as an implementation of Article 2 (c) of the MOU on ASEAN Cooperation Mechanism for Joint Oil Spill Preparedness and Response

the effects of the response itself and the effectiveness of each clean-up measure. Response measures are well described in this literature, the key characteristics being as follows:

Natural recovery is defined herein as the return of natural resources to a dynamic baseline, breaking down the HNS with no active human intervention. In an oil spill, biological communities will generally recover naturally to baseline conditions about as quickly and effectively as they will with active human interventions, and at a lower cost and lower commitment of people, equipment, and materials. Understanding of HNS spill consequences and monitoring of natural recovery are requirements to plan whether a response may be required or not.

Chemical dispersants: are capable of quickly removing significant quantities of HNS spill on the sea surface by transferring it into the water column. This measure can achieve significant environmental and economic benefits. However, this solution has its limitations and accounts must be taken characteristics of HNS, physical conditions, and environmental sensitivities. This requires consensus on the level of influence of the water column based on forecasted exposure profiles, the effectiveness of the solution, and the monitoring of dispersant oil supported.

Containment and recovery of HNS on water measures is to collect and recover the spilled HNS by using response equipment such as booms, and skimmers. The requirement of this measure is a specific decision-making framework to select areas of recovery effectively under present conditions.

Shoreline clean-up: regarding response measures taken at sea, some HNS will reach the coast, and cleaning operations may be necessary to allow the coast to recover its natural state and socio-economic value by using possible manual and mechanical techniques. This requires an evaluation of effective equipment and clean-up teams to choose the best method in each case depending on various factors like the type of coastline, tide range, meteorological conditions, collection rate and effectiveness of available equipment, cost, impact of cleaning operations on the environment, and national policy.

Other response measures: For example, Institu burning measure (HNS on water or shoreline) and bioremediation measure (microbial breakdown of HNS). The secondary response measures are cited in the contingency plans. Interventions must be assessed for effectiveness and feasibility considering the environmental benefit for the spill. Generally, available, and preferred response measures should be designated, with stakeholders before an HNS spill occurs.

As an example, response options to DG/HNS incidents at sea and in ports should be explained by using the factors below.

<Available response options at sea>
*Vessel
– Extinguish fire
– Reduce or contain fire
– Move vessel

– Scuttle vessel
– Monitor
*Cargo
– move cargo
– stop spill
– protect cargo
– destroy cargo
– immobilize cargo
– activate biological degradation
– monitor
*Released HNS
– dispersion
– chemical treatment
<available response options in port>
*Vessel
• Extinguish fire • reduce or contain fire
• Move vessel • scuttle vessel• monitor
*Cargo
• Move cargo • stop spill
• Protect cargo • destroy cargo• immobilize cargo
• Activate biological degradation • monitor
*Released HNS
• dispersion • chemical treatment
• containment • recovery
• Biological treatment • destroy
• immobilize • monitor

Arrangement of organizations and equipment

The organizational arrangement for proper response to the HNS incident should be explained, with the specific roles of the group, tools, and equipment. For example, the monitoring group has the role of reporting, monitoring analyzing situations of spill, and preparing and using remote sensing tools, transceivers, VHF…

Arrangement of Safety Equipment: about importance and drawbacks of personal safety such as heat stress, and claustrophobia should be clearly explained to trainees.

Nine steps of HNS response [IMO]

Basic response steps or activities that describe what the OSC must consider when responding to an HNS incident. As the nine steps will form the basis of the response strategy, which the OSC will be required to consider, lecture shall be made clearly by using videos and slides so that trainees may understand the key essence and requirements of each step as follows:

(a) Step One—Assess the situation

The first step of the HNS response is to assess the situation.

- Report and document accurately all incoming incident information using a standard reporting form.
- Identify the type and source of products involved.
- Quantify the size of the spill.
- Assess the hazard implications, e.g., fire and explosion, human health, shoreline type, and effects on flora and fauna. Ensure that actions are prioritized.
- Identify the threat to people, environment, and facilities.
- The OSC needs to prioritize what actions will be taken.
- An important point is that the OSC identifies the political and economic significance of the spill.
- Gather samples or other supporting data sufficient to allow for the identification of the substances to take place.

(b) Step Two—Activate Contingency Plan

Step two is to activate a contingency plan.

- Inform the senior manager of the organizational procedure and content of the plan.
- Notify appropriate authorities and key personnel according to the plan.
- Activate initial response. Decide how, where, and when. Coordinate identification and mobilization of initial response resources (personnel and equipment) as per the plan.

(c) Step Three—Activate Organisational Response

Step three of the response is to activate the organization's response, according to the agreed strategy plan, including:

- Mobilising and organizing the response team.
- Identifying, assigning, and delegating tasks.
- Establishing both internal and external communication channels.
- Set up a command post in the safe zone (site equipment, safety and security, procedures/resources).
- Implement safety, security, and decontamination procedures.
- Notify, mobilize, and coordinate contracted and subcontracted services.
- Assess needs, define roles, and execute contracts.

(d) Step Four—Prepare Response Action Plan

Step four is to prepare a response action plan strategy. Response organization should.

(i) Identify and prioritize resources at risk e.g., local populace, type of shoreline, local infrastructure. (ii) Identify the resources required, taking into consideration the influence of the sea and weather conditions on the spill properties, the behavior of the HNS, and spread rate and identify additional resources, if necessary (e.g. regional/international agreements), and so on.

(ii) Identify and prioritize response strategies, such as for monitoring, containment and recovery, chemical treatment, shoreline clean up, transfer, and disposal.
(iii) Maintain accurate records as required e.g., log and time-keeping procedures for contractors and subcontractors, rented equipment, consumables.
(iv) Develop a response action plan to cover items such as any resource time available, resources available, containment plan, regular briefings on the response action plans, etc.

(e) Step Five—Activate Operational Response

Step five involves activating the operational response. This involves:

(i) Directing and supervising all monitoring, investigation, recovery, clean up, and decontamination procedures.
(ii) Monitoring adherence to plans.
(iii) Coordinating containment, recovery, and disposal of all pollutants.
(iv) Conducting regular briefings.
(v) Deploying personnel and ensuring logistical support.

(f) Step Six—Managing the Response

Step six is managing the ongoing response. This involves gathering, assessing, and adjusting response information, and interpreting and reviewing reports as follows:

(i) Assessing, maintaining, advising, directing, and controlling response team activities, and plans.
(ii) Monitoring, evaluating, and adjusting strategies as required.
(iii) Monitoring site safety plan and occupational hygiene.
(iv) Enforcing regulations and procedures.

(g) Step Seven—Deactivate the Response

Step seven involves deactivating the response. It should be determined whether the response should be continued, suspended, or terminated, and advise the OSC accordingly. Rank the termination criteria relevant to the operations by using effort/benefit analysis.

(i) Assess the potential for recontamination.
(ii) Coordinate support for scientific monitoring of the area.
(iii) Shut down field operations and conduct an orderly termination of operations.
(iv) Maintain, repair, and replace equipment as appropriate.

(h) Step Eight—Consolidate Costs

As this is the key point for the next stage of the response and follow-up, trainees must be instructed to keep detailed records by taking the following actions:

(i) Verify/certify costs.
(ii) Plan for record and filing maintenance.
(iii) Consolidate records and produce a report of expenditure by category.

(iv) Produce costs documentation report.
(v) Construct a simple coding structure and produce a final cost report.

(i) Step Nine—Debrief the Response

Step nine is to debrief the response. Trainees learn that OSC is required to take the following actions at this step:

(i) Conduct an operational review of the response with the response teams and analyze the strengths and weaknesses of the response.
(ii) Conduct a review with the OSC and analyze the strengths and weaknesses of the response. Identify the lessons learned and the areas that may require improvement.
(iii) Consolidate information, documentation, and evidence. Submit and circulate the final operations report.
(iv) Make and circulate recommendations for improved preparedness.
(v) Implement the lessons learned.
(vi) Update contingency plan.

After explaining the nine basic steps, the trainees must be reminded about the very basic key points so that trainees may take timely actions, effectively and properly during the HNS incident. Trainees are taught to recognize that an OSC and his team can only function effectively if provided with the structure and organization necessary to mount a response in a timely and appropriate manner before the HNS has a chance to affect the local populous or sensitive areas. It should, therefore, be emphasized that:

(i) Sometimes, this is not possible due to the scale of the emergency, the location, or the prevailing weather conditions.
(ii) The probability of small spills can be high, but with a rapid response the damage they cause is often minor and their clean-up costs are relatively low.
(iii) Small spills are ugly and a constant reminder of the risks of much larger spills. A good response to small spills will ensure an effective response to larger spills.
(iv) Regardless of size, marine HNS spills tend to become unmanageable very quickly.
(v) The key to effective response is to have a comprehensive, realistic, and well-exercised contingency plan, a minimum of appropriate equipment, and a response organization of highly trained individuals who will react very quickly and knowledgeably.

Trainees are also reminded that there exist some important interested parties and that the OSC and his team should always pay attention to these parties. Therefore, it is important to explain further that:

(i) During a major incident, there will be a multiplicity of agencies involved that can be of great assistance in the response. These will include government agencies and private groups alike.
(ii) In terms of government groups, most countries have developed joint command centers for spill response or unified command teams that include many agencies.
(iii) In addition to government agencies, there will also be an influx of outside individuals such as media persons, equipment, salespeople, and curious parties of all sorts, who have legitimate interests and some not so legitimate. It is important to recognize that such groups will inevitably be part of the scene and that mechanisms must be established to communicate with them on a fair basis without interfering with the emergency job at hand at the same time.
(iv) The OSC and his team must be allowed to implement the response strategies with the minimum outside interference.

10.7 Management After DG Incident from Ship

DG or HNS transported by ships pose major risks if spilled at sea. ASEAN nations with extensive coastlines and marine resources must be prepared to respond and manage the impacts. When HNS cargoes are spilled, timely assessment and monitoring of contamination levels in seafood is critical. However, HNS encompasses diverse chemicals, requiring customized approaches based on their fate and toxicity.

After a major HNS spill in ASEAN waters, a coordinated regional response should immediately collect information on the type and quantity released. Authorities must rapidly assess risks to human health and marine ecosystems. Spill modeling and monitoring can identify likely transport pathways and impacted areas. Monitoring is challenged by the variety of possible HNS. Sensory analysis may detect tainting in seafood, guiding chemical testing. Targeted sampling and chemical analysis should focus on priority toxic substances of concern based on the spilled HNS. Testing of caged mollusks and farmed fish is crucial as they cannot evade contamination.

To expedite reopening harvest areas, ASEAN nations should agree on regional standards for HNS contamination limits in seafood, referencing Codex Alimentarius and WHO guidelines. Concentrations posing negligible risks, if consumed even by sensitive groups, should be defined. However, localized context and dietary patterns may require modified limits. Benchmark dose and margin of exposure approaches can support evidence-based decisions on safe levels.

Following the spill, a tiered process to reopen fisheries is advised. Initial visual surveys and sensory testing may allow rapid partial reopening of cleaner zones. Progressive chemical sampling would support fully reopening larger areas later. Clear protocols and criteria should balance socioeconomic needs with consumer safety. Communications must assure the public about thorough testing and oversight.

Spill response requires extensive capabilities and infrastructure. ASEAN cooperation can pool expertise and resources through integrated networks of analysts, laboratories, monitoring equipment, and vessels. Common operating procedures, information systems, and training will enable coordinated mobilization. Response planning must engage maritime, fishery, health, and environment authorities.

Lessons on managing seafood contamination should be gleaned from major spills worldwide. Reviews help tailor best practices to ASEAN contexts, considering local fisheries and consumption patterns. However, severe data gaps hinder region-specific HNS risk assessment currently. Priority research can generate key toxicity data for commonly transported HNS relevant to tropical species.

Overall, ASEAN's resilience to HNS marine spills hinges on cross-border coordination. This includes harm mitigation research, monitoring assets, response plans, seafood standards, and communications. No single nation has all the capacities required, but pooling resources can enable ASEAN to better manage HNS impacts on vital marine resources and coastal populations. With proper preparation, ASEAN can respond decisively to safeguard seafood safety and livelihoods after HNS incidents.

10.8 Conclusion

The ASEAN region member states have undertaken cooperative efforts to protect the marine environment since the 1970s on both a bilateral and multilateral basis. There has been considerable focus on the development of oil spill preparedness and response in the ASEAN region. The efforts of the governments, regional bodies, the oil industry, and a range of other stakeholders and international organizations are in addition to the primary focus on the prevention of marine spills. However, many issues to fulfill cooperation have been put on hold as a result of political, and social-economic aspects such as the political inertia of member states, overlapping maritime claims, the wide economic gap, and territorial sovereignty disputes. Moreover, beyond the unclear geographical cooperation of the ASEAN region, it is necessary to develop a concept of the geographical scope of the region within the ASEAN area to include DG/HNS pollution as well as radioactive pollution. Particularly, the ASEAN member states need to strengthen legal instruments at both domestic and regional levels for oil and HNS spill preparedness and response. Finally, the ASEAN member states need to establish the ASEAN trust fund to which the ASEAN members will contribute and support an annual target amount of specific money every year for promoting HNS and the oil spill response capability of the ASEAN region.

References

Alidi AS, Al-Faraj TN (1994) Locating oil spill response equipment and facilities: a cluster analysis approach. GeoJournal 34(4):379–382. https://doi.org/10.1007/BF00813132

Alvarez S, Larkin SL, Whitehead JC, Haab T (2014) A revealed preference approach to valuing non-market recreational fishing losses from the deepwater horizon oil spill. J Environ Manage 145:199–209. https://doi.org/10.1016/j.jenvman.2014.06.031

Behmiri NB, Manso JRP (2013) How crude oil consumption impacts on economic growth of Sub-Saharan Africa? Energy 54:74–83. https://www.cheric.org/research/tech/periodicals/view.php?seq=1191210

Chang Y-T, Park H, Lee S, Kim E (2018) Have Emission Control Areas (ECAs) harmed port efficiency in Europe? Transp Res Part D Transp Environ 58:39–53. https://doi.org/10.1016/j.trd.2017.10.018

Elliff CI, Kikuchi RKP (2015) The ecosystem service approach and its application as a tool for integrated coastal management. Natureza Conservação 13(2):105–111. https://doi.org/10.1016/j.ncon.2015.10.001

EIA U.S. (2016) Annual energy outlook 2017 with projections to 2050, vol 64

Garza-Gil MD, Surís-Regueiro JC, Varela-Lafuente MM (2006) Assessment of economic damages from the prestige oil spill. Mar Policy 30(5):544–551. https://doi.org/10.1016/j.marpol.2005.07.003

IPIECA-IOGP (2015) The global distribution and assessment of major oil spill response resources. Oil Spill Response Joint Industry Project. http://www.oilspillresponseproject.org/wp-content/uploads/2016/02/JIP-9-Response-resources.pdf

ITOPF (2017) Country profile: Vietnam - ITOPF. http://www.itopf.com/knowledge-resources/countries-regions/countries/vietnam/

Reynolds DB (2014) World oil production trend: comparing hubbert multi-cycle curves. https://www.sciencedirect.com/science/article/pii/S0921800913003819

Shaverdi M, Heshmati MR, Ramezani I (2014) Application of fuzzy AHP approach for financial performance evaluation of Iranian petrochemical sector. In: Procedia computer science, 2nd international conference on information technology and quantitative management, ITQM 2014, vol 31, pp 995–1004. https://doi.org/10.1016/j.procs.2014.05.352

Su (2015) Looking ahead the port industry—how Asean Ports respond to the changing global maritime trade.Txt

Chapter 11
Control of Environmental Pollution from Maritime Activities

Tran The Nam, Nguyen Cao Hien, Nguyen Xuan Long, Nguyen Thanh Son, Nguyen Thanh Dieu, Ngo Nhu Tai, and Phan Van Hung

Abstract Along with economic development and scientific and technological progress, maritime activities play an increasingly important role in the global economy. Along with that, maritime activities still cause serious environmental pollution problems such as oil pollution, hazardous and noxious substance pollution, wastewater pollution, ballast water, garbage pollution, air pollution, noise pollution, pollution caused by sinking substances, etc. Therefore, the International Maritime Organization has made efforts to introduce measures to improve safety, security, prevention and control of environmental pollution through measures. Convention, Codes and related guidelines and regulations. And controlling environmental pollution from shipping is a key issue that determines the development of the maritime industry. The chapter focuses on issues such as oil pollution, hazardous and noxious substance pollution, wastewater pollution, ballast water, garbage pollution, air pollution, noise pollution, pollution caused by sinking substances. The content of the book chapter is both theoretical and practical; is the basis for state management officials at seaports, officers, seafarers and related organizations and individuals to understand and apply to prevent environmental pollution in maritime activities as well as be ready to respond when environmental pollution incidents occurred in compliance with statutory compliance conventions. The book chapter is systematically compiled, based on domestic and international legal regulations in the maritime domain, and is composed of 11 sections covering pollution issues in maritime activities: Oil pollution from ship; HNS pollution; Waste water pollution; Ballast water management; Pollution caused by garbage; Dumping Substance pollution at sea; Air pollution and GHG emissions; Antifouling system; Salvage sunken property; Ship breaking; Noise and vibration pollution. This chapter is useful for students, graduate students, scientists, lecturers and experts in the maritime field, especially marine officers. It is one of the important foundations that contribute to perfecting knowledge and skills for practical application.

T. T. Nam · N. X. Long · N. T. Son · N. T. Dieu · N. N. Tai · P. V. Hung (✉)
Vietnam Maritime University, 484 Lach Tray, Le Chan, Haiphong, Vietnam
e-mail: phanvanhung@vimaru.edu.vn

N. C. Hien
Ministry of Transport, 80 Tran Hung Dao, Hoan Kiem, Hanoi, Vietnam

V. Maccarrone and M. Fadzil Akhir (eds.), *Coastal Sustainability*, Coastal Research Library 39, https://doi.org/10.1007/978-3-031-75749-5_11

Keywords National demand · Oil spill recovery · Maximum amount of oil spill · Oil remaining

11.1 Control Oil Pollution from Maritime Activities

11.1.1 National Demand for Marine Oil Spill Recovery

Vietnam is located in the tropical monsoon area, is the oil-producing country with more than 3444 km of shoreline, thousands of islands, rivers and canals. The frequency of exploration and production of offshore oil and gas and the amount of oil and petroleum products transported by vessels that enter and depart from Vietnamese ports and refineries are increasing, resulting in the big risk of a marine oil spill from oil transportation activities around Vietnam (Hung and Kim 2019; Eight Institutions of Europe and Vietnam 2006). However, it is not known how much the amount of marine oil spill response demand should be in Vietnam. Furthermore, the oil spill response demand plays supposedly a vital role in setting up national response capability to combat a marine oil spill incident in the future. Therefore, the criteria of marine developed countries for oil spill response and the current condition based on oil transportation, offshore activities in Vietnam were analyzed. Therefrom the maximum amount of oil spill and oil remaining on the water were calculated in order to estimate Vietnam national demand for marine oil spill recovery from the sea surface. The oil recovery demand of Vung Tau and Thanh Hoa area was estimated about 18,750 tons oil group 3 and 21,563 tons of oil group 4 for 3 days respectively (Nguyen and Nguyen n.d.). The data suggest that both the amount of oil spill and oil remaining are important to calculate oil recovery demand. The results are expected to be science platform to support policy making, and contingency plan making for preparedness and response to marine oil spills.

(1) Introduction

Marine pollution control policy needs to identify the major sources of pollution in the sea area, which can be assumed by pollution incidents along oil transport routes. The worst case discharge of oil tanker can be estimated through the maximum amount of oil spill by hypothetical oil outflow (Phan 2017), thereby the contingency plan was made to ensure equipment, personnel, and additional support for each area, including mobilization of oil spill response (OSR) resources in other areas as well as international assistance. Actually, the advanced countries such as USA, Canada, Japan, and South Korea have assumed the largest oil spill area by region from the target vessel for the estimation of national marine OSR capability. From there, the forces and facilities are built to ensure proper management and flexible mobilization between regions.

Vietnam mentioned the possibility of responding to oil spillages at various levels, such as grassroots, regional and national levels, in legal documents. In cases where the level of oil spill is high or complicated and beyond the capability of the units, the forces and equipment shall be requested to the superior agencies and specialized units in the localities in order to dispatch to spill site. However, there is no specific document or research specifying the criteria for calculating and building oil spill response capabilities such as target vessels, maximum oil spills from ships, the percentage of oil response at sea, duration of response process as well as a model for calculating daily response to equipment such as oil skimmer, oil boom, and oil temporary storage in Vietnam.

Therefore, this study will be analyzed on the criteria of marine developed countries for oil spill recovery and the current condition based on oil transportation, offshore activities in Vietnam, maximum amount of oil spill and oil remaining on water that directly affect the basic standards to estimate national demand for marine oil spill response such as requirements of equipment and facilities in Vietnam.

(2) Concept of National OSRC

Oil spill response capability (OSRC) is defined as the resources to deal with the oil spill incident, including three categories: equipment, response personnel, and additional support (IPIECA-IOGP 2015). Vietnam's Government has built 3 National Oil Spill Response Centers. There are 24/28 of oil spill contingency plans of Coastal Provinces and various Contingency plans in-house level that considering about historical spill have approved. There are some solutions that have been proposed to improve oil spill response capability such as response measures for oil spills in Da Nang City (Phung 2005), enhancing local capacities in oil spill preparedness and response (Nguyen and Nguyen n.d., 200). However, Vietnam is still a lack of scientific criteria related to national demand for marine oil spill response, and the scientific and systematic researches on criteria is necessary to estimate national oil spill response demand in Vietnam. In term of this study, oil spill response demand (OSRD) is defined as the estimating amount of oil remaining on surface of sea for specific response duration. The OSRD can be calculated from the maximum oil discharge (MOD) based on hypothetical oil outflow (HOO) from the largest oil tankers arrival and departure area (IMO 2003).

The OSR capability set by the advanced marine countries is often referred to as the oil recovery capabilities that means the total amount of spilled oil can be recovered from the surface of sea to the maximum. It is named as national OSR capability in nation level, as regional OSR capability in regional level and as local OSR capability in local level. The main OSR equipment are oil recovery devices (oil skimmer), oil boom and temporary oil storage tank. The concepts of national OSR capabilities established in advanced countries are used to mean the total amount of oil that can be recovered from sea surface to the maximum by the oil recovery system in the national level. Therefore, a national OSR capability is sometimes referred to as marine oil recovery capability (OSRC) of a nation.

In the advanced maritime countries, the biggest source of marine oil spill is assumed to be the largest vessel or oil tanker operating in their territorial sea. The major oil transport routes in Vietnam's waters and the largest tankers arrivals and departures ports are considered to select target vessel in oil spill recovery by region.

Vietnam has been carrying out effective exploration and production of oil and gas, about 420,000 barrels per day. Vietnam's gas and oil potential locates in 7 basins: Cuu Long basin, Con Son basin, Song Hong basin, Malay-Tho Chu basin, Phu Khanh basin, Hoang Sa basin, and Truong Sa basin group. There are 7 types of crude oil were made from different oil field such as Bach Ho, Dai Hung, Rang Dong, Bungakekwa/ Cai Nuoc, Su Tu Den.

The oil transport route in Fig. 11.1 and the major oil tankers operating in Vietnamese waters in Table 11.1 have shown that the risk of oil spills from oil tankers of 300,000 DWT exists in all three regions of the North, Central and South. Oil tankers of about 150.000 DWT carry crude oil from oil and gas exploration offshore in Vung Tau area to petrochemical refineries such as Dung Quat oil refinery with high operating frequency. In addition, the East Sea is one of the world's busiest international sea-lanes. More than half of the world's oil tankers passes through the East Sea every year. Over half of the world's merchant fleet sails through the South China Sea every year. More than half of the world's annual merchant fleet tonnage passes through the Straits of Malacca, Sunda, and Lombok, with the majority continuing on to the East Sea. Almost one third of global crude oil and over half of global LNG trade passes through the East Sea, making it one of the most important trade routes in the world. The frequency of operations of oil tankers has been gradually reduced from the South to the North, from the offshore oil field and the oil import of Middle East route to petrochemical refineries. Thus, in order to estimate OSRC, the biggest source of oil spill is assumed to be 300.000 DWT of crude oil tanker operating in the three areas, North–Central–South areas of Vietnam (Fig. 11.1 and Table 11.1).

(3) Estimation Methods of National OSRC and OSRD

Calculating Maximum Amount of Oil Spill

According to IPIECA's recommendations (IPIECA 2000), the largest oil spill from a tanker is the amount of oil contained in two cargo tanks of oil tanker. For 300,000 DWT of target vessels, the volume of central cargo oil tank is 30,000 m^3, so the largest oil spill is $(2 \times 30{,}000) = 60{,}000$ m^3. This calculation is based on oil spill historical statistics before 2000 that was introduced by IPIECA. However, nowadays the structure and design of the oil tanker have been changed significantly. Beside the seafarers are heavily supported from development of science and technology. As a result the incident rate tends to decrease in both the number of cases and the amount of oil spill (ITOPF 2017a).

Therefore, it would be more scientific and practical to apply the Regulation 24 and Regulation 25 of Annex I, MARPOL 73/78. The estimation of maximum amount of oil spill for 300,000 DWT of oil tankers achieve about 28,000 tonnes for single hull and about 45,000 tonnes for double hull (Phan 2017) respectively, as there are differences in the arrangement of the oil tanks. The results of estimation of maximum oil spill in the area are shown on Fig. 11.2.

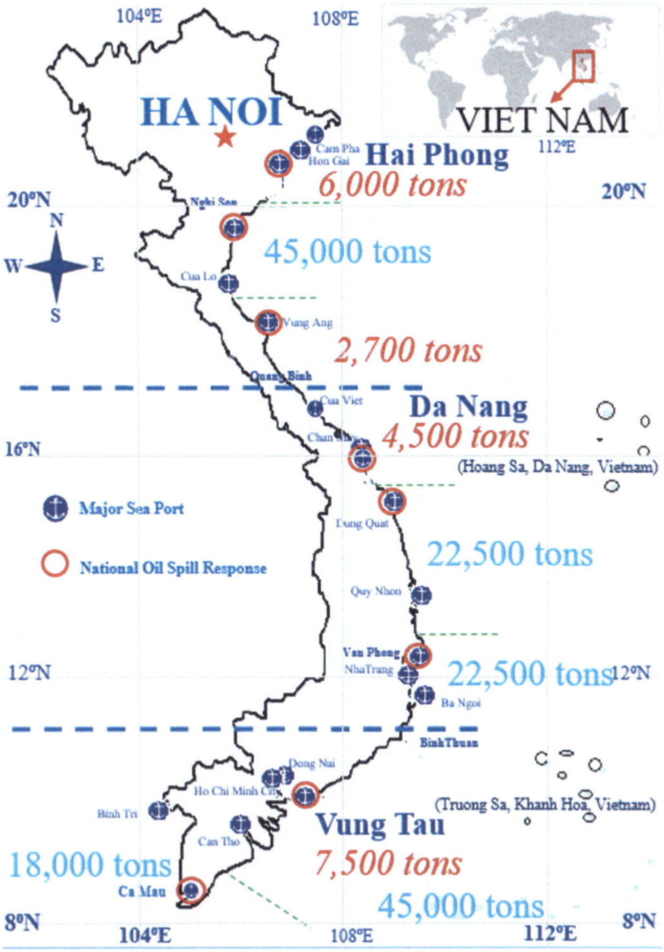

Fig. 11.1 Estimating max. amount of oil spill

Vietnam exploits 17 types of crude oil at domestic oilfields and one from Algeria to provide Dung Quat refinery with the demand for raw material for production. Crude oil from Vietnamese oil fields has a high freezing point of temperature, which fluctuates from 20 to 39 °C, with high paraffin content, its viscosity at 50 °C ranges from 6.05 to 65.72 mm^2/s (DOJ 2015). Nghi Son refinery imports crude oil from Kuwait for production and supply of oil productions for domestic market as well as export. In spite of many ways to classify types of oil, the method of dividing oils into 4 groups is used in this study as shown in Table 11.2 (Fingas 2000; ITOPF 2014).

Oil group 1 is capable of rapid natural dissipation, and completely dissipation on day 6. In Vietnam waters, the most of oil has transported in oil group 2 and oil group

Table 11.1 Vietnam's provinces and port areas which oil tankers are entering and leaving

Province	Port area	Oil tanker (DWT)
Quang Ninh	Cai Lan oil port	40.000
Hai Phong	BC Petec, PTSC Dinh Vu	20.000
Thai Binh	Tien Hai gas field	–
Thanh Hoa	SPM Nghi Son	300.000
Nghe An	DKC oil port	30.000
Ha Tinh	Vung Ang total storage oil	18.000
Quang Binh	Gianh River oil port	1.000
Thua Thien Hue	Thuan An oil port	2.000
Da Nang	My Khe oil port	30.000
Quang Ngai	SPM Dung Quat	150.000
Binh Đinh	Quy Nhon oil port	10.000
Phu Yen	Vung Ro port	10.000
Khanh Hoa	Van Phong port	150.000
	Mui Chut oil port	10.000
Ba Ria-Vung Tau, Dong Nai	Cai Mep oil port	50.000
	PVC port	50.000
	Long Son port	30.000
	PTSC port	10.000
	Phu My Electric Co. oil port	10.000
	K2 oil port	7.000
	09 oil and gas fields (Rong Doi, Rang Dong, Hong Ngoc, Lan Tay, Su Tu Den, Dai Hung, Chi Linh, Ba Vi, Vietsopetro01)	150.000 300.000
Ho Chi Minh City	Nha Be oil port	20.000
	Petechim oil port	25.000
	Sai Gon Petro oil port	25.000
	Cat Lai oil refinery	32.000
Can Tho	Can Tho oil and gas port	5.000
Ca Mau	Song Doc oil field	120.000

– Type of oil group

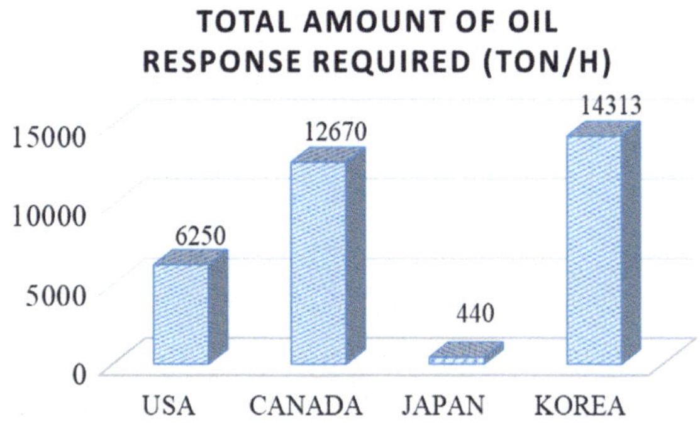

Fig. 11.2 Comparison of recovery requirement between developed countries

3. Therefore, it should be possible to select these two oil groups to estimate oil spill response demand of each area.

In oil spill incident, the maximum oil discharge should be estimated and appropriate response capability is ensured. However, it is unreasonable to set a certain level of response goal and uniformly apply it to all water because the maximum oil discharge depends on the oil spill incidents, the size of the vessel, the type of incident and the place of incident. For this reason, the maximum amount of oil discharge is estimated differently according to the sea area and place in order to allocate the oil equipment and personnel as much as possible near the high risk areas of oil spill. It is difficult to judge the superiority of national oil spill response capability at each country by simply comparing the amount of oil spill response goal. However, comparison of calculating effective response capability per day that is possible to show the rationality and validity of the good calculation method of response capability and the response goal value.

Criteria of USA for Oil Spill Recovery Capacity

In the United State of America, the Oil Pollution Act (USCG 2008a, b) require the ship owner or the ship operator under its contract or other approved method have to prepare sufficient response resources to respond to the worst case discharge (WCD).

Oil or Hazardous Material Pollution Prevention Regulations for Vessels (USCG 2008a), Appendix B suggests in details with respect to the maximum flow at the time of calculation methods.

Accordingly, it is assumed that the oil tankers operating in the sea area are subject to the maximum flow, and all the cargo loaded on the accident vessel is discharged. The oil recovery period at sea is from 4 to 10 days, depending on type of oil and spill site as Inland/Near Shore/Great Lakes, River, Open Ocean, and Offshore, and the daily working time is based on 24 h work. Table 11.2 shows the number of working days per pollution and the percentage of pollution control by type of oil. Marine pollution control plans are set differently depending on the type of oil and location of the accident. The volume of oil for which a vessel owner or operator should plan for should be calculated through the application of factors contained in Tables 11.2 and 11.3.

Table 11.2 Oil groups and natural dissipation by time (Fingas 2000; ITOPF 2014)

Types of oil		°API	Viscosity (10–20 °C)	Specific gravity	Typical example	% Natural dissipation					
						Day 1	Day 2	Day 3	Day 4	Day 5	Day 6
Non-persistent oil	Group I	> 45	< 3 CSt	< 0.80	Gas oil products	80	80	80	80	95	100
Persistent oils	Group II	35–45	4 CSt—semi-solid	0.80–0.85	Light crude oils and diesel-like products	20	32	40	50	75	90
	Group III	17.5–35	8 CSt—semi-solid	0.85–0.95	Medium-grade crudes and intermediate products	5	10	20	30	60	75
	Group IV	< 17.5	1500 CSt—semi-solid	> 0.95	Heavy crude oils and residual products	5	5	5	10	50	50

Table 11.3 Removal capacity planning table (USCG 2008b)

Spill location		Nearshore/inland/Great Lakes			Offshore		Open ocean		
Sustainability of on-water oil recover (days)		4			6		10		
Oil group	% Natural dissipation	% Recovered floating oil	% Oil on shore	% Natural dissipation	% Recovered floating oil	% Oil on shore	% Natural dissipation	% Recovered floating oil	% Oil on shore
I Non-persistent oils	80	20	10	95	[5][a]	/	100	/	/
II Light crudes	50	50	30	75	25	5	90	10	/
III Medium crudes and fuels	30	50	50	60	40	20	75	20	[5][a]
IV Heavy crudes/ residual fuels	10	50	70	50	40	30	50	20	[30][a]

Note Percentage may not sum to 100; reflects enhanced on-water recovery capacity
[a] Included in table for continuity; no planning required

For example, if the oil spill incident occurs in an oil tanker carrying 300,000 tons of Group III crude oil in nearshore area, it will evaporate up to 30%, response on water 50% and shoreline cleanup 50% (4 days, 24 h/day). However, the target value of the control amount exceeds 100% (130%), it means that it can be applied differently depending on the marine and weather conditions and other factors.

$$R = T \times 24 \times E \tag{11.1}$$

$$R = D \times U \tag{11.2}$$

$T = 300{,}000/(4 \times 24 \times 0.2) = 15{,}625$ tons/h

R Effective daily recovery capacity
T Throughput rate in barrels per hour (nameplate capacity)
E 20% efficiency factor (or lower factor as determined by the Coast Guard)
D Average Oil Recovery Rate in barrels per hour (Item 13.2.16 in ASTM F 631; or actual performance data)
U Hours per day that a vessel owner or operator can document capability to operate equipment under spill conditions. Ten hours per day must be used unless a vessel owner or operator can demonstrate that the recovery operation can be sustained for longer periods.

Determining required resources for on-water recovery for each tier using mobilization factors: (ton per hour on-water recovery requirements) = (on-water planning volume as calculated above) × (mobilization factor) = 15,625 × 0.40 = 6250 tons/h (USCG 2008b).

Obviously, the amount of oil spill in the region is required to be allocated, but no specific calculation method is suggested. The amount of the temporary oil storage tank is twice the effective oil return per day (EDRC). The oil treatment agent prescribes the quantity required to use 25% of the EDRC for 3 days. In addition, it does not provide a specific calculation method for the sorbent material and the special manpower assistant device.

Criteria of Canada for Oil Spill Recovery Capacity

The oil recovery system of Canada is generally similar to that of USA. In accordance with the Pollution Control Act, which is based on the Canada Shipping Act (CSA) and the National Pollution Control System, ship owners and user ship must provide necessary measures against pollution accidents, and the pollution control is carried out through the Certified Response Organization. In Canada, marine pollution accidents are classified into stages according to the frequency of occurrences and the extent of previous pollution accidents. According to CCG (1995a, b, 2011), the target vessel is regional largest oil carrier with maximum amount of oil discharge about 10,000 tons. Oil spill recovery capability in Canada was distributed different according to region for 10 operational days (Table 11.4). Effective daily recovery capacity was introduced in the formula (11.3):

Table 11.4 Emulsification factors for petroleum oil cargo groups (USCG 2008b)

Type of oil		Emulsification factor
Non-persistent oil 72G	Group I	1.0
Persistent oil	Group II	1.8
	Group III	2.0
	Group IV	1.4

$$\mathrm{EDRC} = \mathrm{T} \times 24 \times \mathrm{E} \tag{11.3}$$

T is a skimmer's throughput rate
24 h
E is an Efficiency factor set at 20% (or 0.2).

Example A tanker incident spills about 10,000 tons in the sea area of Point Tupper.

Oil recovery by region is required Onshore 4000 tons (40%), on-water 6000 tons (40% of Sheltered, 20% of Unsheltered). Therefore, total amount of effective recovery is 6400 tons (10 days) that includes Shoreline recovery by flotation machine (appropriate 10%) about 400 tons, and On-water (sheltered and unsheltered) total runoff about 6000 tons. Total amount of oil recovery capability per day is 3040 tons/day, namely 50% of on water recovery (3000 tons) and 10% of shoreline recovery (40 tons). As a result, total amount response required can calculate follow this formula (T) = EDRC ÷ (24 × E).

$$\text{Working 24 h a day: } \mathrm{T} = 3040 \div (24 \times 0.2) = 633\,\text{tons/h}$$

$$\text{Working 12 h a day: } \mathrm{T} = 3040 \div (12 \times 0.2) = 1267\,\text{tons/h}$$

Criteria of Japan for Oil Spill Recovery Capacity

In Japan, like the control system in the United States and Canada, it has a regional approach to control. The maximum volume of incident scenarios are set in the regional emergency control plan (RCP) in Japan, and the appropriate control capability is specified. In the case of Tokyo Bay, the volume of accident caused by a collision between a tanker and another vessel in the vicinity of a harbor tanker mooring facility is adopted as a scenario of maximum leakage accident. The maximum oil spill accidents and disposal standards for Tokyo Bay are based on oil tankers loaded with 260,000 DWT, and it is assumed about 23,000 kl due to collision with other ships in the vicinity of the terminal (JCG 2017). The principle of quantity calculation method by Japan is that the maximum oil spill is simply treated at sea for 2–3 days from the occurrence of the pollution accident without consideration of the weathering process of the outflow oil, 1 day is 12 h working.

Japan's pollution control method aims at treating 80% of the effluent with mechanical recovery by oil recovery and oil fence, and 20% of oil remaining and oil cleanup by sorbent. It is judged that the nominal recovered capacity of the recovered wastewater is regarded as the effective recovered capacity as it is because there is no formula and the efficiency or the mobilization rate is not considered. Therefore, the oil recovery capacity of the oil recovery system in Japan can be calculated by the following Eq. (11.4).

$$\mathrm{EDR} = \mathrm{T} \times 12 \tag{11.4}$$

EDR Efficiency recovery capacity (ton/day)
T Rate on name plate
12 1 day working (h)

For instance of worst case discharge, 260,000 DWT of oil tanker incident spills 23,000 kl of crude oil in the Tokyo Bay area. Total response is 18,400 kl (3 days, 12 h/day). Therefore, total amount of oil response required is 511 kl/h (about 440 ton/h).

Criteria of Korea for Oil Spill Recovery Capacity

Korea has experienced a significant number of oil spills such as the Keum Dong No. 5 oil spill incident in 1993, the Sea Prince oil spill incident in 1995. Especially, the Hebei Spirit spilled about 10,800 tons of three different crude oils in 2007. The incident has a wide range of consequences on extensive areas of fisheries, marine and coastal resources as well as socio-economic aspects. In Korea, the target vessel was regional largest tanker (300,000 DWT). Maximum amount of oil discharge is 45,000 kl. During 3 days for oil spill response, a third of oil spillage is assumed to dissipate through natural weathering processes such as evaporation, a third (7500 kl) to be recovered from the sea surface and a third (7500 kl) to be treated through shoreline cleanup.

National recovery capacity (NRC) was calculated with following considerations such as efficiency, mobilization efficiency and operating capability of response personnel after calculating mechanical recovery capability of oil recovery ships and oil skimmers (Lee 2001).

$$\begin{aligned}\mathrm{NRC\ (kl)} = {} & \mathrm{Recovery\ Capacity\ (kl/h)} \times \mathrm{Working\ Hours} \\ & (3\ \mathrm{days} \times 8\ \mathrm{h/day}) \times \mathrm{Mechanical\ Efficiency}\ (0.2) \\ & \times \mathrm{Mobilization\ Efficiency}\ (0.33) \times \mathrm{Operating\ Efficiency}\ (0.65).\end{aligned}$$

And then, the mechanical efficiency factor and operating efficiency factor were grouped into one factor:

$$\mathrm{NRC} = \mathrm{T} \times (3 \times 8) \times 0.2 \times 0.33 \tag{11.5}$$

Therefore, total amount of oil recovery required is calculated according to formulas (11.1) and (11.2), about 16,632 kl/h (14,313 ton/h) and 1894 kl/h (about 1630 ton/h) respectively.

According to the total amount of oil response required on water to be prepared to the worst case of discharge from the largest tanker incident in each country (Fig. 11.2), it was shown that the total amount of oil response required on water in Korea was the highest in 4 countries, peaking 1894 kl/h. Working time per day (8 h/day) considering time for maintenance equipment and charging or refresh personnel, mechanical efficiency, operating efficiency and mobilization are applied to estimation of oil recovery required in Korea. Therefore, the estimation method of Korea can be used as a useful method to set up the response demand of Vietnam.

Calculation Method of Oil Spill Response Demand

Considering methods of advanced marine countries and conditions in Vietnam, authors propose a calculating method to estimate oil recovery capacity by region as shown in Table 11.6.

The amount of oil spill recovery should calculate by using formula (11.5).

$$AORD = Q_{\max} \times P \tag{11.6}$$

$$\mathrm{AORD} = \mathrm{T} \times \mathrm{D} \times 8 \times \mathrm{E} \times \mathrm{M} \tag{11.7}$$

$$\mathrm{T} = \mathrm{AORD}/(\mathrm{D} \times 8 \times \mathrm{E} \times \mathrm{M}) \tag{11.8}$$

where

AORD	is the amount of oil recovery by region
R	is the amount of oil recovery demand
Q_{max}	is the estimation of maximum amount of oil spill by region
P	is percentage of oil spill recovery depending to during response and type of oil spill
T	is recovery capacity
D	is the number of day for response time
8	is the number of time working per day
E	is mechanical efficiency
M	is mobilization efficiency.

Coastal Vietnam can be divided into three regions (NOSRCEN, SOSCEM, NASOS) where the greatest potential for pollution by oil tanker is likely to occur. The NOSRCEN is subdivided into Hai Phong, Thanh Hoa, and Ha Tinh area, the SOSCEM into Da Nang, Quang Ngai, Khanh Hoa, and the NASOS into Vung Tau and Ca Mau area (Figs. 11.1 and 11.2).

Applying Table 11.5 combined the formula (11.7) to estimate the amount of oil recovery demand by region depending time response. Generally, the AORD reduce by time depending on type of oil spill and environmental condition. In Thanh Hoa area, the AORD achieves highest position, peaking 42,750 tons at 1 day, 40,500 tons at 2 days, and 22,500 tons at 6, 10 days. The AORD in Da Nang drops to the lowest level, about 450 tons at 10 days.

Table 11.5 Regional pollution response (%) in Canada (CCG 1995a)

Pollution location	On-water response (%)	
	Sheltered sea	Exposed sea
Point Tupper	40	20
Halifax	30	30
Quebec city	30	10
Vancouver	40	20
Montreal	30	0
Nanticoke	30	20
Holyrood	40	20

By using the formula (11.8), the amount of oil recovery demand (Table 11.6) and mobilization efficiency factor (Table 11.5) to estimate recovery capacity by region and area. The estimation of oil recovery capacity was showed in Table 11.7. These results are science platform to support for policy making, contingency plan making as well as responders to prepare and respond a spill in the future (Table 11.8).

In conclusion, the criteria of advanced marine countries for oil spill recovery and the current condition based on oil transportation and offshore activities in Vietnam were analyzed in order to suggest oil recovery demand on water. The target vessel and the maximum amount of oil spill were identified and calculated such as 300,000 DWT of an oil tanker in NORCEN and 45,000 tons of maximum amount of oil spill in Thanh Hoa area. In addition, the estimation of oil recovery demand by region and area was conducted through the target vessel, the maximum amount of oil spill, and

Table 11.6 Percentage of oil spill response demand

Spill location	High priority protected areas	Medium protected areas	Low priority protected areas	Offshore	Open ocean
Sustainability of on-water oil recovery (days)	1	3	4	6	10
Mobilization factor	0.60	0.40	0.33	0.21	0.12
Oil group	% Recovery on-water				
1 Non-persistent oil	20	20	20	5	0
2 Light crudes and diesel products	80	60	50	25	10
3 Medium crudes and intermediate products	95	80	70	40	25
4 Heavy crudes and residual products	95	92	90	50	50

Table 11.7 The amount of oil recovery demand

Region	Area	Oil group	High priority protected areas	Medium protected areas	Low priority protected areas	Offshore	Open ocean
			1 day	3 days	4 days	6 days	10 days
NOSRCEN	Hai Phong	2	4800	3600	3000	1500	600
	Thanh Hoa	4	42,750	41,400	40,500	22,500	22,500
	Ha Tinh	2	2160	1620	1350	675	270
SOSRCEM	Da Nang	2	3600	2700	2250	1125	450
	Quang Ngai	3	21,375	18,000	15,750	9000	5625
	Khanh Hoa	3	21,375	18,000	15,750	9000	5625
NASOS	Vung Tau	3	42,750	36,000	31,500	18,000	11,250
	Ca Mau	3	17,100	14,400	12,600	7200	4500

Table 11.8 Estimation of oil recovery capacity

Region	Area	Oil group	High priority protected areas	Medium protected areas	Low priority protected areas	Offshore	Open ocean
			1 day	3 days	4 days	6 days	10 days
NOSRCEN	Hai Phong	2	5000	1875	1420	744	313
	Thanh Hoa	4	44,531	21,563	19,176	11,161	11,719
	Ha Tinh	2	2250	844	639	335	141
SOSRCEM	Da Nang	2	3750	1406	1065	558	234
	Quang Ngai	3	22,266	9375	7457	4464	2930
	Khanh Hoa	3	22,266	9375	7457	4464	2930
NASOS	Vung Tau	3	44,531	18,750	14,915	8929	5859
	Ca Mau	3	17,813	7500	5966	3571	2344

the group of oil. According to these results, the oil recovery demand of Vung Tau area was estimated 18,750 tons for 3 days, 5859 tons for 10 days.

The results of this study are expected to be science platform to support policy making, and contingency plan making for preparedness and response to marine oil spills in Vietnam and ASEAN.

11.1.2 Enhancing National Measure Plans on Marine Oil Spills in Vietnam Through the Analysis of the Present States and National System

Introduction

Marine oil spill incident is serious environmental pollution problem in Vietnam because it is prevalent in Vietnamese waters and has a wide range of impacts and long-term consequences on marine ecosystems, coastal resources, human health and socio-economy. Improving national capacities of management (prevention, preparedness, response and recovery) to combat oil spills at sea seems best way to minimize health risks and mitigate economic losses due to oil spill incidents. Vietnam is in the process of improving the policy and legislation on oil pollution, so some solutions supposed to improve Vietnamese law on oil pollution (Hung and Kim 2019; Nguyen et al. 2017; Nguyen and Nguyen n.d.; Bui and Nguyen 2013), enhancing local capacities in oil spill preparedness and response (Bui and Nguyen 2013), and oils as an assistant tool were proposed. Economic loss valuation on resources and marine systems caused by oil spills and recovery measures showed theory and international experiences (Bui and Nguyen 2013). Besides, Vietnam has not signed on OPRC Convention 90, OPRC/HNS 2000, Fund Protocol 1992 and 2003 yet, while the Republic of Korea is a contracting party to the above international conventions and Protocols (IMO 2004b).

Generally, previous studies of marine oil pollution have made important contributions to theoretical basis and practical at the same time creating a premise to continue research. However, it is not clear whether impacts of oil spills are an independent national capacities for management of oil spill incidents. We hypothesized that national capacities for management play a key role in minimizing effects due to oil spill events. In addition, it is very hard to find book or papers of researchers related marine oil spills between Vietnam and Korea. In this paper, overviews of oil spill incidents in Vietnam and Korea such as the number and volume of oil spills, national system are presented and the improvements in Vietnamese national capacities to manage oil spill incidents at sea around Vietnam, especially, through a comparison between Vietnam and Korea are suggested.

The Present State of Oil Spills in Vietnam and Korea

(1) *The number and volume of oil spills in Vietnam*

Oil spills at sea in Vietnam from 1996 to 2015 occurred from tanker vessels, from facilities, another kind of vessels, from pipelines, from unknown sources. Particularly, unknown spills in 2007 had effects on the shorelines of many coastal provinces in 2007. These incidents can be divided into 4 periods as below: Firstly, from January 28 to February 7, 2007, oil slicks drifted and stranded on shorelines of 6 central provinces involving Ha Tinh, Quang Binh, Thua Thien Hue, Da Nang, Quang Nam, Quang Ngai, peaking 1867 tonnes of oil spilled. Secondly, from March 1 to 31 March 2007, it affected 12 central and southern provinces. Thirdly, from April 11 to 19 April 2007, it made an impact on 4 provinces including Hai Phong, Phu Yen, Khanh Hoa and Ninh Thuan. Finally, on 20 May 2007, it produced an effect on Tien Giang and on 4 May 2007, oil appeared in Ca Mau province. Further, the cases of oil leaks appear in the dry season 1994 and 1995 in Ben Tre and Tien Giang, on May, 1994 in Do Son and Hai Hau, on June, 1995 in Quang Ninh, Hai Phong and Nam Dinh, on April, 1997 in Soc Trang, on May, 1997 in Khanh Hoa and June, 1997 in Da Nang causing huge economic losses and serious pollution and long-term effects in the environment. Annual numbers and quantities of oil spilled in Vietnam for twenty years from 1996 to 2015 were showed in Table 11.9 (VINASARCOM 2007; MONRE 2016; VASI 2016).

A few large-scale oil spills have been recorded in Vietnam. In 2009, the incidents of Army's Petrol and Ocean store in Da Nang released 1600 tonnes of fuel A80 and 1000 tonnes of DO respectively. Hence it showed a tendency to increase in the volume of oil spilt, around 2096 tonnes oil spills per year on average from 2006 to 2010, peaking at 3773 tonnes (about 3321 kl) in 2007. The average of oil spill accidents and incidences was approximately 4 cases per year and 79 cases of the total. As a result, the total volume of an oil spill was about 13,773 tonnes (about 15,648 kl) and an average quantity of oil spill per year was about 782 tonnes (689 kl).

There was a tendency to decrease in the volume of oil spilt between 2010 and 2015, around 334 tonnes (about 380 kl) of oil spills per year on average.

The bar chart, combined with the line chart, offers a glimpse of the trend annual number and volume of oil spilt in Vietnam between 1996 and 2015 (Fig. 11.3). According to the first bar chart, it is apparent from the information supplies that, in 2010 the number of oil spilt reached the highest level, peaking 14 cases because of QD-TTg (2010). Interestingly, there were not oil spill accidents occurred in 2002.

(2) *Comparison of trend in annual number and volume of oil spill between Korea and Vietnam*

Korea and Vietnam are marine nations with thousands of kilometers of shoreline and islands, geographic proximity, cultural similarities. Most of exports and imports in Korea and Vietnam have been undertaken by shipping industry. Nevertheless, marine transport activity around the Korea and Vietnam coast produces a continual risk of pollution of the seaborne environment by a host of forms of oil spills such as fuel oils, cargo and operational waste.

Total number and total volume of oil spills in Korea for 20 years from 1996 to 2015 were 6300 cases and about 28,489 kl (about 25,100 tons) respectively, and

Table 11.9 Annual number and volume of oil spills in Vietnam from 1996 to 2015

Year	Annual number of oil spills	Annual oil spillage	
	(Case)	Volume (kl)	Quantity (M/T)
1996	3	82	72
1997	3	341	300
1998	2	47	41
1999	1	130	114
2000	2	5	4
2001	1	1136	1000
2002	0	0	0
2003	3	1022	900
2004	1	34	30
2005	4	700	616
2006	3	1763	1552
2007	6	3773	3321
2008	5	1235	1087
2009	9	3358	2956
2010	13	351	309
2011	6	482	425
2012	5	101	89
2013	4	499	439
2014	3	550	484
2015	5	39	34
Total	79	15,648	13,773
Average	3.95	782	689

average annual number and volume of oil spills in Korea were 315 cases and 1425 kl, respectively (Kim et al. 2022, 2; KCG 2008). There was a fluctuation of the annual number of incidents from year to year in Korea. A few large spills are responsible for high volume of oil spills that have occurred since 1996 are shown below.

Particularly, on 7th December 2007, the VLCC Hebei Spirit (146,848 GT), was hit by a crane barge whilst at anchor off Taean, South Korea, approximately 10,900 tonnes (about 12,547 kl) of Iranian Heavy, Upper Zakum and Kuwait Export crude oils were released to the sea (ITOPF 2017b). The spill incident impacted significantly three provinces and several hundred kilometers of coastline, on the mainland and on numerous islands, along the western coast of South Korea. Generally, it showed a tendency to decrease in a number of cases and the volume of oil spills in Korea.

Obviously, the numbers of oil spill incidents between Vietnam and South Korea are big different. Because of the fact that oil spill report and monitoring systems in Vietnam is incomplete.

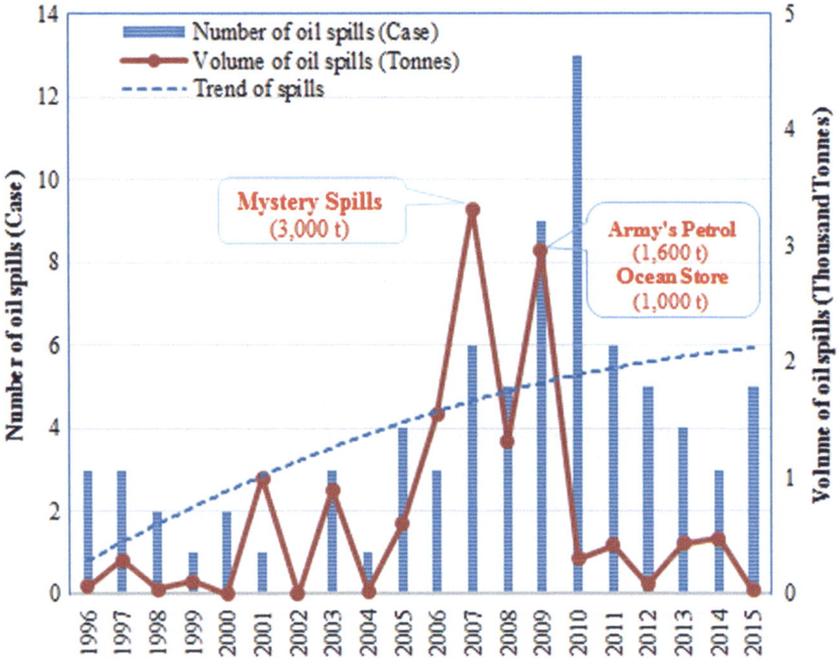

Fig. 11.3 Trend of annual number and volume of oil spills in Vietnam from 1996 to 2015

Looking on Fig. 11.4, the 5-year average for number of oil spill incidents in Korea was a downward trend, around 11.73%, 20.12%, 21.16% less proportion of the number of oil spill incidents in last 5-years than in the previous 5-year, respectively. While it reflected an upward trend in Vietnam, rapid increase 300% in 5-year (2006–2010) period versus previous 5 years. Figure 11.2 shows the amount of oil spilt in incidents in Korea and Vietnam over the last 20 years.

It is notable that a few large spills are responsible for a high percentage of the total oil spilt. For instance, in the five-year period 2006–2010 there were 1413 spills in Korea, totalling 12,476.24 tonnes, but 11,450.710 tonnes (91.78%) was spilt in VLCC Hebei Spirit incident. Most recently for example, in 2006, a total of spillage in Vietnam was 1552 tonnes, but 1500 tonnes (96.65%) was spilt in tanker La Palmas incident.

The volume of oil spilt in five-year (2006–2010) averages of Vietnam and Korea were the significant increase comparison with the previous 5 years (2001–2005), around 362.4% and 250% respectively (Fig. 11.5).

Generally, over the two decades the number and the volume of oil spills in South Korea shown a downward trend as opposed to increasing trend in Vietnam (Figs. 11.5 and 11.6). Therefore, improving national capacities for management of marine oil spill incidents around Vietnam through Korean experiences seems to be necessary.

National Framework of Vietnam

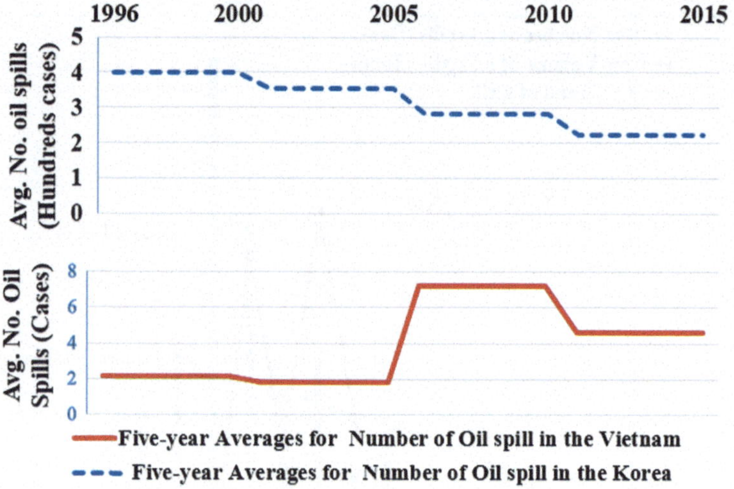

Fig. 11.4 Five-year averages for number of oil spill in Vietnam and Korea

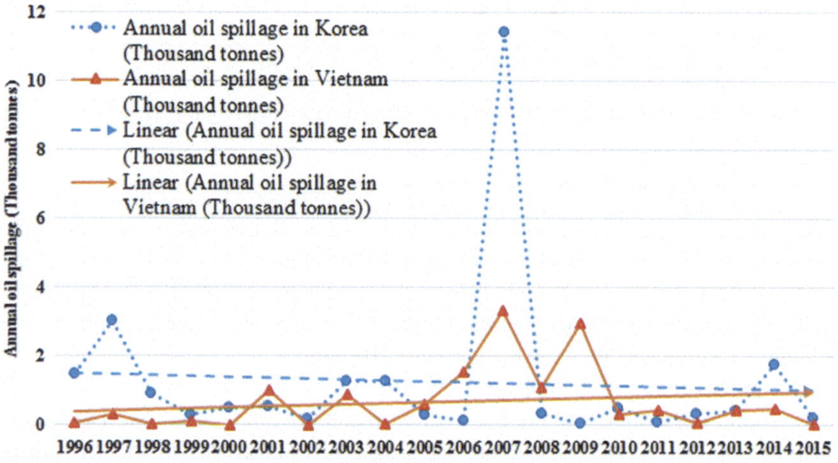

Fig. 11.5 Trend of annual volume of oil spill in Vietnam and Korea

(1) *National framework for oil spill management in Vietnam*

Currently, Vietnam's policy has been closely related to a vast number of the economy in the world, leading to the fact that national economy has been developing rapidly. Besides, it has been exerting the negative impacts on the environment and human health. Establishing laws and regulations on environmental protection has became interest a key focus on working programme of the Government and National Assembly, specifically in protection of the marine environment and oil spills responses that are shown via legislative documents in Table 11.10. National

Contingency Plan on oil spill response in 2001–2010 period (QD-TTg, 2001) and Prime Minister Decision 02 (QD-TTg, 2013) promulgating the oil spill response regulation divided Vietnam into 3 areas (Fig. 11.7).

The National Southern OSR Center (NASOS) in Vungtau, the National Northern OSR Center (NANOS) in Haiphong and the Central OSR Center (NACOS) in Danang (Fig. 11.7). These centrals shall be organized and operate under the statute of public-utility State enterprises with professional response forces acting as core units fully qualified and ready to perform the task of combating to tier 1 and tier 2 level of oil spills. All three OSR centers are under direction of Vietnam National Committee for Search and Rescue (VINASARCOM) for those tasks related to oil spill response (Fig. 11.7). According OSR regulation states that National OSR system is divided into 3 level: grassroots, area and national level, corresponding to governmental, provincial and private industry resources. The provinces have jurisdiction to the 20 km coastal waters limit; outside the 20 km coastal waters as the National OSR Centers and VINASARCOM designated authority (Fig. 11.7).

Table 11.10 Major laws and regulations related oil spill response in Vietnam

Law, regulations	Major contents
Cir. 2262/TT-MTg of Ministry of Science, Technology and Environment. Circular on overcoming oil spills, 29 December 1995 (MSTE 1995)	Regulation cases are considered oil spill incidents The measures to prevent and response the occurrence of oil spills, compensation procedures remedy environmental damage are gave
Decision No. 129/2001/QD-TTg, dated 29 August 2001 of Prime Minister. National Oil Spill Contingency Master Plan (QD-TTg, 2001)	The plan was distributed to oil spills into 3 levels Level 1: below 100 tonnes Level 2: 100–2000 tonnes Level 3: 2000 tonnes On organization response is divided into 03 levels: local level, area level and national level
Decision No. 103/2005/QD-TTg of Prime Minister. Regulation on activities to response to oil spill incidents (QD-TTg, 2005)	Provisions on the subjects and scope of oil spills and responsibility of organizations of management of oil spills (VINASARCOM, NASOS, NACOS, NANOS) Additional positions, roles and responsibilities of provincial level in response to oil spills
Decision No. 43/2010/QĐ-TTg of the Prime Minister. On promulgating the national statistical indicator system (QD-TTg, 2010)	The first time, statistics criteria-related oil spill at sea, including criteria No. 2105–2109 were provided
Decision No. 02/2013/QD-TTg of Prime Minister dated 14 Jan 2013, Oil Spill Response Regulations (QD-TTg, 2013)	Provides national plan, including three levels of responses Tier 1: small local spill, less than 20 tonnes of oil Tier 2: medium spill, 20–500 tonnes of oil Tier 3: large spill, greater than 500 tonnes of oil
Law No. 82/2015/QH13 of National Assembly. Law on Natural Resources and Environment of Sea and Islands (Law 2015)	It has dedicated Chapter IV to show: pollution control and response to oil spills, toxic chemicals and engulfed at sea

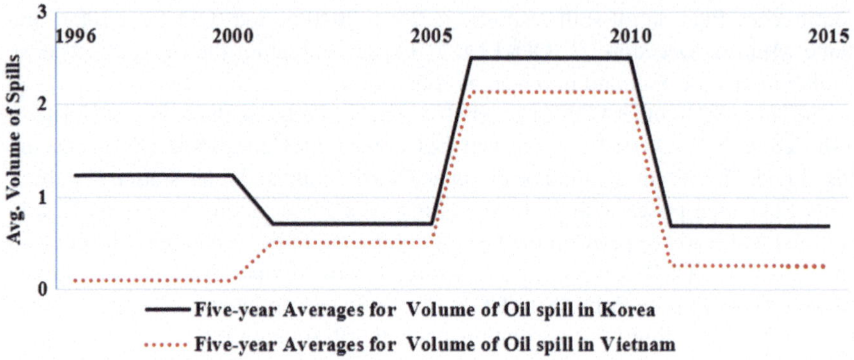

Fig. 11.6 Five-year averages for volume of oil spill in Vietnam and Korea

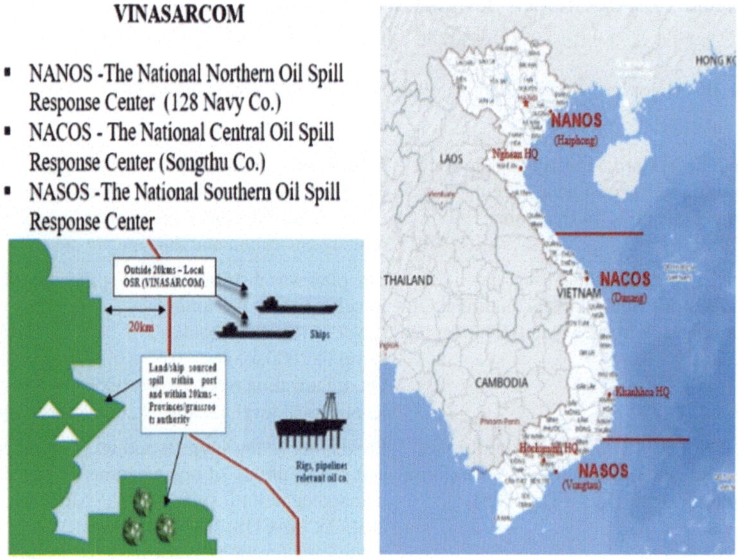

Fig. 11.7 Area oil spill response in Vietnam (QD-TTg, 2001, 2005)

Provincial resources: There are 28 coastal provinces with its own contingency plan. In terms of provincial OSR, provinces sets up province's Search and Rescue and Flood Prevention Steering Board (under the authority of People's Committee of the province) to response oil spill accidents. Elaborate and organize the implementation of OSR plan at provincial level on the basis of mobilizing forces, means, facilities and equipment for responding to oil spills caused by organizations located in the localities under their respective management. Direct and inspect organizations and enterprises under their respective management, which are likely to cause oil spills, in working out and organizing the implementation of their OSR plan. However, there

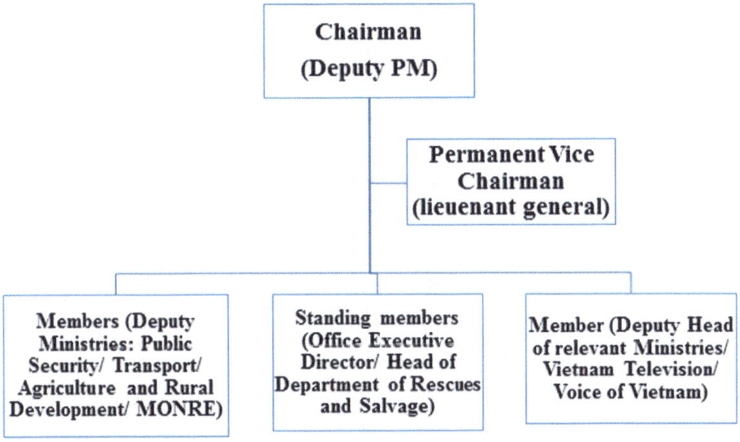

Fig. 11.8 Vietnam national OSR system

are no provinces in Vietnam that has OSR equipment because of the lack of financial and experts in using and maintaining the equipment (Fig. 11.8).

Admittedly, in the grassroots level, draw-up of OSR plan, investment in facilities or sign of contract with grassroot's OSR service supplier or Regional OSR center (Fig. 11.9) to response to Tier 1 of oil spills. According legal requirement, those organization that can cause oil pollution have to sign service contract with OSR service providers or to procure adequate equipment and training personnel. However an OSR center may contract various grassroots, it will not response requirement of these grassroots when many incidents occur simultaneously.

Interestingly, Law (2015) was established by the National Assembly in Hanoi, June 25, 2015. It will be enforced from July 1, 2016. Especially, this Law has a Chapter VI about Control Pollution, Response to oil, chemical toxic spills and engulfing at sea, involving 22 articles, from Article 42 to Article 63.

Moreover, Vietnam has cooperated closely with countries in the region to oil spill response such as Thailand, Cambodia, Philippines through joint statements as follow: Joint Statement of Cambodia, Thailand and Vietnam on partnership in Oil Spill Preparedness and Response in the Gulf of Thailand (GOT) signed on 12 January 2006; Joint Statement of Philippines and Vietnam on partnership in Oil Spill Preparedness and Response in the sea signed on October 2010; Regional Programme for the Prevention and Management of Marine Pollution in the East Asian Seas with the ASEAN countries, China, Cambodia, PDR Korea and Vietnam.

(2) *Comparison between Vietnam and ROK in national system for preparedness and response to oil spill incidents at sea*

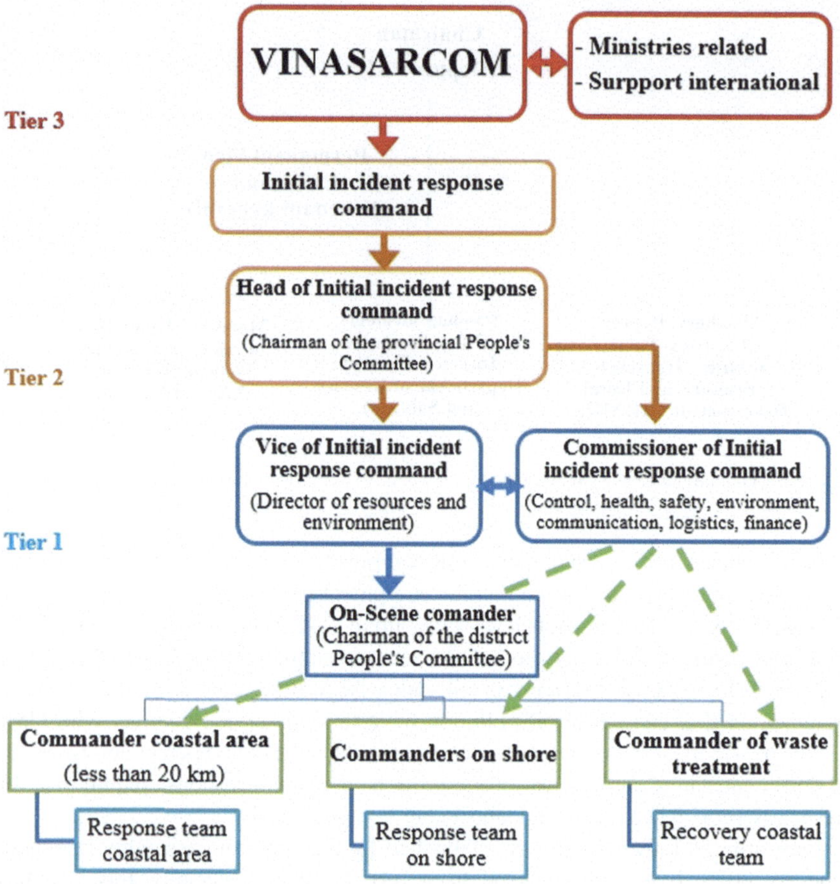

Fig. 11.9 OSR structure in Vietnam

The ROK is a country with plenty of experience for preparedness and response to the large oil spills such as Keum Dong No. 5, Sea Prince, Honam Sapphire and Hebei Spirit. Korean oil spill legal system is relatively complete which was shown through ratified and implemented most of international conventions and agreements regarding oil spills. While Vietnam has not signed on OPRC 90, FUND Protocol 1993, FUND Protocol 2003 and OPRC/HNS 2000 yet (Table 11.11). There is only one responsible leading authority in charge of national response system in both Vietnam and Korea (Table 11.11). National Contingency Plan of Vietnam and ROK are also combined oil and HNS spills response. However, Vietnam is really a lack of HNS detail plan. National Education and Training System, National Report and Monitoring System of ROK is fairly efficient, training various course for manpower. Beside, the number of modern equipments and materials in ROK will be a good shape in order to study and develop in Vietnam.

Table 11.11 Comparison between Vietnam and Korea in national system for management of combating oil spill incidents at sea

No.	Element	Vietnam	ROK
1	Length of shoreline	3444 km (WA 2016)	2413 km (WA 2016)
2	National Contingency Plan	Oil and HNS spill	Oil and HNS spills
3	Contracting state of convention	MARPOL 73/78 CLC Protocol 1992 Bunkers Convention 01	MARPOL 73/78 CLC Protocol 1976 CLC Protocol 1992 FUND Protocol 1992 SUPP FUND 2005 OPRC Convention 90, OPRC/HNS 2000 Bunkers Convention 01
4	National report and monitoring system	VNINASARCOM, ministries related, local Lack of specific regulations limit reporting	KCG Report oil spills more than 100 L
5	Agency or organization	VINASARCOM (leading agency), MONRE	KCG (leading agency), KOEM
6	Local response	Area Contingency Plan (ACP) 1. NANOS (Haiphong and Nghe An HQ) 2. NACOS (Danang and Khanh Hoa HQ) 3. NASOS (Vung Tau) Coastal provinces contingency plan	Area Contingency Plan (ACP) 1. South area KCG 2. East area KCG 3. West area KCG 4. Jeju area KCG KOEM center basic CP and 12 area branches plans
7	Response level	Tier 1: < 20 tonnes Tier 2: 20–500 tonnes Tier 3: > 500 tonnes	Tier 1: 30–100 kl (25–85 tonnes) Tier 2: 100–1000 kl Tier 3: over 1000 kl
8	National education and training system	NASOS, NACOS, SOSMoitruong Several awareness course	KCG and KOEM Various courses
9	Equipments and response vessels	Governmental, provincial and private industry resources (2016.03) Oil spill combat vessel: 30 vessels Oil boom: 20.71 km Oil skimmer: 34 units (Recovery capacity: ~ 1700 m^3/h) Dispersant: 12,000 L Sorbent: 4.5 tonnes Personal protcctive Clothes: 8 units	KOEM and KCG (2015.05) Oil spill combat vessels: 94 vessels Oil boom: 76.77 km Oil skimmer: 276 (Recovery capacity: 15,408 m^3/h) Dispersant: 288,11 kl Sorbents: 217.77 tonnes Personal protective Clothes: 139 units

(continued)

Table 11.11 (continued)

No.	Element	Vietnam	ROK
10	Man power	VINASARCOM 281 persons Others 235 persons Total: 516 persons	KOEM 579 persons KCG 313 persons Total: 892 persons
11	Recovery experience	NEPTUNE ARIES (1994) spilt 1500 tons of gas oil FOREMOSA ONE (2001) spilt 615 tons of gas oil	KEUM DONG No. 5 (1993) spilt 1300 tons of Bunker C. SEA PRINCE (1995) spilt 7000 tons of oil HONAM SAPPHIRE spilt over 1000 tonnes of crude oil HEBEI SPIRIT (2007) spilt 10,500 tons of oil

Obviously, an upward trend of the number and volume of oil spill incidents as well as import and export oil through 8 refineries combined with poor response system required that improving capacities for management of oil spill incidents or disasters such as more equipments, improving the legislative system, reinforcing the organization, providing national response training program are suggested below.

Enhancing National Measure Plans on Oil Spill Incidents

Considering the present states, national system as well as comparison with Korea, the challenges that Vietnam is facing and recommendation followings are suggested:

(1) *Enhancing monitoring programme*

Vietnam has made important strides in marine surveillance and monitoring oil spill programmes such as the coastal information system of Vietnam including 30 coastal communications, 01 ground communications satellite-HPLES, 01 satellite information station-VNLUT/MCC, 01 long range identification and tracking system-LRIT; The very-small-aperture-terminal (VSAT) fixed satellite communication network; Vessel traffic services-VTS systems in Haiphong and Hochiminh areas; The national remote sensing agency has used radar satellite imagery combined with GIS technology to tracking, monitoring and contaminating position of oil trails. However, these monitoring oil spill system were limited as follow: Lack of information monitoring oil spill from vessel less than 500 GT and/or non AIS; Lack of cooperation monitoring between organizations in a oil spills; Lack of standardisation of procedures for conducting monitoring programmes following a spill. In order to enhance more effective monitoring programme, the marine monitoring guidelines should be established with specific standardisations, including the definition of a monitoring programme (what, where, why and how to monitor, frequently) and the implementation of a monitoring programme (survey design, sampling strategies and standard methodologies). It is noted that small spills control and statistics measures should be effectively implemented.

(2) *Training and exercising programme*

Annually, each area centers remain at least one training course for provincial officers in the designated area. The course normally lasts 3 days in which 02 days in class and 01 day of practical training with equipment. A course mainly focused on overview of OSR and its environmental impact on the environment and shoreline clean-up activities that VINASARCOM has organized for military officers who participate in OSR. The authorities of coastal provinces also remain awareness OSR training course. However, the structure and major contents of the training programme have not been standardized yet. Therefore, the Vietnam should be researched and developed the standard contents of courses in training programme at different levels for expert in response oil spill incidents. These response training and education program should be initiated interested both in quantity and quality of response manpower. As part of its management of the national plan, OSR organizations conduct regular training programs and exercises for personnel likely to be involved in a spill response. These programs should be designed to provide Vietnam with numbers of trained personnel to have an effective response to an oil spill incidents.

In order to raise capability of response adaptation, have been providing response personnel with various training programs such as operational level courses for scene responders, administrative level courses for on-scene commanders and managers, oversea training courses, etc. It has been raising adaptation capabilities to the National Contingency Plan and the Local Contingency Plans through the joint exercises conducted by Civil and Authorities, and has been continuously improving problems coming from such exercises. The government should give regulations on business incentives to private companies trained OSR experts meet standard training programme and providing oil spill response services.

(3) *Enhancing legal liability and compensation*

According to Vietnam legislations, only vessels over 1000 GT are obligatory to buy civil liability insurance but the insured amount is not clearly defined. Beside, Vietnam is party to Civil Liability Convention 1992 and Bunker Convention 2001, not yet a party to 1992 Fund and the Supplementary Fund. Hence, the lack of financial guarantee to ensure compensations for damages and loss caused by oil spills as well as costs of cleanup and recovery.

It is necessary to build a roadmap to accede to the 1992 Fund as a legal basis for the process of claiming compensation for damages caused by oil pollution in Vietnamese waters. The participation of 1992 Fund contributes to the development and finalization of the compensation regime for marine environment suitable to the region and the world; to create a legal ground for promoting international cooperation in marine environmental protection to meet the requirements of integration. The decisions and judgments in force of the competent Vietnamese Court in respect of compensation for oil pollution from ships shall be automatically recognized by the other member States.

(4) *Strengthening cooperation with foreign forces*

In recent years, Vietnam has been cooperating actively in preparing for oil spill response to regional countries such as GOT program (got training, ESI mapping). In particular, Vietnam has issued the Decree on licensing and cooperation with foreign search and rescue forces in Vietnam, which the specific responsibilities of relevant government agencies. However, there is still lacks of detailed procedures and detailed form to mobilize foreign forces into Vietnam to assist in search and rescue in general and oil spill response in particular. Beside, it should be actively researching and cooperating with experienced oil spill response countries such as the United States, Canada, Korea, Japan etc. in human resource training, risk assessment, developing contingency plans, studying and cooperating in transferring technologies to manufacture equipment in services of oil spill response.

In this section, the analysis of oil spill system shows that Vietnam has the actually high potentiality of oil spills in the sea and coastal waters as well as in inland waterways. Through overviews of Vietnam national system and comparison between Vietnam and ROK in national system, enhancing national measure plan on oil spill incidents involving monitoring programme, legal liability and compensation, and strengthening with foreign forces such as effective monitoring guideline, ensuring financial guarantee for damages and loss caused by oil pollution, and enhancing cooperation with other countries in OSR are suggested.

11.2 Control HNS Pollution from Maritime Activities

Contingency Plan and Response

Purpose, legal basis, area covers.

The contingency plan serves to describe the emergency reporting network, provide procedural guidelines for HNS spill responders, and outline the policy framework (as defined by the OPRC-HNS Protocol 2000) regarding the role of agencies and organizations and involvement in an accidental spill of HNS in environmental emergencies.

This section should include an introduction to the HNS response system's components, the legal foundation, and the plan's subject areas. Organizational structures, planning specifications, monitoring and reporting, clearly defined operational processes, training and exercise, media, and financial liability arrangements are a few of the variables that need to be taken into account.

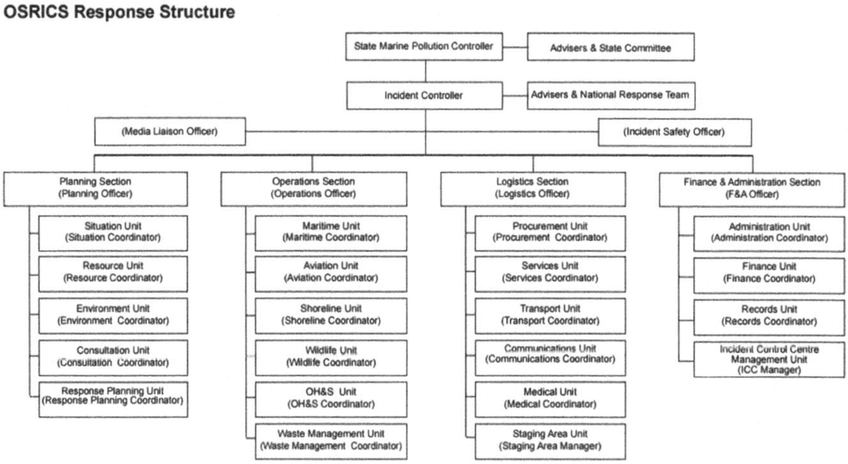

Fig. 11.10 Oil spill response incident control system (OSRICS) in Australia (Jeevan et al. 2016)

Setting up the national response unit or response organizations.

The specific party's tasks, responsibilities, and skills should be made clear in case of an emergency. In order to clearly identify the individual who bears the responsibility for reacting to an incident. Development, implementation, and maintenance must also be taken into account and delegated to other parties for several tasks.

Four broad categories can be used to group these tasks: response, recuperation, readiness, and prevention (Fig. 11.10).

(1) **Prevention**

Critical phases in emergency response include incident prevention and effect minimization.

The entirety of the currently in effect laws, rules, regulations, and other material, including national, regional, and international documents.

These documents will be compared and evaluated in order to determine the necessary requirements for compliance.

(2) **Preparedness**

It is expected of trainees to understand that: The purpose of preparedness exercises is to improve one's capacity for emergency response. This includes several supporting components, including:

Level of service: Specifying the needs for emergency response plans, as well as the readiness and response capacity standards for facilities, vessels, and port/harbor authorities.

Create an emergency plan: The following sections should be included in any applicable guidelines, policies, and backup plan that are prepared:

(i) The introduction will contain the distribution plan and any updates that have been made. It will also include the purpose, scope, and environmental policy of the plan.
(ii) The strategy section describes the scope of the plan. The geographical coverage, the perceived risks, roles and responsibilities of those charged with implementing the plan, and the proposed response strategy.
(iii) The action and operations section will set out the emergency procedures, which will allow rapid assessment of the spill and mobilization of appropriate resources.
(iv) The data directory will contain the relevant maps, resource lists, and data sheets to support the HNS spill response effort.

There are seven key steps to developing a contingency plan (IMO)

Define the scope of the plan
Conduct risk assessments
Develop your strategy
Decide on the structure and layout
Procure appropriate equipment
Conduct training and exercises
Update your plan as circumstances change with feedback from incidents or exercises.

Marine risk indicator: Sources of pollution, vessel traffic, risks to marine navigation, current weather patterns, environmental sensitivity, places used for cultural purposes, and socioeconomic activity are a few examples of marine risk indicators. It considers the different scenarios related to likely spills.

Personnel training and exercise: This includes industries, local communities, and anybody who might have to collaborate in the event of an accident. Information gathering and dissemination have to be the goal. locating chemical reaction units, educating their staff, and liaising with the International Maritime Organization (IMO), national and international organizations, and information sources in both ordinary and emergency circumstances (Fig. 11.11).

Response resources and equipment: Readiness; establish protocols for routine inspections and audits to make sure that the right tools, methods, and processes are in place and being followed.

Media: Creating channels for alerts and communication, including connections to the media, among other things.

Monitoring and reporting of incidents: Reporting protocols, requirements for reporting incidents, and general coordination should be provided; information databases about shipping details (such as bills of lading and stowage plans), chemical and physical properties, and response protocols should be established.

NATIONAL PLAN TRAINING FRAMEWORK

Courses	IGA Level	Status	Level	Notes
Online learning (Introduction to Pollution Response)	All	Available for use	N/A	1
Incident Management Team Course (Core)	2/3	Developed	AQF 5	2
Marine Pollution Controller	3	Same as Incident Controller	AQF 5	3
Incident Controller Course	2	Developed	AQF 5	
Planning Course	2	Under development	AQF 5	
Operations course	2	Under development	AQF 5	
Logistics Course	2	Developed	AQF 5	
Finance and Admin Course	2	States/NT	AQF 5	4
Aerial Observers Course	1	Under Review		
Wildlife Course	1	States/NT		
Shoreline Course	1	States/NT		
Health and Safety Course	2	To be developed		
Exercises (Desktop)	All	States/NT		
Exercises (Practical)	All	States/NT		
Exercises (Combined)	All	States/NT		

1 *Online learning must be completed prior to attending any of AMSA management courses. This free course is available to anyone. www.amsa.gov.au (Industry Training tab).*

2 *Incident Management Team Course is AMSAs Core and will be offered four times each financial year.*

3 *At this stage Marine Pollution Controllers are encouraged to participate in the Incident Controllers training for familiarisation.*

4 *Many States already have established Finance and Admin courses. It was agreed that owing to differing finance and procurement guidelines States/NT would be responsible for this.*

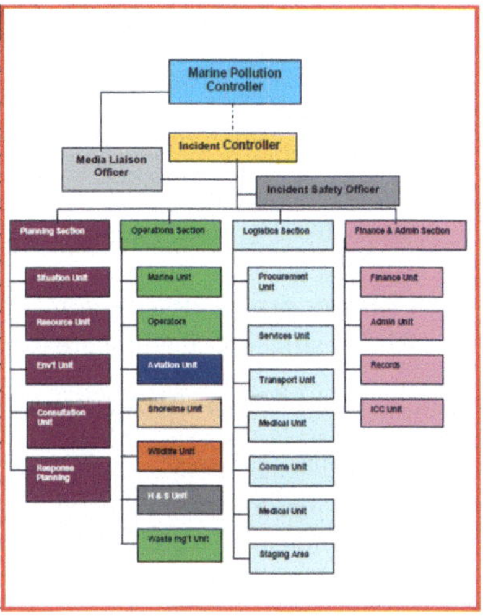

Fig. 11.11 Australia's plan training framework

It is essential that establishing an advisory group, consisting of individuals with expertise in the following areas:

navigation
ship technology
naval architecture
chemical hazards
marine biology
physical and chemical oceanography
dispersion modeling
spill response
fire lighting.

(3) **Risk Assessment**

The first step of a risk assessment is a correct identification of all hazards and possible scenarios. The risk is assigned a category and to be managed.

According ROPME SEA AREA Regional Contingency Plan to Combat Pollution of the Sea By Hazardous and Noxious Substances (HNS), the risk possibilities for chemical spills and chemical releases include:

Collision
Grounding
Transfer spill and equipment failure
Structural failure
Severe weather events.

Sensitive Area at risk, this will include:

- Environmentally sensitive areas
  - Mangrove swamps
  - Salt marsh
  - Muddy shores
  - Turtle nesting sites
  - Dugong areas
  - Fisheries especially spawning areas and shell fisheries
  - Bird nesting and feeding areas
- Socio-economic sensitivity
  - Desalination plants
  - Power station water intakes
  - Other industrial water intakes
  - Mariculture facilities
  - Tourist facilities
  - Tourist beaches …

(4) **Mapping**

Mapping the shoreline type and its general environmental sensitivity: For the various types of shoreline (and riverine or lacustrine ecosystems), the widely accepted Environmental Sensitivity Index (ESI) can be adapted for each country. The ESI, ranging from 1 (low sensitivity) to 10 (very high sensitivity), integrates the:

- shoreline type (grain size, slope) which determines the capacity of oil penetration and/or burial on the shore, and movement;
- exposure to wave (and tidal energy) which determines the natural persistence time of oil on the shoreline; and
- general biological productivity and sensitivity.

Figure *Colour code of the Environmental Sensitivity Index*

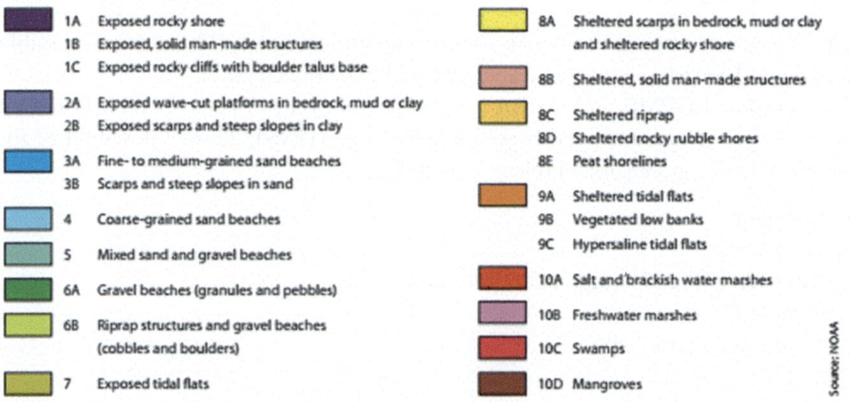

Mapping the biodiversity-sensitive elements of the areas: Mapping both protected areas and areas of biodiversity importance should include the coastal species, habitats and natural resources that could be affected by accidental oil pollution. The ESI (index 1–10, see above) does not consider the broader context which might be influenced by the oil spill. In particular, this includes:

- protected areas and important areas of biodiversity that can be identified using the UNEPWCMC Integrated Biodiversity Assessment Tool (IBAT);
- different types of coastal habitats/ecosystems; and
- endangered species that can be identified using the IUCN Red List.

Mapping the sensitive socio-economic features: Sensitive socio-economic features to be mapped should include: non-living resources that may be directly injured by incident; managed areas that may suffer economically.

These features can be grouped into various categories:

- subsistence, artisanal and commercial fishing, and fishing villages;
- aquaculture;
- water intakes (salt marsh plant, desalinization plant, aquaculture and salt production, industrial use);
- tourism and recreation areas (hotels, restaurants, marinas, beaches, recreational fishing, diving, etc.);
- port (including the activities and infrastructures);
- industrial activities (relying on maritime transport);
- infrastructures related to oil and HNS exploration, production and transport activities;
- cultural sites (archaeological, historical, religious, etc.).

Mapping oil and HNS spill response operational and logistical features: General logistic and operational resources include:

- detailed information to assist the On-Scene Commanders and operators on site for the response operations (without overloading the map);
- location of incident command posts and their geographical limits;
- existing stockpiles of equipment;
- pre-approved dispersant areas and their geographic limit(s);
- environmental recommendations (optional) to limit the impact of the pollution and of the cleaning operations, particularly in the sensitive areas.

Mapping potential sources of accidental pollution: To support the contingency plan, and identify High Risk Areas, the identification of the potential sources of accidental pollution is aided by the mapping of:

- offshore oil activities (oil fields, and exploration and production installations including platforms, pipelines, floating production, storage and offloading vessels (FPSOs), single point mooring buoys (SPMs), etc.);
- onshore oil and HNS activities (exploration, production, storage, refining and transport installations, etc.);

- maritime transport activities (traffic lanes, and port infrastructures including loading/offloading, bunkering, passenger traffic, etc.); and
- other potential sources of oil and HNS pollution (sunken vessels, etc.).

(5) **Response and Recovery**

When HNS is spilled at sea, measures must be taken to minimize physical damage to and pollution of marine resources and the environment. The measures of HNS spill response are relatively limited and are often dependent on various factors such as physical characteristics of the HNS, meteorological conditions (sea state, sunlight, air temperatures), characteristics of the seawater, distance from shore and considering safety. The measure is selected must also consider the protection of environmental sensitivity to marine and coastal resources, and balancing against effects of the response itself and the effectiveness of each clean-up measures. Response measures are well described in this literature, the key characteristics being as follow:

Natural recovery: Is defined herein as the return of natural resources to a dynamic baseline, breaking down the HNS with no active human intervention. In an oil spill, biological communities will generally recover naturally to baseline conditions about as quickly and effectively as they will with active human interventions, and at a lower cost and lower commitment of people, equipment, and materials. Understanding of HNS spill consequences and monitoring of natural recovery are requirements in order to make a decision whether a response may be required or not.

Chemical dispersants: Are capable of quickly removing significant quantities of HNS spill on the sea surface by transferring it into the water column. This measure can be achieved significant environmental and economic benefits. However, this solution have their limitations and account must be taken characteristics of HNS, physical conditions and environmental sensitivities. This requires consensus of the level of influence of water column based on forecasted exposure profiles, effectiveness of solution and monitoring of dispersant oil supported.

Containment and recovery of HNS on water measures: Is to collect and recover the spilled HNS by using response equipments such as booms, skimmers. Requirement of this measure is a specific decision-making framework to select areas of recovery effectively under present conditions.

Shoreline clean-up: Regarding response measures taken at sea, some HNS will reach the coast, and cleaning operations may be necessary to allow the coast to recover its natural state and socio-economic value by using possible manual and mechanical techniques. This requires an evaluation of effective equipment and clean-up teams to choice of the best method in each case depending on various factors like type of coastline, tide range, meteorological conditions, collection rate and effectiveness of available equipment, cost, impact of cleaning operations on the environment, national policy.

Other response measures: For example, In-situ burning measure (HNS on water or shoreline); Bioremediation measure (microbial breakdown of HNS). The secondary response measure are cited in the contingency plans. Interventions must be assessed effectiveness and feasibility considering environmental benefit for the spill. Generally, available and preferred response measures should be designated, with stakeholders before HNS spill occurring.

As an example, response options to HNS incident at sea and in ports should be explained by using factors as below.

< Available response options at sea >

*vessel

– Extinguish fire
– Reduce or contain fire
– Move vessel
– Scuttle vessel
– Monitor

*cargo

– move cargo
– stop spill
– protect cargo
– destroy cargo
– immobilize cargo
– activate biological degradation
– monitor

*released HNS

– dispersion
– chemical treatment

< Available response options in port >

*vessel

• extinguish fire • reduce or contain fire
• move vessel • scuttle vessel
• monitor

*cargo

• move cargo • stop spill
• protect cargo • destroy cargo
• immobilize cargo
• activate biological degradation • monitor

*released HNS

• dispersion • chemical treatment
• containment • recovery
• biological treatment • destroy
• immobilize • monitor

Arrangement of Organizations and Equipment

The organizational arrangement for proper response to the HNS incident should be explained, with specific role of group, tools and equipment. For example, monitoring group has role reporting, monitoring analyzing situation of spill and preparing and using remote sensing tools, transceivers, VHF …

Arrangement of Safety Equipment: explaining about important and drawbacks of personal safety such as heat stress, claustrophobia should be clearly explained to trainees.

(6) **Nine Steps of HNS Response [IMO]**

Basic response steps or activities which describe what the OSC must consider when responding to an HNS incident. As the nine steps will form the basis of the response strategy, which the OSC will be required to consider, lecture shall be made clearly by using videos and slides so that trainees may understand the key essence and requirements of each step as follows:

(a) **Step One—Assess the Situation**

The first step of the HNS response is to assess the situation.

- Report and document accurately all incoming incident information using a standard reporting form.
- Identify the type and source of products involved.
- Quantify the size of the spill.
- Assess the hazards implications, e.g. fire and explosion, human health, shoreline type, and effects on flora and fauna. Ensure that actions are prioritised.
- Identify the threat to people, environment, and facilities.
- The OSC needs to prioritise what actions will be taken.
- Important point is that the OSC identifies the political and economic significance of the spill.
- Gather samples or other supporting data sufficient to allow for the positive identification of the substances to take place.

(b) **Step Two—Activate Contingency Plan**

Step two is to activate contingency plan.

- Inform the senior manager as per the organisational procedure and content of the plan.
- Notify appropriate authorities and key personnel according to the plan.
- Activate initial response. Decide how, where, and when. Coordinate identification and mobilisation of initial response resources (personnel and equipment) as per the plan.

(c) **Step Three—Activate Organisational Response**

Step three of the response is to activate the organisation's response, according to the agreed strategy plan, including:

- Mobilising and organising the response team.
- Identifying, assigning and delegating tasks.
- Establishing both internal and external communication channels.
- Set up a command post in the safe zone (site equipment, safety and security, procedures/resources).
- Implement safety, security and de-contamination procedures.
- Notify, mobilise, and co-ordinate contracted and sub contracted services.
- Assess needs, define roles, and execute contracts.

(d) **Step Four—Prepare Response Action Plan**

Step four is to prepare a response action plan strategy. Response organisation should;

(i) Identify and priority resources at risk e.g. local populace, type of shoreline, local infrastructure.
(ii) Identify the resources required, taking into consideration the influence of the sea and weather conditions on the spill properties, the behaviour of the HNS, spread rate and identify additional resources if necessary (e.g. regional/international agreements), and so on.
(iii) Identify and prioritise response strategies, such as for monitoring, containment and recovery, chemical treatment, shoreline clean up, transfer, and disposal.
(iv) Maintain accurate records as required e.g. log and time keeping procedures for contractors and sub contractors, rented equipment, consumables.
(v) Develop response action plan to cover items such as any resource time available, resources available, containment plan, regular briefings on the response action plans, etc.

(e) **Step Five—Activate Operational Response**

Step five involves activating the operational response. This involves:

(i) Directing and supervising all monitoring, investigation, recovery, clean up and decontamination procedures.
(ii) Monitoring adherence to plans.
(iii) Coordinating containment, recovery and disposal of all pollutants.
(iv) Conducting regular briefings.
(v) Deploying personnel and ensuring logistical support.

(f) **Step Six—Managing the Response**

Step six is managing the on-going response. This involves gathering, assessing, and adjusting response information, and interpreting and reviewing reports as follows:

(i) Assessing, maintaining, advising, directing and control response team activities, and plan.
(ii) Monitoring, evaluating, and adjusting strategies as required.
(iii) Monitoring site safety plan and occupational hygiene.
(iv) Enforcing regulations and procedures.

(g) **Step Seven—Deactivate the Response**

Step seven involves deactivating the response. It should be determined whether the response should be continued, suspended, or terminated and advise the OSC accordingly. Rank the termination criteria relevant to the operations by using effort/benefit analysis.

(i) Assess the potential for recontamination.
(ii) Coordinate support to scientific monitoring of the area.
(iii) Shutdown field operations and conduct an orderly termination of operations.
(iv) Maintain, repair, and replace equipment as appropriate.

(h) **Step Eight—Consolidate Costs**

As this is the key point for the next stage of the response and follow-up, it is essential that trainees are instructed to keep detailed records by taking following actions:

(i) Verify/certify costs.
(ii) Plan for record and filing maintenance.
(iii) Consolidate records and produce a report of expenditure by category.
(iv) Produce costs documentation report.
(v) Construct simple coding structure and produce final cost report.

(i) **Step Nine—Debrief the Response**

Step nine is to debrief the response. Trainees learn that OSC is required to take following actions at this step:

(i) Conduct operational review of the response with the response teams, and analyse strengths and weaknesses of response.
(ii) Conduct a review with the OSC and analyse strengths and weaknesses of response. Identify the lessons learned and the areas that may require improvement.
(iii) Consolidate information, documentation, and evidence. Submit and circulate the final operations report.
(iv) Make and circulate recommendations for improved preparedness.
(v) Implement the lessons learned.
(vi) Update contingency plan.

After explaining the nine basic steps, it is imperative that the trainees are reminded about the very basic key points so that trainees may take timely actions, effectively and properly during the HNS incident. Trainees are taught to recognize that an OSC and his team can only function effectively if provided with the structure and organisation necessary to mount a response in a timely and appropriate manner, before the HNS has a chance to affect the local populous or sensitive areas. It should, therefore, be emphasized that:

(i) Sometimes, this is not possible due to the scale of the emergency, the location or the prevailing weather conditions.

(ii) The probability of small spills can be high, but with a rapid response the damage they cause are often minor and their clean up costs are relatively low.
(iii) Small spills are ugly and a constant reminder of the risks of much larger spills. A good response to small spills will ensure an effective response to larger spills.
(iv) Regardless of size, marine HNS spills tend to become unmanageable very quickly.
(v) The key to effective response is to have a comprehensive, realistic, and well exercised contingency plan, a minimum of appropriate equipment and a response organisation of highly trained individuals who will react very quickly and knowledgeably.

Trainees are also reminded that there exist some important interested parties and that the OSC and his team should always pay attention to these parties. Therefore, it is important to explain further that:

(i) During a major incident, there will be multiplicity of agencies involved that can be of great assistance in the response. These will include government agencies and private groups alike.
(ii) In terms of government groups, most countries have developed joint command centers for spill response or unified command teams that include many agencies.
(iii) In addition to government agencies, there will also be an influx of outside individuals such as media persons, equipment, sales people, and curious parties of all sorts, who have legitimate interests and some not so legitimate. It is important to recognize that such groups will inevitably be part of the scene and that mechanisms must be established to communicate with them on a fair basis without interfering with the emergency job at hand at the same time.
(iv) The OSC and his team must be allowed to implement the response strategies with the minimum outside interference.

11.3 Waste Water Pollution from Ship

Black water and gray water are the two separate wastewater streams produced by the ship. Black water is sewage with a high concentration of contaminants that comes from animal cages, ship toilets, and the hospital. Grey water, on the other hand, has a lower pollutant content than sewage and is produced by general kitchen and cleaning waste from sinks, baths, laundries, and the ship's restaurant. Passenger ships typically use treatment systems that gather and filter greywater and sewage in the same tank using integrated procedures.

Marine ecosystems are seriously threatened by ship-based effluent, which contains contaminants such organic matter, nutrients, suspended particles, and coliforms. Even though organic compounds are frequently biodegradable, during their breakdown, they can remove dissolved oxygen, which can cause an oxygen shortage in the marine environment. This procedure can have negative impacts when put into semi-enclosed and warmer waters. These effects can include harm to natural life, the discharge of harmful gasses, and unpleasant scents because of the aerobic conditions.

Overabundance of some nutrients, such as phosphorus and nitrogen, which are necessary for the flourishing of marine organisms, can cause eutrophication. When there is insufficient water flow and circulation, eutrophication occurs and turns bodies of water into marshes. In ship-based wastewater, proper nutrient discharge control is essential to preventing such environmental problems.

Coliform is essential for marine life and public health since it is a harmful microorganism indicator. Although coliform is not intrinsically harmful, its existence in marine settings indicates the possibility of waterborne infections.

A key factor in water quality is pH, which indicates how acidic or alkaline the water is. The pH range of seawater is generally 7.5–8.4, with the content of carbon dioxide having a considerable impact. Abnormally high or low pH levels can have detrimental effects on aquatic life by changing the toxicity of compounds such as ammonia, sulfides, and cyanides.

The oxygen needed to oxidize every material in water is measured by the chemical oxygen demand, or COD. The natural equilibrium in aquatic habitats is upset by the introduction of biodegradable organic materials, which may lower oxygen levels and have an impact on biological processes.

International conventions like the International Convention for the Prevention of Pollution from Ships (MARPOL 73/78) have been formed to address these concerns. Wastewater discharge into the ocean can be hazardous to human health, deplete oxygen in the water, and aggravate coastal areas' visual pollution, all of which can have an adverse effect on countries that depend heavily on tourism.

In order to address this problem, the MARPOL convention's Annex IV lays forth rules pertaining to ship sewage discharge into the ocean. These rules address the sewage control equipment and systems on ships, the availability of sewage port reception facilities, and the certification and survey requirements.

The common consensus is that raw sewage can be organically handled by microbes in the open ocean. Consequently, unless certain requirements are satisfied, Annex IV forbids the dumping of sewage into the sea within a given radius of the closest land.

It is the responsibility of governments to guarantee that ships arrive at ports with enough space for reception.

Annex IV was put into action on September 27, 2003, revised on April 1, 2004, and come into force on August 1, 2005. The updated Annex is applicable to ships approved to carry more than 15 people or those on international journeys with a gross tonnage of 400 or more. Annex IV stipulates that ships must be outfitted with a sewage holding tank, an approved sewage treatment plant, or an approved sewage comminuting and disinfection system.

Only under certain circumstances is it acceptable for a ship to dump sewage into the ocean; for example, the ship may run an authorized sewage treatment plant or discharge comminuted and disinfected sewage more than three nautical miles from the closest shore. According to resolution MEPC.157(55), untreated sewage may be released over 12 nautical miles from land as long as the ship is en route and moving at least 4 knots. However, this is subject to administration approval.

Resolution MEPC.157(55) was another recommendation made by the MEPC about guidelines for the pace at which untreated sewage is released from ships.

Technology-Based Sewage Treatment and Solutions

A holding tank is usually the most basic on-board sewage management system aboard a ship. The size of the tank needs to take into account various criteria, including the number of people on board, the ship's operations, and the capacity to retain sewage. The holding tank also needs to include a visible indicator to show the volume of its contents.

When a ship is within three nautical miles of the closest shore, it must have accommodations for temporary sewage storage if it is using an approved sewage comminuting and disinfecting system. This system uses a physical/chemical method that includes chlorination and sewage size reduction.

There are several different kinds of authorized sewage treatment plants. An on-board treatment system that uses aerobic or biological digestion is a common choice. There are three divisions in this system, which includes primary and secondary treatment. Similar to traditional septic tanks, the first chamber helps sediments settle to the bottom while enabling scum to float to the top. To help bacteria break down sediments even more, partially cleared wastewater is combined with air in the second compartment, known as the aeration chamber. The third section allows for more sediments to settle and is where the final disinfection step of chlorination takes place.

Most of the ships inspected for this study had one of the systems previously described installed on them, or a combination of them. On the other hand, in one particular case, a ship did not have a holding tank and chose to release the sewage straight from the sewage treatment facility.

Options for managing sewage on board a ship include holding tank collection, comminuting and cleaning the sewage, and treating the sewage in a sewage treatment plant. Either a storage tank or the ocean is used to receive the treated sewage plant effluent. Black water from ships without sewage treatment plants is collected in a holding tank. Grey water and/or galley water can also be collected using this holding tank. Grey water, on the other hand, is occasionally kept in separate holding tanks rather than being diverted to the holding tank. Grey water can be recycled into the toilet flushing system, mixed with sewage for treatment, or dumped straight into the sea.

About 25% of cruise ships have Advanced Wastewater Treatment Systems (AWTS) installed. These systems combine and process black and grey water to produce sewage sludge, also known as bio-residual, which must be discharged ashore. Cruise ships frequently have separate galley water tanks that are released in accordance with food waste laws. Water on cruise ships is usually mixed, agitated, and sanitized prior to being released into the ocean.

According to ship audits, the vast majority of vessels have on-board treatment systems; the handful that choose to disinfect sewage are usually smaller, older vessels. The treated wastewater is typically released into open waters.

Elements Affecting the Generation of Sewage

The amount of sewage produced is influenced by several factors, including:

- Number of Passengers and Crew Members: The number of people on board, including both crew members and passengers, affects the overall amount of sewage produced.
- Toilet Type: The total amount of sewage produced is impacted by water toilets, which produce more sewage than vacuum toilets.
- Length of Voyage: How much sewage is produced depends in part on how long the trip takes.
- Type of Treatment: Different amounts of waste are produced by the existence of sewage treatment plants or comminuting and disinfection systems.

After undergoing a variety of treatment processes, sewage can be safely dumped at sea. The number of people on board and the kind of system being used are two examples of the variables that affect the amount of sewage discharge generated. The range of wastewater output per person per day is estimated to be between 0.04 and 0.45 m^3. Of this volume, black water makes up about 0.01–0.06 m^3, and grey or galley water makes up the remaining portion.

11.4 Ballast Water Pollution from Ship

Maritime shipping plays a vital role in the transportation services (Hung and Kim 2019). According to statistics (UNCTAD 2020) with more than 98,000 commercial vessels, with a total tonnage of 2.069 billion DWT, sea transport is a method of transporting more than 80% of the world trade volumes every year. Ballast water is taken in and discharged by ships to maintain stability and maneuverability, and to mitigate hull stress during rough sea conditions, loading/unloading operations, or changes in fuel and water levels, control trim (Aridgides et al. 2004; Altug et al. 2012; Chen et al. 2010). However, this water can contain a variety of organisms, including phytoplankton, zooplankton, bacteria, viruses, and macro-fauna. The spread of potentially invasive alien species (approximately 7000–10,000 different species) of marine microbes, plants and animals globally each day results an economic loss about tens of billions US dollars per year (Drillet et al. 2023; Neill and Arim 2019) and impacts on Aquatic Ecosystems. The International Maritime Organization (IMO 2004b) lists many unwanted species and invasive species in ballast water operations (Table 11.12). The introduction of non-indigenous species through ballast water discharge can have serious impacts on local ecosystems. Invasive species can outcompete native species for resources, altering the structure and function of the ecosystem. They can also introduce diseases and parasites, which can harm native species and disrupt the food web. Some invasive species can also alter water chemistry, potentially leading to eutrophication and the proliferation of harmful algal blooms. The impacts of ballast water discharge on aquatic ecosystems can also have significant economic consequences. The introduction of invasive species can harm commercial and recreational fisheries, reducing their yields and affecting the livelihoods of those who depend on them. Invasive species can also damage infrastructure and property, increasing maintenance costs and reducing property values. Additionally, the costs associated with controlling and removing invasive species can be substantial, and they can persist for many years.

Specifically, the International Convention for the Control and Management of ship's ballast water and sediments (Ballast Water Management Convention) was eventually adopted by an IMO Diplomatic Conference in February 2004. From 8 September 2017, the BWM Convention entered into force globally, ships must manage their ballast water so that aquatic organisms and pathogens are removed or rendered harmless before the ballast water is released into a new location. This will help prevent the spread of invasive species as well as potentially harmful pathogens. Ships may be subject to port State control in any port or offshore terminal of a Party to the BWM Convention. This inspection may include verifying that there is onboard a valid Certificate and an approved ballast water management plan; inspection of

Table 11.12 Species in ballast water listed IMO (IMO 2004a, b; Aridgides et al. 2004)

Species	Original habitat water
Asian kelp *Undaria pinnatifida*	Asia
European green crab *Carcinus maenas*	Europe
North Pacific seastar *Asterias amurensis*	North Pacific
North American comb jelly *Mnemiopsis leidyi*	North American
Zebra mussel *Dreissena polymorpha*	Black and Caspian Seas
Cholera *Vibrio cholerae* (various strains)	India
Cladoceran water flea *Cercopagis pengoi*	North America
Mitten crab *Eriocheir sinensis*	East Asia
Toxic algae (red/brown/green tides) (various species)	
Round goby *Neogobius melanostomus*	Black and Caspian Seas
Skelton shrimp (*Caprella mutica*)	North Pacific Ocean
Leathery sea squirt (*Styela clava*)	Northwest Pacific Ocean
Asian sea weed—green fingers (*Codium fragile*)	Asia
Asian sea weed—a japweed (*Sargassum muticum*)	Asia
Chinese mitten crab (*Eriocheir sinensis*)	China
Japanese kelp (*Undaria pinnatifida*)	Japan
Japanese kelp (*Undaria pinnatifida*)	Japan
Squirt (*Perophora japonica*)	Northwest Pacific Ocean
Majid spider crab *Hyas araneus*	North Atlantic Ocean
Enteric bacteria	
Cholera	Indonesia

the ballast water record book; and/or sampling of the ship's ballast water, carried out in accordance with the Guidelines for ballast water sampling (G2) that meets the D-1, D-2 standard. Ballast water management systems (D3) must be approved by the Administration taking into account IMO Guidelines. Revised Guidelines for approval of ballast water management systems (G8) were adopted in 2016 and have now been reworked as a draft mandatory Code for approval of ballast water management systems (BWMS Code), the Procedure for approval of ballast water management systems that make use of Active Substances (G9). Particularly, all ships must meet D2 Standard by 8 September 2024 (IMO 2004b).

The physico-chemical parameters such as temperature, color, turbidity, salinity, dissolved oxygen conductivity, suspended solids and radioactivity are some of the water quality parameters which are vital for the proper management of the marine environment (Briton et al. 2007; Liu et al. 2021; Salleh et al. 2021).

Up to now, multiple ballast water treatment systems or combinations of treatment systems have been developed and applied in practice to control ballast water compliance with the standards of the BWM Convention. Therefore, this study analyzes ballast water treatment technologies to find an effective treatment technology for the ballast water to be D-2 compliant that is the most energy efficient, cost effective, low power consuming.

Ballast Water Management Treatments

Ballast Water Exchange

Ballast water exchange is the most commonly practiced and trusted method of ballast water management to minimize the risk of ballast water intrusion. As a supplementary guidance for ships, IMO 2006c has produced guidelines for the Design and Construction Code for Ballast Water Exchange (G11). According to the Ballast Water Convention, ballast water exchange must take place at least 200 nautical miles from the nearest land mass and at a depth of at least 200 m (IMO 2004b). Furthermore, regulation D-1 of the BWM Convention states that "ships carrying out ballast water exchange shall carry out this work with a volumetric exchange efficiency of at least 95%" (IMO 2004b).

Various methods are used for ballast water exchange, including the sequential method, flow-through method, and dilution method. It investigated the effectiveness of these methods and found that the flow-through method could remove 95% of organisms in ballast tanks. Various security issues arise during these exchange processes. Safety aspects such as stability, longitudinal stress, hull vibration due to waves, fore and aft drought, and visibility of bridges must be considered when performing replacements at sea (IMO 2023). Rigby and Hallegraeff (2001) found that continuous ballast water flushing does not significantly affect ship loads and bending moments. Furthermore, exchange tests showed that approximately 4% of the original water remained after exchanging three tank volumes. They found that 5% of the original port's dead plankton remained in the tank after replacement investigated the dilution efficiency of ballast water exchange using the marker dye rhodamine WT. The study concluded that 90% efficiency was observed for ships with a tank capacity of 114 m^3 and 99% efficiency was observed for ships with a tank capacity of 1435 m^3 (Rigby et al. 1999). From these studies it can be concluded that the efficiency of ballast water exchange in diluting pollutants, salinity, phytoplankton and zooplankton increases with increasing ballast water tank capacity and exchange frequency.

However, the Ballast Water Convention sees ballast water exchange as a short-term measure as ballast water exchange requirements do not apply to new buildings built after 2009 with a ballast water capacity of 5000 m^3 (IMO 2004a). Ballast water exchange will be allowed until 2021, but vessels built before 2009 will have to carry her BWMS to meet the standards of the BW Convention (IMO 2004b). Shapoori and Gholami (2014) reported that during mid-ocean exchange, factors such as water depth, season, and exchange method influence the effectiveness of the exchange process. Sample analyzes of ballast water after central ocean exchange show different results each time (Shapoori and Gholami 2014). The variability in results indicates that this is highly unlikely to be an effective ballast water treatment method that yields consistent results. Care must be taken when using ballast water exchange methods in local seas to reduce the risk of introducing alien species (Shapoori and Gholami 2014). To conduct BWE safely; the weather, sea conditions, time and place must be optimal for the exchange to take place. Furthermore, the efficiency of BWE depends on several factors, including species and behavior of organisms in the tank, tank design and construction, mixing within the tank, sediment properties and behavior (Liu et al. 2021; Rigby and Hallegraeff 2001; Salleh et al. 2021). Numerous studies have been conducted using various combinations of replacement and treatment systems to reduce the degree of invasion. Briski et al. (2015) found that ballast water replacement combined with ballast water treatment contained fewer risk organisms that were less likely to survive in the recipient freshwater ecosystem than ballast water treatment alone. Various alternative designs, including piping arrangements and different pumping arrangements, have been proposed to improve the efficiency of various forms of ballast water exchange, but this further increases the operating costs of the vessel (Shang et al. 2019; Bailey and Rajakaruna 2017).

Treatment Technology

Filtration

Filtration is physical separation treatment to remove larger organisms. Popular filter pore/mesh size of $\geq 50\ \mu m$ are used (Shafiquzzaman 2021; Yang et al. 2021). In order to remove large organisms and sediments from the ballast water, a filtration step is combined with gravity separation (Lakshmi et al. 2021). The capacity of filtration is influenced by various factors such as salinity, temperature, turbidity, organics, vibration, and other variables. These factors can significantly impact the subsequent processes in the ballast water treatment system, leading to alterations in its overall effectiveness (Gollasch et al. 2019; Briski et al. 2015). The hydrocyclone proved ineffective in the removal of zooplankton, whereas screen filters require lower pump pressure for operation. However, it was discovered that its efficiency levels are significantly lower compared to filtration methods (Rigby and Hallegraeff 2001; Burkholder et al. 2007; Baldrighi et al. 2019; Aridgides et al. 2004; Kim et al. 2022).

Mechanical Treatment

There are variuos mechanical treatment such as electric pulse, electro-ionization magnetic separation, hydrodynamic cavitation; heat treatment; micro wave; ultra violet radiation, and ultrasound.

Hydrodynamic cavitation is an efficient physical treatment method for ballast water treatment. It is developed by the National Institute of Oceanography in India and is easy to handle, environmentally friendly, and effective in destroying microorganisms and micro pollutants. The method requires no expensive parts or chemical storage, has a small footprint, and is energy efficient. However, it can cause problems at high pumping rates, and hull integrity may be compromised by repeated exposure to high waves.

Ultrasound technology can be used as a mechanical secondary treatment option. It has been shown to remove green algae, zooplankton, and bacteria. The effectiveness of ultrasound treatment depends on factors such as growth stage, flow rate, and transducer amplitude. Intermittent or pulsative treatment is suggested for onboard application to reduce power consumption and increase the lifespan of ultrasonic devices. The positioning of ultrasound transducers and the material of the tank can also affect its effectiveness. Further research is needed to understand the influence of organism size on ultrasound treatment.

Microwave technology, electric pulse, heat treatment, and ultraviolet radiation are also effective methods for ballast water treatment. Microwave heating, electric pulse, and heat treatment can achieve inactivation of various organisms, but they require careful control of operational parameters. Ultraviolet radiation treatment is proven to be a bactericide and virucide, but its effectiveness depends on the size and morphology of organisms. UV treatment systems can be combined with filtration and hydrocyclones for enhanced results. However, they require energy consumption and may have limitations in treating water with high turbidity.

Electro-ionization magnetic separation is a sequential process that introduces ionized gas into the water, which coagulates contaminants and allows for magnetic separation filtration. It has shown promising results in removing red phytoplankton.

Chemical treatment with chlorine is another option, but its effectiveness depends on factors such as temperature, residual chlorine, and reaction time. Chlorine concentrations above 3 ppm are not recommended due to the formation of high values of Trihalo methanes. Low concentrations of chlorine in continuous flux are suggested for improved efficiency.

Each treatment method has its advantages and limitations, and further research is needed to optimize their effectiveness and determine their impact on different organisms and the environment.

Chemical Treatment

There are various chemical treatments used for ballast water treatment. Chlorine is a commonly used biocide, and its effectiveness depends on factors like temperature, residual chlorine, and reaction time. Chlorine concentrations above 3 ppm are not recommended due to the formation of Trihalo methanes. Chlorine dioxide is another strong disinfectant that can effectively inactivate various microorganisms. Peraclean, a biocide solution containing peroxyacetic acid, is effective against bacteria, yeasts, molds, and viruses. Sea Kleen, a mixture of naphthoquinone, showed effectiveness against different organisms. Other biocides like Vibrex, copper ions, hypochlorite, glutaraldehyde, acrolein, ozone, and de-oxygenation are also discussed. Each biocide has its advantages and limitations. Electrolytic chlorine generation can produce sodium hypochlorite on-site. Overall, different chemical treatments have shown varying degrees of effectiveness in removing bacteria, phytoplankton, zooplankton, and other microorganisms from ballast water.

Ballast Water Treatment Systems

Ballast water treatment systems are essential for maintaining the ecological balance of marine ecosystems and preventing the spread of pests by ballast water discharges. A variety of treatment methods are used to effectively remove organisms and contaminants from the water.

Primary treatment plays an essential role in removing larger organisms and sending them back to their natural environment. It improves the effectiveness of subsequent mechanical and chemical processes. A number of treatment processes are often combined for better results.

A variety of treatment combinations were tested and found to be effective. For instance, a combination of filtration, hypoxia and high CO_2 levels showed significant mortality of the organisms tested. Mechanical treatment with special pipes, with or without ozone injection, has proven to be very effective at eliminating phytoplankton and zooplankton.

Gravity separation, filtration, and the use of oxidizing agents like chlorine-free oxidizers or ferrate proved successful in removing test organisms. Thermal treatment, UV treatment, ultrasound, ozone, and other advanced oxidation methods were also found effective against phytoplankton, zooplankton, and bacteria.

Various companies have developed ballast water treatment systems with a variety of characteristics. Filtration is a common step in most designs. Companies such as Alfa Laval, EcoChlor, Hyde Marine Lamor Corporation, and Marenco Technology have developed their own systems based on factors such as space use, energy consumption, costs, and processing efficiency.

The system of each company uses various filtration methods, such as screens with automatic washing, carbon steel or stainless steel filters, and filters capable of handling high or medium flow rates. Other treatment processes, such as electrolysis disinfection, UV or ozone, are also integrated in the plans.

Technical specifications and capabilities differ among the systems offered by different companies. Factors like flow rate, power consumption, and footprint vary accordingly. For instance, EcoChlor and Ocean Guard systems offer high flow rates with relatively low power consumption and small footprints, while Clean Ballast and NK-03 Blue Ballast systems have higher power consumption and larger footprints.

Ballast water treatment systems use various treatment processes, including filtration and disinfection methods, to ensure that harmful organisms are removed. Different companies have developed their own systems, considering factors like efficiency, space utilization, and power consumption. These systems play a critical role in protecting the marine environment from the potential hazards associated with the release of ballast water.

Ballast water management has become a significant issue as a result of the transfer and introduction of invasive species, which impacts the environment, the economy, and human health. It was found that a combination of ballast water exchange and treatment can be an efficient approach to ballast water management. However, most existing treatments mainly target larger organisms (> 50 μm) and have drawbacks such as the production of toxic by-products, high energy and space requirements, and expensive operation.

Among the various treatment methods, it has been observed that the combination of filtration and strong hydroxyl radical treatment shows the most effectiveness in treating the physico-chemical parameters of ballast water and inactivating smaller organisms (< 50 μm). However, further research is required to explore the potential and effectiveness of two-tier filtration laser treatment and to determine the optimum dose for inactivation of organisms in ballast water (Thach and Van Hung 2023, 2024).

It is essential to conduct new studies that explore different variations in existing systems and to develop innovative methods of treating ballast water. This research is essential before the implementation of D-2 compliance, which requires all ships to meet the ballast water management standards by the end of 2020. By addressing these research gaps and developing more effective treatment approaches, we can ensure better ballast water management and mitigate the negative impacts of invasive species on marine ecosystems, economies, and human well-being.

11.5 Pollution Caused by Garbage from Ship

Marine life may be harmed by ship-generated trash just as much as by chemicals or oil. Plastic is a big problem since it can linger for years and endanger fish and marine life. Sometimes plastics are mistaken for food by marine life, and they can also become tangled in plastic ropes, nets, bags, and even seemingly harmless items like the plastic rings used to close beverage cans.

Passing ships provide a big contribution to shoreline debris, even though the majority comes from onshore activities such as beachgoers and fisherman throwing trash into rivers or the ocean. Ships that choose to dispose of waste overboard rather than properly in ports are the main source of finding garbage in some areas.

It was once thought that anything thrown into the water would be absorbed by it, but this idea has been debunked by growing environmental consciousness. Many things will ultimately deteriorate in the sea, although it may take months or even years.

Education is necessary to address this problem and deter people from utilizing the oceans as landfills. While the antiquated notion that the ocean is capable of handling anything persists, stricter enforcement of laws—especially those pertaining to MARPOL Annex V—is required.

The goal of MARPOL Annex V is to reduce the amount of trash that ships release into the ocean. With more than 150 nations having ratified, the Annex went into force on December 31, 1988, and it is applicable to all kinds of ships operating in the maritime environment, including platforms, yachts, pleasure craft, and cargo ships.

Annex V, which forbids the general discharge of trash into the sea, makes exceptions for food waste, residues from cargo, cleaning supplies, additives, animal corpses, and safety-related matters pertaining to the ship and its occupants. A ship's regular operations produce a variety of waste kinds that are included in Annex V's definition of garbage. The most recent version of the living guidelines for implementing MARPOL Annex V, developed by the MEPC to aid in compliance, is resolution MEPC.295(71).

Approved as a major impediment to complete compliance with MARPOL, the perceived deficiency of port welcome facilities was the subject of an Action Plan approved by MEPC 55 in October 2006. The Plan was created by the Sub-Committee on Flag State Implementation (FSI) to support the efficient administration and shipping of MARPOL.

The Action Plan included a range of tasks intended to improve the availability and application of sufficient port welcome facilities. These tasks included examining the types and quantities of waste generated on board in relation to the type and capacity of port reception facilities; revising the IMO Comprehensive Manual on Port Reception Facilities; identifying technical challenges during waste transfer between ship and shore; standardizing garbage segregation requirements and containment identification; and creating a guide to good practice for port reception facility providers and users.

Regional Arrangements: In March 2012, MEPC 63 endorsed resolution MEPC.216(63) amending MARPOL Annex V to permit Small Island Developing States (SIDS) to satisfy the requirements for relevant reception facilities through regional arrangements when, due to special circumstances, this is the only feasible way to comply.

A common Advance Notification Form was created as part of the Action Plan to ensure uniform application and seamless execution, reducing the possibility of ship delays. To guarantee consistent record-keeping worldwide, a standard Waste Delivery Notification form was also developed.

In the course of working on the Action Plan, FSI created the Guide of Good Practice for Port Reception Facility Providers and Users, which includes a compilation of relevant rules and guidelines as well as guidance and an easy-to-refer to best practices for the use and provision of port reception facilities.

The Guide to good practice for port reception facility providers and users (MEPC.1/Circ.671/Rev.1) and four other circulars pertaining to port reception facilities (MEPC.1/Circ.469/Rev.2, MEPC.1/Circ.644/Rev.1, MEPC.1/Circ.645/Rev.1, and MEPC.1/Circ.470/Rev.1) were combined to form MEPC.1/Circ.834/Rev.1,

Revised Consolidated Guidance for port reception facility providers and users, which was adopted by MEPC in March 2018.

Rubbish Management Procedures: A rubbish management plan must be carried on board by all ships with a gross tonnage of 100 or more, as well as any vessel certified to carry 15 or more people. Written protocols for reducing, gathering, storing, processing, and discarding waste should be included in this plan, together with information on using on-board equipment (regulation 10.2). It is necessary to name the person in charge of the plan, and it needs to be written in the crew's working language. The 2012 Guidelines are provided by Resolution MEPC.220(63) for creating waste management strategies.

Garbage Record Book: The application and enforcement of waste management are the main topics of Regulation 10.3. It requires every fixed or floating platform, ships with a gross tonnage of 400 or more, and ships certified to transport 15 or more people on trips to ports and offshore terminals under the jurisdiction of another Party to keep a Garbage Record Book. The date, time, ship's position, a description of the waste, and the approximate amount burned or released must all be documented for every disposal and incineration operation. After the last entry, the Garbage Record Book is to be kept for two years. This rule makes it easier to follow trash restrictions, guaranteeing precise tracking and responsibility. MARPOL Annex V's Appendix 2 offers a template for a garbage record book.

Regulation 1.2 of Annex V defines cargo residues as residues that are not covered by other Convention Annexes and that are left on deck or in holds after loading or unloading. The only residues that can be discharged are those that are not recoverable using standard unloading techniques. It is not permitted to discharge cargo residues containing compounds hazardous to the marine environment (HME) at sea; instead, port receiving facilities must be utilized. Whether or if non-HME cargo residues are in wash water determines their specific requirements.

The shipper must determine whether or not a solid bulk cargo is hazardous to the marine environment by applying the standards outlined in MARPOL Annex V's Appendix 1.

Shipboard Incinerator: The design, production, performance, operation, and testing of incinerators intended for rubbish and shipboard waste incineration are described in detail in the Standard Specification for Shipboard Incinerators (resolution MEPC.244(66)).

Verification of Compliance: Parties are required to employ the Code for Implementation and submit to the IMO Member State Audit Scheme (IMSAS) in order to verify compliance and implementation, as per Chapter 2 of MARPOL Annex V. The required program, IMSAS, went into effect on January 1, 2016.

Many writers draw attention to the particular environmental risk that cruise ships pose. Cruises are one of the tourism industry's fastest-growing areas; 27.2 million passengers were carried in 2018 alone. The largest ships in this class have the capacity to accommodate nearly 2000 crew members and 6500 passengers, making them into floating cities on the sea. As a result, it is imperative to find long-term solutions to lessen the negative environmental effects of this type of tourism. There are two possible outcomes that are taken into consideration: either the adoption of environmentally friendly technology to reduce the detrimental effects of cruise ships, or the investigation of alternatives to the typical cruise ship experiences.

11.6 Control Air Pollution and GHG Emissions from Ship

The transportation of goods by sea is one of the crucial activities in the global economy, offering significant advantages in terms of volume and distance covered. According to statistics from the United Nations Conference on Trade and Development, with over 90,000 commercial vessels totaling 1.86 billion deadweight tons (DWT), maritime transport accounts for more than 80% of the annual global trade volume (Gritsenko 2017).

Alongside this process, maritime activities release a substantial amount of greenhouse gases and other harmful emissions, causing adverse effects on the environment and global climate. According to studies on greenhouse gas emissions in 2014, maritime transport emitted approximately 1 billion tons of CO_2 annually, constituting 3% of the global greenhouse gas emissions from fuel combustion. Projections indicate that by 2050, greenhouse gas emissions from maritime activities could increase by 50–250%. Furthermore, maritime activities contribute to 12% of sulfur oxide (SO_x) emissions and approximately 13% of nitrogen oxide (NO_x) emissions globally (Su 2015).

To address the negative impacts of emissions from maritime activities, the International Maritime Organization (IMO) has undertaken extensive efforts over many decades to establish legal and technical frameworks for controlling and reducing emissions from ships. Specifically:

Through Annex VI of the International Convention for the Prevention of Pollution from Ships (MARPOL) and its amendments (in effect since January 1, 2013), limits have been set for NO_x emissions, the use of low-sulfur fuels (reducing SO_x emissions), and mandatory technical and operational measures for all ships with a gross tonnage of 400 or more to enhance energy efficiency.

Through Resolution MEPC.304(72) in 2018 on the IMO Initial Strategy for reducing greenhouse gas (GHG) emissions from ships, the goal is to reduce the total annual GHG emissions by at least 50% by 2050 compared to 2008.

Issuing guidelines for establishing a legal framework to control emissions from ships and ports (Ship and Port Emission Toolkit), guiding member states to review their specific conditions and circumstances, assess the current situation, integrate Annex VI of the MARPOL Convention into national law, and develop a national strategy for reducing emissions from maritime activities.

In recent years, with the economic development of the country, Vietnam's maritime industry has experienced strong recovery, boasting a fleet of nearly 1300 ships with over 11 million DWT. With an increasing role and responsibility in the international community, Vietnam is gradually participating in and realizing collective commitments to minimize emissions from maritime activities.

The Vietnam Prime Minister has issued a decision approving the plan to implement Annexes III, IV, V, VI of the MARPOL Convention, outlining tasks such as reviewing the national legal system, assessing the pollution status caused by maritime activities, researching control measures, reducing emissions from ships, enhancing enforcement capabilities, and other tasks towards full implementation of the MARPOL Convention annexes (VAWR 2008).

Following the signing of the Paris Agreement, under the nationally determined contributions (NDC) program for reducing greenhouse gas emissions, Vietnam has committed to achieving a target of an 8% reduction in total GHG emissions by 2030, which could increase to 25% with international support through bilateral and multilateral cooperation, and the implementation of new mechanisms in the global climate agreement.

In recent years, the International Maritime Organization (IMO) has made considerable efforts to introduce regulations promoting energy-efficient practices on ships, aiming to control the generation of greenhouse gases. These efforts are specified through technical standards and measures such as:

- Energy Efficiency Design Index (EEDI)
- Energy Efficiency Operational Indicator (EEOI)
- Ship Energy Efficiency Management Plan (SEEMP)
- Carbon Intensity Indicator (CII).

The Energy Efficiency Design Index (EEDI) can be considered one of the technical standards aimed at minimizing CO_2 emissions from ships. The IMO's Marine Environment Protection Committee (MEPC) has continuously refined these regulations since 1997 to apply to ship fleets. In July 2011, mandatory regulations for EEDI and SEEMP were adopted and officially incorporated into Annex VI of the International Convention for the Prevention of Pollution from Ships (MARPOL 73/78), applying to all existing and newly built ships with a capacity of 400 gross tonnages (GT) or more. Subsequently, the amended Annex VI of MARPOL 73/78, effective from January 1, 2013, introduced Chapter 4, "Regulations on the Energy Efficiency of Ships" (IMO 1995).

The Energy Efficiency Design Index (EEDI) was adopted at the 62nd meeting of the Marine Environment Protection Committee (MEPC, IMO) as the first and most crucial tool for controlling and limiting CO_2 emissions in the maritime sector.

EEDI consists of technical provisions designed to reduce CO_2 emissions from ships by promoting the efficient use of energy, applying to newly built ships with a total capacity of 400 GT or more (details on ship types and applicable ranges are outlined in Chapter 4 of this manual). Accordingly, EEDI for new ships must undergo validation by an authorized agency and be issued an International Energy Efficiency Certificate (IEEC).

These regulations mandate that newly built ships must achieve energy efficiency levels of over 10% from 2015, 20% by 2020, and 30% by 2025. If implemented according to the specified timeline, by 2030, CO_2 emissions are projected to be reduced by 263 million tons. Despite the relatively high initial investment costs for research, modifications, and technology development, the efficiency gained from

EEDI is substantial. Annual fuel consumption is estimated to decrease by about 75 million tons, resulting in fuel cost savings equivalent to approximately USD 52 billion.

The expected effectiveness of increasing economic efficiency and reducing CO_2 emissions, based on applied technological changes and ship design modifications, is illustrated. The horizontal axis of the graph represents the timeline, while the vertical axis on the left indicates the total CO_2 emissions from maritime activities in cases without applying the Energy Efficiency Existing Ship Index (EEDI, baseline), applying EEDI from 2015 (no delay), and applying EEDI from 2019 (4-year delay). The right vertical axis shows the expected fuel savings if applied from 2015 or 2019, corresponding to the shaded areas on the graph (Perera and Mo 2016).

From this graph, it can be observed that CO_2 emissions from maritime activities are still dangerously increasing, posing a significant factor in accelerating the greenhouse effect and, consequently, the Earth's climate change. The implementation of appropriate solutions to reduce CO_2 emissions demonstrates the maritime industry's responsibility to the community in preserving a sustainable and environmentally friendly living environment.

To mitigate environmental pollution from diesel engine exhaust gases, the International Maritime Organization (IMO) has established mandatory standards outlined in Annex II and Annex VI of MARPOL 73/78, specifying limits on the concentration of harmful substances in diesel engine exhaust gases, particularly NO_x. For fuels obtained through non-petroleum refining methods, compliance with the following sulfur content limits is required:

- Before January 1, 2012: 4.5% m/m;
- After January 1, 2012: 3.5% m/m;
- After January 1, 2020: 0.5% m/m.

In practice, maritime transport is considered the transportation sector with the lowest fuel costs per ton-mile and has a competitive advantage due to its ability to transport large quantities of cargo over long distances with minimal fuel costs (Van Hung 2022). However, ship operations face several challenges, including high fuel costs and relatively low fuel efficiency, as well as loose technical management and ship operations that are nearly passive.

The mentioned issues negatively impact the operational efficiency of ships, particularly when fuel prices and related transportation costs continue to rise while freight rates decrease. Therefore, fuel-efficient usage and management have become a particular concern. In Vietnam, the National Assembly has approved the Law on Energy Conservation and Efficiency on June 17, 2010, which includes specific regulations for the Transportation sector in Chapter IV. Subsequently, on October 2, 2012, the Government issued Decision No. 1427/QD-TTg approving the national target program on "Energy Conservation and Efficiency for the 2012–2015 period," with a prominent objective being "Developing a transportation system that meets high-quality transportation needs, is fuel-efficient, and limits environmental pollution. Promoting the application of new technologies and using renewable energy to replace traditional fuels in transportation." To realize this objective, the Government focuses on “Applying new technologies, renewable energy in transportation,” with the main contents:

- Applying innovative technical and new technology initiatives in transportation to save energy and reduce the emission of environmental pollutants.
- Implementing the use of renewable energy and alternative energy sources (CNG, LPG, LNG, biofuels, electric energy, and other potential energy sources) for vehicles and transportation equipment to enhance energy efficiency.

The impact on the operational characteristics and fuel efficiency of sea vessels involves various factors, and some key groups of factors include:

Ship Design

Ship design aims to enhance the efficiency of vessel transportation by optimizing hull designs to reduce hydrodynamic resistance, decrease the weight of the empty vessel, and increase cargo load.

Some practical solutions with high applicability include:

- Increasing cargo load: The load of the vessel can be increased by modifying the structure or using lighter materials to reduce the weight of the empty vessel. However, ensuring the durability of the vessel is crucial.
- Optimizing hull design: Hull resistance accounts for up to 70% of engine power consumption. Therefore, improving hull design can significantly impact fuel efficiency, especially when applied to vessel types with suboptimal hydrodynamic designs. Additionally, designing a bulbous bow or adding devices to enhance hydrodynamic efficiency can be considered.
- Optimizing aerodynamic shape: When facing strong headwinds, wind resistance can contribute up to 10% of the total resistance of the vessel. Applying appropriate design technologies to the superstructure or installing additional structures to alter the direction of the airflow can reduce wind resistance, thereby increasing energy efficiency.
- Optimizing the flow pattern: Changing the flow pattern of the vessel can have a significant impact on efficiency. This can be achieved by modifying the vessel’s steering mechanism or installing devices to adjust the flow around the vessel.

Wind Energy

Wind energy, if effectively utilized, can contribute substantially to reducing CO_2 emissions. To use wind energy efficiently, investments in research and the installation of wind energy utilization systems, such as sails or wind deflectors for vessels, are necessary. However, considerations must be made to ensure the vessel’s stability when wind affects sails or wind deflectors. Additionally, wind stability should be assessed to guarantee the effectiveness of using wind for vessels.

Solar Energy

Solar panels are a renewable energy source that can be used for vessels, helping to reduce CO_2 emissions and other pollutants. However, the installation cost of these solar panels is high, and a large surface area is required for effective energy efficiency. As a result, solar panels currently serve as a supplementary measure rather than a complete replacement for traditional energy sources.

Nuclear Energy

Nuclear energy, if utilized in maritime vessels, completely eliminates CO_2, NO_x, and SO_x emissions. Another advantage of nuclear energy is its ability to save engine room space while providing a large energy source to meet vessel operation requirements. However, safety and nuclear waste management need thorough consideration.

Engine Technology

Ship engine manufacturers have proposed various solutions to modify design and engine technology to increase efficiency and reduce the harmful emissions of ship engines.

Additionally, there are numerous technological solutions that can be applied to vessels, such as changing the material of the vessel's hull coating to reduce drag, optimizing the design of the propeller system, and efficiently recovering waste heat.

Vessel Operation

Reducing a vessel's speed can help decrease fuel consumption and CO_2 emissions. However, reducing speed may lead to a decline in the competitiveness of maritime transportation or require an increase in alternative transportation methods such as road or air transport.

Moreover, increasing vessel size allows for more efficient exploitation. A larger vessel, capable of carrying a greater cargo load, helps reduce fuel consumption per unit of cargo. However, the effectiveness of this solution depends on the receiving capacity of ports and harbors.

Specific factors affecting energy efficiency in vessel exploitation can be analyzed as follows:

Vessel Speed

Analyzing the correlation between speed and fuel consumption of Container vessels provides the following example: A 9000 TEU Container vessel's fuel consumption at various speeds is as follows:

Speed 18 knots: 85 tons/day;
Speed 25 knots: 230 tons/day.

Therefore, for a voyage of 1728 NM, the corresponding fuel consumption is:

Speed 25 knots (3 days): 690 tons;
Speed 18 knots (4 days): 340 tons.

Through this example, the author notes that when operating vessels at lower speeds, fuel consumption is lower, resulting in a reduced greenhouse gas emission from maritime vessels.

So, in the process of vessel operation, based on the exchange of complete and accurate information among the parties involved, operators can precisely determine the time the vessel needs to be present at the destination port. At that point, they can calculate the optimal sailing speed to ensure fuel efficiency, reduce greenhouse gas emissions by running the vessel at a lower speed, and still have the vessel arrive at the destination port within the specified timeframe. This is the method of optimizing vessel speed through the Just In Time technique.

Advantages: Using this method will maximize fuel savings for maritime vessels, reducing greenhouse gas emissions from vessels to a minimum. As part of the program to increase energy efficiency and reduce greenhouse gas emissions from maritime vessels, the consideration of reducing vessel speed is being seriously examined due to the advantages it brings (Van Hung et al. 2023).

Disadvantages: This method requires continuous and transparent information exchange among the parties involved in vessel operation to calculate appropriate berth times, avoiding penalties for late arrivals.

Weather Information

Daily weather information is transmitted to vessels via email, including comprehensive weather factors such as wind, waves, currents, and sea ice.

Through a list of required information, maritime officers can choose to display all or some specific weather information necessary for the voyage. This information can be displayed as weather maps on the screen or printed on paper. Weather information display modes include:

- Weather information at a point (Display weather information in the area around a preselected point).
- Weather information at the vessel's location.
- Weather information at the port.
- Tropical cyclone information.

Notifications

Regular weather information is sent to vessels in the form of grid point data, indicating the values of weather factors (wind direction, wind speed, wave direction, etc.) at predefined points on a grid. This data is processed and displayed by the on-board SPOS software. Information to be displayed in the form of lines (such as fronts, isobars in cyclones, anticyclones) is often sent as a series of points on that line, connected to provide information about the required object.

Maritime Route Calculation

To calculate and select the sailing route, the following information is used:

- Vessel characteristics (or characteristics of speed variation due to the influence of waves and wind), fuel consumption characteristics.
- Forecast weather information, including waves, wind, and currents.
- Information about fuel consumption at different speeds in calm waters.

Based on the actual operational characteristics or experience of vessel operation, as well as the current stowage status, maritime officers can modify this data before starting calculations. Additionally, the system allows the captain or maritime officer to choose limiting weather conditions, such as:

- Maximum latitude limit.
- Limit on the coverage area of ice (expressed as a percentage).
- Wave/wind limit.
- Minimum distance to the center of an anticyclone.

These factors are selected based on the captain's or maritime officer's understanding and practical experience. The optimized route is calculated and displayed on the screen. For comparison, this route is drawn alongside the shortest maritime route (great circle route). Waypoints, estimated time of arrival at the destination port, etc., are also appropriately displayed.

It should be noted that the optimal route is only a recommended route, and weather forecasts are always subject to errors. Therefore, the safe and efficient operation of the vessel depends heavily on the captain's experience, understanding, skill, and reasonable analysis of the available information, as the captain has the highest responsibility for the safety of the vessel.

Fuel Usage for Ship Engines

This solution group focuses on using new fuels for vessels, such as LNG or biofuels. In theory, these types of fuels can significantly reduce the emissions of CO_2, NO_x, SO_x. However, there are practical issues related to supply and safety. Additionally, the environmental impact of biofuel production is not yet entirely clear.

Currently, maritime vessels primarily use petroleum-based fuels (Diesel Oil, Freight Oil), which release a large amount of air pollutants and are the main contributors to air pollution, including CO_x, NO_x, SO_x, VOC, CFCs. Due to increasing concerns about climate change and its impacts on human life, there is a clear awareness of the need to control greenhouse gas emissions and the use of fossil fuels. In its role as the regulatory body for global cargo shipping activities in recent years, the IMO has invested a considerable amount of time and effort in effectively controlling the energy use of maritime activities.

The Impacts of Emissions from Ships

Climate change affects marine ecosystems, causing disruptions in biodiversity, populations, and marine resources, directly impacting the livelihoods of coastal communities. The phenomenon of mass coral bleaching over the past 20 years is attributed, in part, to rising sea temperatures. The potential effects of climate change on the environment and socio-economic systems can be assessed through the sensitivity, adaptability, and vulnerability of these systems.

Meanwhile, the International Maritime Organization (IMO) has committed to reducing greenhouse gas emissions from ships in Annex VI—"Regulations for the Prevention of Air Pollution from Ships" of the International Convention for the Prevention of Pollution from Ships, 1973, as amended by the 1978 Protocol (MARPOL 73/78), effective from May 19, 2005.

The purpose of Annex VI is to control emissions of ozone-depleting substances, nitrogen oxides (NO_X), sulfur oxides (SO_X), volatile organic compounds, and the incineration of waste on ships.

The application scope of Annex VI MARPOL 73/78 is specified for all ships, irrespective of the year of construction, operational areas, and ship size. All seagoing ships with a gross tonnage of 400 GT and above constructed on or after May 19, 2005, must be surveyed and certified in accordance with the requirements of Annex VI. For seagoing ships with a gross tonnage of 400 GT and above constructed before May 19, 2005, they must be surveyed and certified not later than the first survey after May 19, 2005, but in any case not later than May 19, 2008. Diesel engines with a power output of more than 130 kW installed on ships constructed on or after January 1, 2000, must be tested for NO_X emissions and have an International Air Pollution Prevention Certificate for Engines (EIAPP) in accordance with the requirements of the NO_X Technical Code.

In Vietnam, there is currently a lack of comprehensive studies on the impact of ship emissions on the marine environment and biodiversity. Greenhouse gases are gases capable of absorbing long-wave radiation (infrared) reflected from the Earth's surface when illuminated by sunlight. The main greenhouse gases include H_2O, CO_2, CH_4, N_2O, O_3, and CFC gases.

The greenhouse effect, originating from the term "effet de serre" in French and named by Jean Baptiste Joseph Fourier, refers to the effect when the radiant energy of sunlight passing through windows or glass roofs is absorbed and scattered back into heat for the interior space, leading to warming the entire interior space rather than just the illuminated areas. In maritime activities, the primary source of CO_2 emissions is the direct use of fuel for ship engines, power generators, and a small part from refrigeration activities to preserve products—food and ice production.

Since the 1800s, fossil fuels have been used for steam-powered ships and have continuously increased during the twentieth century. The use of fossil fuels has become a characteristic feature of transport fleets and fishing vessels. An assessment study of the greenhouse gas emissions from the U.S. fishing fleet with vessels longer than 22.9 m (1050 vessels) revealed that the daily emissions of CO_2, NO_X, CO, and CH_4 were 16,995 tons, 306 tons, 40 tons, and 13 tons, respectively. In 2012, researchers Peter Tyedmers and Robert Parker from Dalhousie University conducted a study on fuel consumption and greenhouse gas emissions from the global tuna fishery. The study showed that products from tuna fishing consumed more energy than aquaculture and livestock farming products. The total fuel consumption of the global tuna fishery in 2009 was about 3 billion liters, equivalent to emitting 9 million tons of CO_2.

Furthermore, the toxic emissions from various types of ships are estimated to cause approximately 39,000 deaths per year in Europe, with the UK being the most heavily affected. This result comes from a study commissioned by the European Union, which plans to establish low-emission sea areas, minimizing pollution from thousands of cargo ships passing through these areas each year. Some super-heavy ships emit pollutants equivalent to millions of cars each year. Scientists have long demonstrated that the sulfur compounds and sulfur emissions from engines pose a threat to human health, leading to heart disease, lung disease, and cancer. These emissions come from factories, vehicles, but only recently have scientists calculated the specific impact of ship emissions, causing 39,000 deaths per year in Europe. Additionally, these emissions reduce the average lifespan of Europeans by a few months. The head of the research team, Janusz Cofala of the International Institute for Applied Systems Analysis in Austria, stated that the acceleration of international trade and the increasing number of ships, mostly originating from China, are making the environment more polluted. The UK and Ireland are most affected because they have long coastlines and are also bustling trade routes with busy shipping traffic. According to the study, the average lifespan of residents in the western coastal areas of the UK will decrease by 20–30 months from 2020. These results were once again confirmed by a study from the Stockholm University's Department of Environmental Science published in the journal Atmospheric Environment, showing that particles and dust from all harmful sources cause 546,000 deaths per year in Europe. Ship emissions have mostly not been included in pollution control laws in various countries. This means that about 289 million tons of fuel consumed by 100,000 ships worldwide each year could be the most significant source of serious pollution. It is 2000 times the allowable SO_x in diesel fuel for cars, along with various solid metals and other pollutants.

Environmental experts in the United States believe that ship transport is a major source of pollution to the environment, especially in port cities and coastal areas, as they use low-quality asphalt fuel, emitting high levels of nitrogen oxides and sulfur dioxide. Additionally, these emissions have also created acid rain and tiny particles in the air. The U.S. Environmental Protection Agency (EPA) announced initiatives earlier this week to prevent around 8300 deaths each year in the United States and Canada caused by inhaling smoke and toxic emissions from ships. According to U.S. government statistics, ships are responsible for two-thirds of sulfur dioxide emissions in the transportation sector in 2002, and the lack of control measures could increase this rate to 98% by 2020. In addition, the U.S. and Canadian governments have set new standards for emissions from large ships. From 2015, new ships must reduce sulfur dioxide (SO_2) emissions by 96% compared to today. Similarly, ships built after 2016 will have to cut nitrogen oxide (NO_x) emissions by 80%.

Domestically, according to the research results of KidiTech at Nam Trieu Fishing Company, Hai Phong (2009), replacing the traditional fishery exploitation lighting system with an LED lighting system will reduce costs and energy consumption, leading to reduced greenhouse gas emissions. For a ship using 40 high-pressure lamps, the lighting system weighs 400 kg, consuming 200 L of diesel per day. When replaced with 100 LED lamps, the weight of the lighting system is reduced to only 125 kg, and the fuel consumption drops to 30 L per day. With an offshore travel time of 20 days per month, the fuel consumption decreases from 4000 L per month to 600 L per month thanks to LED technology. This means that each sea trip, a large-capacity ship can save nearly $3000 in fuel costs by using LED technology and reduce emissions by about 3026 tons of greenhouse gases.

Vietnam has committed to implementing measures to address climate change, including signing the Paris Agreement in 2015 under the United Nations Framework Convention on Climate Change (UNFCCC) and submitting its Intended Nationally Determined Contributions (INDC). In the INDC, Vietnam set a target to reduce greenhouse gas emissions by 8% by 2030 compared to the business-as-usual scenario with domestic resources and envisaged increasing this target to a 25% reduction with international support.

The UNFCCC requires member countries to regularly update and adjust their Nationally Determined Contributions (NDC), and Vietnam has revised its goal to achieve a 9% reduction in greenhouse gas emissions by 2030 compared to the business-as-usual scenario with domestic resources, aiming for a 27% reduction with international support (Gritsenko 2017; Zhen et al. 2018; Russo et al. 2018).

In recent years, Vietnam has implemented various policies, programs, and plans actively addressing climate change and promoting sustainable development. Notable actions include:

- The 11th Central Committee of the Communist Party of Vietnam passed Resolution No. 24-NQ/TW on June 3, 2013, emphasizing proactive responses to climate change, enhanced natural resource management, and environmental protection.
- The National Assembly of the Socialist Republic of Vietnam enacted the Environmental Protection Law No. 55/2014/QH13 on June 23, 2014, which includes specific provisions regarding coping with climate change in Chapter IV.
- The National Assembly of the Socialist Republic of Vietnam passed the Meteorology and Hydrology Law No. 90/2015/QH13 on November 23, 2015, addressing issues related to monitoring, assessing impacts, and responding to climate change.
- The Vietnamese Government approved Resolution No. 120/NQ-CP on November 17, 2017, on sustainable development in the Mekong Delta to adapt to climate change, with a comprehensive action plan outlining specific tasks, actions, and deadlines for various sectors and localities.
- The Prime Minister issued Decision No. 2053/QD-TTg on October 28, 2016, approving the Action Plan to implement the Paris Agreement on climate change.
- The Ministry of Transport issued the Action Plan for Coping with Climate Change and Green Growth for the 2016–2020 period, as stated in Decision No. 1456/QD-BGTVT on November 15, 2016. The plan focuses on developing sustainable, environmentally friendly transportation, reducing greenhouse gas emissions from transportation activities.

Vietnam, being a significant contributor to international maritime transport and shipbuilding, has developed an action plan in alignment with global trends, adapting to climate change and fostering sustainable development in the maritime industry. The country's maritime fleet comprised 1756 ships in 2011, contributing 0.348% to the world's total commercial shipping tonnage and emitting approximately 2.803 million tons of CO_2. In 2020, Vietnam's maritime fleet had 1276 ships with a total tonnage of 8716 DWT, emitting 4.029 million tons of CO_2 in that year.

Despite the challenges posed by the COVID-19 pandemic, Vietnam's economy is forecasted to grow in 2021, with estimated GDP growth rates of 5.49% (baseline scenario), 6.9% (high scenario), and 3.48% (low scenario). However, the current energy consumption in Vietnam is projected to be three times the GDP growth rate. The scenarios for CO_2 emissions from Vietnam's maritime fleet until 2050 are depicted in the figures. According to the baseline scenario in 2008, the emissions amounted to 2.80 million tons of CO_2.

The National Strategy on Reducing Greenhouse Gas Emissions from Vietnam's maritime activities focuses on specific issues, including controlling emissions from Vietnamese ships, regulating emissions from ships in ports, enhancing the energy efficiency of ships, monitoring and controlling CO_2 or greenhouse gas emissions from ships, and managing air pollution from ships (Dang et al. 2022). The strategy aligns with the International Maritime Organization's (IMO) guidelines on minimizing greenhouse gas emissions from ships. The IMO's strategy aims to reduce the intensity of carbon emissions (e.g., CO_2 emissions per transport work) by at least 26% compared to 2019 levels, achieve an annual total reduction of at least 30% in greenhouse gas emissions from international maritime transport by 2050 compared to 2019 levels (equivalent to a 70% intensity reduction), and pursue zero emissions from international shipping as early as possible in the twenty-first century.

Moreover, the IMO's specific measures to achieve these goals are categorized into short-term, medium-term, and long-term solutions, covering technology and operational efficiency measures applicable to both existing and new ships, low-carbon fuel solutions, and eventually, zero-carbon fuel solutions.

In conclusion, Vietnam's commitment to addressing climate change and reducing greenhouse gas emissions is evident through various legislative and strategic measures. The country is actively aligning its policies with global efforts and is implementing specific actions to contribute to sustainable development and climate resilience.

11.7 Dumping Pollution from Maritime Activities

Significant environmental toxins have been introduced into marine ecosystems during the past century by the fast urban and industrial expansion along rivers and coastlines. Because of its potential for toxicity and bioaccumulation, heavy metal pollution has become a significant problem in marine settings. Dredging and disposal operations have demonstrated that minute variations in the water and sediment conditions can have an impact on the mobility and bioavailability of metals.

Microplastics were released from three ocean dumping sites in South Korea's East Sea (ES) and Yellow Sea (YS) between 1988 and 2015. These sites gathered terrestrial waste, mostly sewage sludge and industrial effluent. The amount of historical ocean dumping was associated with the historical pattern of microplastic pollution in age-dated core sediments from the East Sea dumping site (EDP) and Yellow Sea dumping site (YDP). Microplastics became less common as ocean dumping declined over time, peaking in 2015 when it was outlawed.

Sediment is disposed of into seawater through global coastal and offshore operations such as dredging, isolating contaminated sediments, shoreline development, and land reclamation. Sediment contamination, a major problem related to dredged materials, impacts roughly 10% of the 190–230 million m^3 of dredged material that are dredged each year in the US and contains organic compounds and heavy metals. Sediment build-up and increasing loss during disposal operations raise concerns because they affect the ecology, marine life, and water quality.

Protecting these settings is essential since important cities are situated in coastal regions and a significant section of the earth's surface is submerged in water. Coastal regions are growing because of the development of ports and beaches. Because it is more expensive to transfer materials from the sea to land, marine structure engineers, which includes port engineers, play a crucial role in dredging operations and the selection of disposal sites for dredged material.

Sites for dredging and disposal, Dredge and dump objectives, material settling, and induced energy regimes must all be thoroughly understood before deciding on the best way to dispose of dredged materials. Preliminary material destiny predictions require a mathematical model to clarify the complex interactions between dredging, transportation, and sedimentation as well as hydrodynamic understanding of the system. The river, lake, estuary, and ocean are the four aquatic environments that are allowed to receive dredged debris. Dredging has three main purposes: decontaminating (cleaning the seabed), aiding civil operations (removing sand from the bottom), and facilitating shipping (maintaining estuary depth). Dredged materials are typically disposed of in three different settings: on land, in the maritime environment, and at particular locations.

Dredged-Sediment Dumping Phases Dedicated barges are frequently used in dredging operations to discharge marine sediments. under this field, a number of experimental, numerical, and analytical initiatives have been finished or are under progress over time.

Material release aims to limit possible contaminants and stop them from spreading into the surrounding area by confining the materials in a certain area. With this strategy, pollution can diffuse under control and stay within tolerable bounds. The word "release" can also refer, more broadly, to the process of processing materials in order to separate or stabilize contaminants and turn them into innocuous compounds.

The ecology surrounding the dumping site is one of the most important considerations when it comes to the transfer of marine sediments from one location to another. Turbidity in water is a serious environmental issue that endangers both local ecosystems and marine life. To limit environmental harm and find the best solutions, a thorough investigation of all parameters and factors affecting the environment is required.

Despite these factors, a study gap persists and the entire range of problems that dumping sites impose on the marine environment have not been adequately explored. By providing a thorough analysis and simulation of the issue using the system dynamics technique, this research seeks to close that gap. Although specific negative effects of dredging operations on marine life and ecosystems have been studied in the past, thorough modeling and analysis taking into account all aspects of the issue are required. The report also makes suggestions for possible remedies to the problems with marine dumping sites' effects on the environment.

Sites for Dredging and Dumping: Environmental Effects and Issues

Large amounts of seabed material must be excavated, moved, and transported as part of dredging projects, which has an impact on the marine ecology both directly and indirectly. An overview of worldwide research on the effects of dumping sites on the environment is provided in this section. Dredged sediments are often classified as either non-polluted or polluted, with hydrocarbons and heavy metals being the most common contaminants, especially in the center of large ports.

Through drilling mud and excavated debris, oil and gas drilling activities at sea release a variety of chemicals into the environment. Marine life is greatly impacted by drilling material discharge, particularly in tropical regions, despite efforts to reduce toxicity and biodegradability. For ecological restoration, quantifying the environmental effects of drilling waste both during and after disposal is essential.

Effects of the Environment on Marine Life

Collisions: When a vessel is moving, dredging puts marine animals at danger of collision. The chance of a collision depends on a number of variables, including the type, speed, location, species, and behavior of the vessel.

Noise Pollution: Dredge-related anthropogenic noise impairs marine mammals' hearing systems, resulting in transient or permanent threshold shifts and behavioral modifications.

Turbidity and Sediment Plumes: Dredging affects the topography and concentrations of sediment, which changes the habitats and feeding strategies of marine species.

Changes in Depth and Topography: Dredging-related large-scale, recurring changes have the ability to positively or negatively impact biodiversity throughout the marine food web.

Concerns concerning environmental deterioration and irreparable harm to marine ecosystems are persistent among governments and society. The ability of marine ecosystems to self-purify is hampered by the widespread usage of the sea, fluctuating discharge rates, and contributions from coastal businesses. In order to effectively tackle marine pollution, national and international initiatives are required to address these challenges.

11.8 Control Antifouling System

Ships with submerged surfaces are painted with anti-pollution paint to stop marine life from adhering to the hull, including mollusks and algae. By reducing drag, this increases fuel economy for the ship. Before the current chemical industry created more efficient anti-pollution paints employing metallic compounds, materials like lime and arsenic were utilized for this purpose.

These substances "drag" into the ocean over time, making it more difficult for hyenas and other marine life to cling to the ship. Studies have revealed that these substances are persistent, damaging the environment, destroying marine life, and getting ingested into the food chain. Tributyltin (TBT), a particularly potent biocide, has been demonstrated to cause sex changes in the shell and deformity in oysters, with long-term effects resulting from persistence and accumulation. bioaccumulator.

In 1989, the International Maritime Organization (IMO) took up this issue after realizing its influence on the environment. The Marine Environment Protection Committee (MEPC) of the International Maritime Organization (IMO) suggested actions in 1990 to get rid of TBT-containing anti-pollution paint on some ships. Then the AFS Convention, also known as the International Convention for the Control of Ships' Anti-Pollution Systems, was ratified in 2001.

The AFS Convention forbids the use of hazardous organic compounds in anti-pollution paints on ships and sets up safeguards against the possible use of other hazardous materials. It came into effect on September 17, 2008. This pertains to all sizes of ships traveling internationally and guarantees adherence by means of certification and paperwork specifications.

The addition of cybutryne in 2021 shows that the Convention may be revised to include control of new drugs. It is recommended of the Parties to the Convention to put the resolutions and recommendations pertaining to the AFS Convention into effect and to keep improving them. Included in them are recommendations for standard and worldwide use in terms of sampling, testing, and surveying anti-infringement systems.

Workshops on capacity building are held to assist developing nations in ratifying and putting the AFS Convention into practice. Additionally, in order to promote ecologically conscious practices in the maritime sector, the Convention fosters collaboration in scientific and technological research on the impacts of anti-pollution systems.

The marine environment presents ongoing challenges for mechanical constructions like ships, marine platforms, offshore platforms, and boats. It's critical to shield these structures from environmental factors including seawater, biological dangers, and temperature changes. Protection techniques need to take into account surface expansion and contraction, water repellency, and controlled diffusion in addition to the use of injectable biocides in closed systems. In order to increase system life and boost dependability, protective organic coatings are commonly used in the shipping sector.

11.9 Salvage Sunken Property

Ships, cargo, and other items that sink or drift in port regions, marine zones, inland waterways, or wash up on a coastal nation's beaches are considered sunken assets.

Sunken assets are dangerous because they can damage resources, impede or jeopardize inland and marine waterway transit, harm human life and health, and pollute the environment.

Receiving information, locating the sunken asset, notifying others, identifying the owner of the item, recovering it, transferring it, receiving it, storing it, selling it, and disposing of it are all part of handling sunken assets.

There is frequently a significant risk of oil contamination and the release of toxic and hazardous materials from accidents involving submerged assets, particularly maritime vessels. Predicting the environmental effects of oil and hazardous substance dispersion is difficult since many vessels transport a variety of oils, including lubricating oil, diesel oil, and fuel oil (FO). Exploration, building and executing floating plans, moving, demolishing, and destroying sunken items are all included in salvage operations.

The location and kind of a sunken asset must be promptly reported to the appropriate authorities, such as the local military authorities, the inland waterway port authority, the maritime port authority, or the local police, upon discovery by the owner or other pertinent entities or individuals. Notifying the closest Customs authority is also necessary for coordination in the event of submerged assets involved in import or export in accordance with Customs law. A variety of channels, including direct conversation, the phone, fax, email, and other suitable means, can be used to provide information.

When information about a sunken asset is received, the responsible authority must confirm receipt of the information, verify the information received, and coordinate or organize the sunken asset's preservation. If the asset is located outside of a port or an inland waterway, the provincial People's Committee should take the lead and coordinate with relevant agencies to organize asset preservation. To guarantee the security of maritime operations, they must also set up warning signals, notifications for inland waterways, maritime notifications, and traffic regulations, if needed.

Sunken assets posing a danger are classified into two levels:

Level 1 Hazardous Sunken Assets: Items that could endanger inland and marine canal operations but haven't yet created traffic jams or need the implementation of traffic restrictions. These assets could include up to 100 tons of petroleum or petroleum products, or up to 50 tons of poisonous and hazardous chemicals. They could also present a risk of disease to people and the surrounding environment.

Level 2 Hazardous Sunken Assets: Items that impede the flow of transportation on inland or marine waterways or necessitate limitations on such movement. They could leak dangerous substances, be a source of disease for people and the environment, hold more than 100 tons of petroleum or petroleum products, or more than 50 tons of hazardous and toxic chemicals.

If the owner of the submerged asset is found, the port authority has to get in touch with them right away. The port authority must notify the Department of Transport, the Vietnam Inland Waterway Administration, or the Vietnam Maritime Administration for further action if the owner cannot be located or the asset is owned by the state. In the event that the submerged asset contains international components, English must also be used for communication. The owner is required to submit the recovery plan and notify the competent authorities of the anticipated start and finish timings of the recovery operation within 30 days of the notification date or sinking date.

The competent authority sets the owner's deadline for finishing the recovery operation and authorizes the recovery plan based on the circumstances and real conditions.

Regarding dangerous submerged assets:

- The recovery plan must be submitted for approval by the owner within one day after the notice or sinking date.
- From the date of permission, the owner shall start the recovery operation within 24 h for Level 2 hazardous assets or 72 h for Level 1 hazardous assets. If it is not possible, the owner is required to provide a documented report outlining the reasons.

Responsibilities for drafting the recovery plan:

- The owner is personally in charge of creating and submitting the recovery plan to the appropriate authority, or they may contract with groups or individuals to do so. The responsible authority may designate other organizations or individuals to construct the plan if the owner cannot be located or does not submit it within the required time frame.

For a plan that the owner designed and implemented, a recovery cost estimate is not included.

Keep in mind that the following actions should be done if the found property belongs to someone else.

- The person recovering property must notify the appropriate local authorities right away if it is unintentionally found in inland waterways, territorial seas, or when it is brought into inland waters or territorial seas of another nation. In Vietnam, the Vietnam Maritime Code mandates that it be notified to the appropriate state agency. Notifying the property owner if feasible and safeguarding the property until it is turned over to them or the appropriate state agency are all included in the notification content, along with the time, place, and other pertinent events.
- The person recuperating is required to safeguard the property until it is turned over, get payment for their efforts, and compensate any associated costs in accordance with the relevant rescue guidelines. The person recovering has the right to auction the item after covering the costs of recovery, preservation, auction organization, and other associated charges, if the recovered property is damaged or the cost of preservation is too high. The leftover sum needs to be put in a bank account and given back to the property owner after subtracting the costs associated with recovery, preservation, holding an auction, and other charges.

The Vietnam Maritime Code also stipulates the following:

- The person recuperating must turn over the item to the appropriate state agency within 15 days of the notice date if the property owner does not want to claim it back or pay off obligations.
- If the property owner does nothing to safeguard their rights within sixty days of the notice date, the appropriate state agency may manage the property in accordance with the law.

The Vietnam Maritime Code stipulates that anyone who finds, rescues, or assists in rescuing another person's property floating at sea is entitled to compensation in accordance with comparable rescue standards for floating assets at sea or washed ashore in Vietnam.

Please refer to Decree No. 05/2017/ND-CP, issued by the Government on January 16, 2017, for more information. This decree governs the management of sunken assets on Vietnam's inland waterways, port areas, maritime zones, and seas. It has been amended and supplemented by Decree No. 69/2022/ND-CP, issued by the Government on September 23, 2022, which amends and supplements certain provisions of related decrees on maritime business activities. This decree is effective October 30, 2022.

11.10 Ship Breaking and Recycling

11.10.1 Overview of Shipbreaking

Usually, a ship's operating life ends after 22–30 years. Because to high labor costs, safety measures being prioritized, and strict environmental restrictions in developed countries, most ship recycling operations take place in less developed Asian countries. Currently, South Asian nations dismantle more than 80% of ships, which puts worker and environmental safety at risk. Furthermore, it is a controversial worldwide issue to establish a single policy and laws for recycling end-of-life vessels.

Ship owners use a variety of techniques in the disposal process, including as breaking, abandoning, decommissioning, demolishing, disassembling, recycling, scuttling, and scrapping. Notably, China, Bangladesh, India, Pakistan, and Turkey are the main locations for recycling end-of-life vessels. Just 0.3% of the gross tonnage of recycled vessels (GTRV) was recycled in other nations as of 2016, with the remaining 35, 30, 22, 9, and 3.7% being handled in facilities located in these five countries. During vessel dismantling, hazardous materials such asbestos, glass fiber, waste oils, polychlorinated biphenyls (PCBs), and solid foam present serious dangers to the environment and public health.

The 2009-adopted Hong Kong International Convention for the Safe and Environmentally Sound Recycling of Ships acknowledges the global necessity for sustainable shipbreaking and recycling facilities (sustainable SSRI), however it hasn't yet gone into effect. A emphasis on appropriate safety and environmental conditions is required for green ship recycling (GSP), which calls for higher expenditures in the planning, building, and management of ship recycling facilities. Alongside the application of environmentally sound management practices, more costly equipment and technology are used. Many academics stress how important it is to continue developing financial tools in order to enable safe and sound ship recycling.

Shipbreaking, which includes the dismantling processes involved, is the central and principal endeavor in ship recycling. Around 5.8 million tons of steel are recycled each year through shipbreaking worldwide. Although 82 nations engage in steel shipbreaking and recycling activities globally, the shipbreaking sectors located in Bangladesh, India, Pakistan, China, and Turkey account for 97–98% of the total amount of recycled steel worldwide.

After first building up in the land, the dangerous byproducts from shipbreaking operations eventually make their way into the sea, contaminating the water and lowering its quality. Mangroves and other vegetation are gradually disappearing as a result of this process. Beyond the environmental damage, shipbreaking operations put workers' health and emotional well-being at risk due to poor sanitary conditions in shipbreaking yards, extended exposure to heat, dust, hazardous chemicals, and contaminated smoke.

Given the current loose regulatory framework, the situation in shipbreaking yards, especially in developing nations, makes it difficult for authorities to implement stronger environmental standards. While the protection of fundamental human rights, safety gear, and clothing could all improve the well-being of shipbreaking laborers in these nations, putting precautionary measures in place to stop hazardous materials from contaminating the soil would be extremely expensive.

Because of more stringent international restrictions, the shipbreaking industries, which mostly use the beaching method, are actively looking for more economical, efficient, sustainable, and environmentally friendly manufacturing processes. In order to create a sustainable and ecologically friendly shipbreaking method, an interdisciplinary joint research effort has been launched, with a focus on efficient management of hazardous material mixing in shipbreaking yards. It is expected that the suggested method will be less expensive as well as easier to execute.

In light of this, the study results discussed in this section provide shipbreaking companies with sophisticated guidelines that offer insights into attaining cleaner raw material production in an environmentally sustainable manner, improving social welfare and bolstering the blue economy.

11.10.2 Controlling the Synthesis of Hazardous Materials from Modern Shipbreaking Activities

Hazardous compounds produced by shipbreaking pose a threat to human health and the environment if improperly managed. However, hazardous waste management is challenging due to a number of transdisciplinary issues, including:

(1) Absent or opaque "Green Passport": According to IMO regulations, ships sold for recycling must have a "Green Passport" that lists any hazardous materials carried on board. Nevertheless, ship owners frequently fail to give or maintain this information. Inaccurate information about hazardous compounds makes it challenging for surveyors to recognize and measure dangers.
(2) Non-idealized testing techniques: There are a number of ways to find hazardous elements like asbestos and PCBs, but no one ideal technique that combines recommendations from several organizations exists. Accuracy testing poses a challenge because approaches vary.
(3) Absence of testing facilities: Specialized tools, software, and labs are needed for testing. Such in-house skills are costly for shipbreaking yards in poor nations. Because government facilities are limited, they are unable to test for every potentially dangerous item.
(4) Absence of facilities for managing hazardous waste: It also costs a lot to maintain and dispose of hazardous wastes properly, requiring large infrastructure investments. The majority of shipbreaking yards rely on central government or private facilities, many of which are either under construction or in the planning stages.
(5) Inconvenient logistics and transports: It can be difficult to store and move hazardous materials when facilities are located distant from shipbreaking yards. It is challenging to contain during transit in order to prevent environmental release.
(6) Unintentional spills: Although regulations require containment structures, hazardous substance spills nonetheless happen often in shipbreaking yards, contaminating the land and water. Mudflat cleanup after an accident is particularly challenging.

In conclusion, the management of hazardous waste is a complicated interdisciplinary task due to incomplete information, inconsistent testing procedures, inadequate infrastructure, challenging logistics, and unintentional releases. In order to limit the effects on the environment and human health, addressing these concerns will require innovative solutions in the areas of regulation, infrastructure development, transport management, and process controls. Addressing these complex concerns is necessary for the global implementation of sound ship recycling techniques.

11.10.3 Sustainable Shipbreaking

Hazardous wastes cannot be completely contained by current shipbreaking techniques, which pollutes the land and water. Mudflat cleanup after activities is difficult. Shipbreaking on a bed made up of layers of sand, gravel, and pebbles with different diffusivity and porosity is the suggested approach.

The fluctuating permeability of the bed can prevent the dilution and diffusion of dangerous contaminants. Lower layers' dense packing and large pebbles' poor permeability prevented the movement of dangerous materials downward. Prior to reaching lower levels, diffusion into pebble pore spaces was made possible by smaller, more porous pebbles in the higher layers. This prevented underlying soil contamination by lowering the overall solution concentration. With this design, the amount of hazardous material transported to the base layer was reduced to zero.

The bed's natural building elements are renewable resources that may be reused. The method's rearranged layers eliminate the need for a fixed infrastructure. It provides easier future material recycling along with reasonably priced pollution management. This approach is economical, flexible, and sustainable. It might make it possible to recycle ships more sustainably without incurring large implementation costs. Additional physical testing can validate its feasibility and accurate design specifications.

11.10.4 Conclusion

Hazardous compounds produced by shipbreaking and ship recycling have the potential to harm both human health and ecological systems. These hazardous pollutants cannot be completely contained by the shipbreaking techniques used today, which pollutes the environment.

The findings were in favor of building a multi-layered shipbreaking bed with decreasing pebble sizes to prevent the diffusion of hazardous materials into the underlying soil. The pollutants in the top layers were trapped by the varying permeability of pebbles with varying sizes. Concentrations were kept from getting to the bottom layer by the suggested bed, which had larger pebbles toward the base.

The method is sustainable because it uses natural, reusable materials to build the bed. The shipbreaking bed makes material recycling and future rearrangement easier by providing inexpensive pollution control without the need for permanent infrastructure. With the help of this innovative method, it is possible to produce raw materials from shipbreaking in a cleaner manner.

To ensure that the input data is accurate and to determine the ideal bed parameters, more physical testing is required. With more verification, this strategy may provide shipbreaking companies around the world with cutting-edge instructions for implementing responsible recycling processes. The bed concept can advance the IMO's shipbreaking requirements by minimizing ecological damage in a cost-effective manner. Overall, there is a good chance that the suggested method will increase sustainability and safety in shipbreaking.

11.11 Noise and Vibration Pollution

Over the past century, there has been a noticeable rise in human-induced underwater noise, with ship traffic playing a major role in raising noise levels in marine habitats. Over 60,000 medium- to large-sized commercial vessels, including tankers, ferries, bulk carriers, cargo ships, and cruise ships, are constantly navigating the oceans each year. Over the past forty years, there has been a notable increase in the amount of noise produced by ship traffic across the world's oceans due to the notable expansion of global maritime trade and the corresponding increase in vessel speed. Underwater noise substantially doubled every decade from the 1960s, when the first noise level data were recorded, until the 1990s. While some recent research suggests that low-frequency ocean noise levels are gradually declining at different marine areas throughout.

The purpose of this chapter is to examine how ship traffic affects marine life and how it contributes to ambient noise. It will also look at whether international legislation is heading toward a management plan that works to solve the problem of "ship noise." These factors will offer a thorough summary of ship-generated noise, its consequences on marine environments, and a synopsis of significant global and national regulation programs designed to lessen the effects of ship noise.

The main source of anthropogenic noise in the undersea environment is clearly ship traffic. Ship noise typically affects the 50–150 Hz frequency band, and it may even extend to several tens of kHz. Sound attenuation at lower frequencies (below a few hundred hertz) is mostly caused by geometrical spreading, as the impact of seawater's salt-induced absorption can be disregarded.

Propeller spinning is the primary source of what is widely known as "ship noise" underwater. Noise from propellers can be classified as either cavitating or non-cavitating. Propeller noise in non-cavitating situations is caused by vibrations in the blades, which result in characteristic narrowband tones at extremely low frequencies and a broadband element at higher frequencies. The broadband component is caused by inflow turbulence and edge effects, whereas the tone frequencies are determined by the number of propeller blades and shaft speed.

Propeller noise is considerably increased by cavitation. It is the main reason why ships make noise. Due to pressure decreases on the suction side, propeller blades cause cavitation noise as they move across the uneven wakefield behind the ship, creating bubbles packed with steam. 'White noise' that can reach upwards is created when these bubbles burst.

Variations in inflow speed, propeller loading, propeller submergence, and propeller design quality all affect how quickly cavitation begins. With typical pressure maxima ranging from 2 to 10 kPa for the first harmonic and generally decreasing with higher harmonics, blade sheet cavitation introduces tonal noise at harmonics of the blade passing rate. The stern and propeller design, as well as the use of wakefield-enhancing appendages, can have a substantial impact on the tonals at multiples of the blade rate.

Cavitation for fixed pitch propellers is directly related to the speed at which the shaft rotates and tends to decrease as ship speed decreases. When blades of controllable pitch propellers depart from the design condition, especially when the propeller is moving at low speeds, cavitation becomes more intense. Because the propeller is usually positioned a few meters below the surface of the ocean, Lloyd's mirror effects affect how propeller noise spreads. Dipole interference is caused by direct waves and their reflection off the sea surface, which determines the low-frequency cut-off which is based on the propeller's depth.

Based on the propeller depth, Lloyd's mirror effects affect the horizontal direction of radiated noise; lower effects are seen at shallower source depths. As a result, the amount of noise that a ship radiates depends on its load; empty ships produce less noise since their propellers are lower in depth.

Propeller spinning is not the only source of ship noise; winches, engines, generators, and other machinery can also produce noise that pierces the hull and enters the ocean. Propellers are usually driven by low-speed, firmly mounted, two-stroke engines. The propeller and engine rotate at the same speed, which adds to the total amount of noise that is transmitted. Medium-speed propulsion diesels are frequently used in smaller ships; they revolve at a steady speed, and the ship's velocity is controlled by varying the pitch of the propeller blades. This propulsion system is noisier at lower speeds when the pitch is suboptimal.

Shipboard diesel generators are another source of underwater noise, generating tones at a range of frequencies up to kilohertz. Ship noise is usually asymmetrical, with far-field acoustic output in the bow and stern aspects being modified by wake bubble absorption and hull blockage. Stern aspect noise levels on contemporary commercial ships are 5–10 dB greater than bow aspect noise levels. The sound signature of a ship is produced by a variety of sources, including engines, generators, and propellers; the main factors influencing noise levels are size and speed of operation.

Noise from commercial ships is mostly heard between 10 and 50 Hz. Noise spectral levels in this band range from 140 dB re 1 $\mu Pa^2/Hz$ at 1 m for small fishing vessels to 195 dB re 1 $\mu Pa^2/Hz$ at 1 m for fast-moving supertankers. Small boats have lower overall noise levels, but they contribute more spectrally in the 1–5 kHz range. Fast ferries produce narrow peaks in the 1–2 kHz range, and their broadband hydrodynamic noise can reach up to 10 kHz. Based on acoustic measurements of 2809 isolated transits, a research conducted in the Salish Sea, Washington State, and British Columbia estimates mean broadband (20–40,000 Hz) source levels for the ship population at 173 $\pm$ 7 dB re 1 μPa. These consequences include changes in behavior as well as physiological and auditory repercussions. An extremely interdisciplinary approach is needed to monitor the effects of ship noise on marine species, including techniques and instruments from a variety of fields, including bioacoustics, ecology, ethology, physical and biological oceanography, among others. The characteristics of the species under consideration, the anticipated consequences, and the characteristics of the noise source all have a significant role in the choice of monitoring approach.

Surveys can be carried out prior to, during, and following exposure for loud offshore operations such as seismic monitoring campaigns and drilling. However, because it can be difficult to determine cumulative exposure levels and because there are uncertainties in the physiology, behavior, distribution, and hearing sensitivity of different species, knowing the consequences of low-frequency diffuse ship noise presents considerable obstacles.

The majority of research on marine mammals' reactions to noise in recent decades has focused on cetacean species. Studies looking into the impacts of noise on fish have just recently come under more scrutiny because to growing concerns about the consequences of ship noise. Sound, usually in the frequency range of 100 Hz–3 kHz, is used by more than 50 fish families for communication and environmental perception.

One of the most common effects that come from ship noise entering the maritime environment is acoustic masking. This phenomena, which is typified by the masking of physiologically significant sounds by ship noise, compromises predator–prey interactions, the perception and interpretation of other environmental or human-induced noises, and the underwater auditory communication of animals. In particular, "communication masking" is the lowering of "communication space," which is the volume of space where it is expected that conspecifics will communicate acoustically, due to the presence of an external noise that interferes with communication. This makes it harder for a receiver to pick up important signals that are in the same frequency range as the noise.

There is evidence of physiological impacts associated with noise exposure in fish and invertebrates, as well as in a number of species of cetaceans. Stress from exposure to ship noise is the main cause of physiological alterations in marine organisms. Long-term noise exposure can have a variety of physiological impacts, most notably immune system suppression, which can jeopardize the health of those who experience it.

Heart rate, auditory response sensitivity, TTS, and PTS are a few examples of markers that can be assessed to evaluate the physiological and auditory state of a particular specimen in connection to ship noise. Furthermore, tracking the levels of stress hormones or the metabolites of stress hormones offers insightful information. Stress hormones (dopamine, adrenaline, and norepinephrine) increased in response to increasing noise levels in captive cetaceans (*Delphinapterus leucas* and *Tursiops truncatus*) subjected to impulsive sounds from a water cannon. In endangered North Atlantic right whales (*Eubalaena glacialis*), It found a connection between baseline levels of stress-related faecal hormone metabolites (glucocorticoids) and lower ship noise levels. In a Canadian port, low-frequency noise levels (< 150 Hz) were tested, and the results showed a 6 dB reduction due to a considerable decrease in ship activity in the days that followed the 9/11 events.

The establishment of worldwide and regional recommendations to address and mitigate noise pollution is a result of the ongoing increase in anthropogenic noise in marine habitats and the growing understanding of its possible effects on marine life.

To reduce noise from commercial shipping, the International Maritime Organization (IMO) approved non-mandatory guidelines in 2014 that were directed towards designers, shipbuilders, and ship operators (IMO 2014). These principles suggest that the initial ship design has the greatest chances for decreasing underwater noise. Propeller diameter, number of blades, pitch, skew, and sections are some of the elements that need to be properly planned or chosen to guarantee consistent water flow in order to reduce cavitation. A homogeneous wakefield should also be the goal of the ship's hull design, which is attained by reducing propeller cavitation in the wakefield the hull creates. When choosing inboard machinery, vibration control mechanisms must be taken into account, and their impact on radiated noise must be assessed. Considerations for operation and maintenance are also covered, such as cleaning propellers and maintaining hull surfaces. Ship speed selection is highlighted as an effective measure to reduce radiated noise from cavitation, and the guidelines suggest the possibility of adjusting routes to minimize ship noise impact on specific marine areas.

Shipyards, shipowners, and ship surveyors are advised by the International Organization for Standardization (ISO) to quantify ship noise according to specified protocols. In particular, ISO 17208-1 is recommended for usage and can be used to any surface vessel that is in motion under specified operating circumstances. Ships traveling faster than 50 km/h are not allowed to utilize this method. The standard specifies conditions for a deepwater ocean test site.

It's crucial to remember that although the IMO rules offer potentially useful ship noise reduction strategies, their implementation is entirely optional and limited to commercial ships. They do not cover other forms of noise introduction, such as that caused by seismic and sonar activity, nor do they apply to the navy or warships (IMO 2014). The use of models to predict future ship noise scenarios can be very helpful in the low-impact strategic design of new shipping routes and traffic patterns, such as those planned for the future European Motorways of the Sea. Moreover, noise hotspots can be identified by comparing modeled noise distributions and trends with animal density data. This is especially important for the most vulnerable species, such marine mammals.

11.12 Conclusion

Maritime activity is becoming more and more significant to the global economy, along with economic growth and advances in science and technology. Furthermore, significant environmental pollution issues related to oil, toxic and noxious substance, wastewater, ballast water, rubbish, air, noise, pollution from sinking materials, and other issues are still brought on by maritime activities. As a result, the International Maritime Organization has worked to implement policies that will enhance environmental pollution prevention, control, and safety. Convention, Codes, and associated rules and laws. Furthermore, limiting shipping's impact on the environment is a crucial factor in determining how the marine sector develops.

The chapter addresses topics such air pollution, wastewater pollution, ballast water pollution, toxic and poisonous material pollution, oil pollution, air pollution, noise pollution, pollution caused by sinking substances.

The book chapter's content is both theoretical and practical; it serves as a foundation for understanding and application of environmental pollution prevention in maritime activities by state management officials at seaports, officers, seafarers, and related organizations and individuals. It also prepares them to respond appropriately when environmental pollution incidents occur in accordance with statutory compliance conventions.

The book chapter, which covers environmental issues in maritime activities, is divided into 11 sections and is methodically collected based on national and international legal laws in the maritime domain: Ship-related oil pollution, HNS pollution, waste water pollution, ballast water management, garbage-related pollution, and dumping Substance pollution at sea; ship breaking; antifouling system; air pollution and greenhouse gas emissions; salvage sunken property; noise and vibration pollution.

Students, postgraduate students, scientists, instructors, and professionals in the maritime field—especially marine officers—will find this chapter beneficial. It is one of the key pillars that helps refine information and abilities for real-world application.

References

Altug G, Gurun S, Cardak M, Ciftci PS, Kalkan S (2012) The occurrence of pathogenic bacteria in some ships' ballast water incoming from various marine regions to the sea of Marmara, Turkey. Mar Environ Res 81:35–42. https://doi.org/10.1016/j.marenvres.2012.08.005

Aridgides LJ, Doblin MA, Berke T, Dobbs FC, Matson DO, Drake LA (2004) Multiplex PCR allows simultaneous detection of pathogens in ships' ballast water. Mar Pollut Bull 48(11):1096–1101. https://doi.org/10.1016/j.marpolbul.2003.12.017

Bailey SA, Rajakaruna H (2017) Optimizing methods to estimate zooplankton concentration based on generalized patterns of patchiness inside ballast tanks and ballast water discharges. Ecol Evol 7(22):9689–9698. https://doi.org/10.1002/ece3.3498

Baldrighi E, Semprucci F, Franzo A, Cvitkovic I, Bogner D, Despalatovic M, Berto D et al (2019) Meiofaunal communities in four Adriatic ports: baseline data for risk assessment in ballast water management. In: Marine pollution bulletin, vulnerability to bioinvasions: current status, risk assessment and management of ballast water through a regional approach—the Adriatic Sea, vol 147, pp 171–184. https://doi.org/10.1016/j.marpolbul.2018.06.056

Briski E, Gollasch S, David M, Dallas Linley R, Casas-Monroy O, Rajakaruna H, Bailey SA (2015) Combining ballast water exchange and treatment to maximize prevention of species introductions to freshwater ecosystems. Environ Sci Technol 49(16):9566–9573. https://doi.org/10.1021/acs.est.5b01795

Briton GH, Yao B, Ado G (2007) Evaluation of the Abidjan lagoon pollution. J Appl Sci Environ Manag 11(2). https://doi.org/10.4314/jasem.v11i2.55030

Bui TTH, Nguyen CH (2013) Building oil spill response contingency plan according to geographical area: one approach to marine spatial planning. In: International conference of union for conservation of nature and nature resources. Gland, pp 45–60

Burkholder JM, Hallegraeff GM, Melia G, Cohen A, Bowers HA, Oldach DW, Parrow MW et al (2007) Phytoplankton and bacterial assemblages in ballast water of U.S. military ships as a

function of port of origin, voyage time, and ocean exchange practices. Harm Algae Ballast Water 6(4):486–518. https://doi.org/10.1016/j.hal.2006.11.006
CCG (1995a) Canadian coast guard, oil spill response field guide
CCG (1995b) Oil spill response field guide, 1995
CCG (2011) Marine spills contingency plan national chapter. http://www.ccg-gcc.gc.ca/folios/00025/docs/national-response-plan-2011-eng.pdf
Chen J, Lin Y, Huo JZ, Zhang MX, Ji ZS (2010) Optimization of ship's subdivision arrangement for offshore sequential ballast water exchange using a non-dominated sorting genetic algorithm. Ocean Eng 37(11):978–988. https://doi.org/10.1016/j.oceaneng.2010.03.012
Dang QV, Nguyen MD, Van Hung P (2022) Assessment of potentially cutting GHG emissions from shipping in relation to energy consumption trends using Fuzzy Analytic Hierarchy Process. Aust J Marit Ocean Aff 14(3):189–203. https://doi.org/10.1080/18366503.2021.1997172
DOJ (2015) U.S. and five Gulf states reach historic settlement with BP to resolve civil lawsuit over deepwater horizon oil spill. https://www.justice.gov/opa/pr/us-and-five-gulf-states-reach-historic-settlement-bp-resolve-civil-lawsuit-over-deepwater
Drillet G, Gianoli C, Gang L, Zacharopoulou A, Schneider G, Stehouwer P, Bonamin V, Goldring R, Drake LA (2023) Improvement in compliance of ships' ballast water discharges during commissioning tests. Mar Pollut Bull 191:114911. https://doi.org/10.1016/j.marpolbul.2023.114911
Eight Institutions of Europe and Vietnam (2006) The risks and consequences of oil pollution in Vietnam & guidelines on bioremediation of oil spills. http://homepages.hs-bremen.de/~office-ikrw/biorem/Download/Handbook_BIOREM%201.0.pdf
Fingas MF (2000) Use of surfactants for environmental application. In: Schramm LL (ed) Surfactants: fundamentals and applications to the petroleum industry. Cambridge University Press, pp 461–539
Gollasch S, Hewitt CL, Bailey S, David M (2019) Introductions and transfers of species by ballast water in the Adriatic Sea. Mar Pollut Bull 147:8–15. https://doi.org/10.1016/j.marpolbul.2018.08.054
Gritsenko D (2017) Regulating GHG emissions from shipping: local, global, or polycentric approach? Mar Policy 84:130–133. https://doi.org/10.1016/j.marpol.2017.07.010
Hung PV, Kim K-S (2019) The estimation of regional oil recovery capacity based on marine oil spill response scenario in Vietnam. J Int Marit Saf Environ Aff Ship 2(2):97–111. https://doi.org/10.1080/25725084.2018.1562512
IMO (1995) MEPC.110(49) revised interim guidelines for the approval of alternative methods of design and construction of oil tankers under regulation 13F(5) of Annex I of MARPOL 73/78. http://rise.odessa.ua/texts/MEPC110_49e.php3
IMO (2003) Amendments to the Annex of the protocol of 1978 relating to the international convention for the prevention of pollution from ships, 1973 (revised Annex I of MARPOL 73/78), resolution MEPC.117(52). https://www.state.gov/documents/organization/102371.pdf
IMO (2004a) Globallast guidelines for development of a national ballast water management strategy, 1997. In: Guidelines for preventing the introduction of unwanted organisms and pathogens from ships ballast waters and sediment discharges. Resolution A.868 (20)
IMO (2004b) International convention for the control and management of ships' ballast water and sediments. International Maritime Organization, London. http://www.imo.org/en/about/conventions/listofconventions/pages/international-convention-for-the-control-and-management-of-ships%27-ballast-water-and-sediments-(bwm).aspx
IMO (2014) Building partnerships to assist developing countries to reduce the transfer of harmful aquatic organisms in ships' ballast water. Globallast partnerships
IMO (2023) Status of conventions. Available from http://www.imo.org/en/about/conventions/statusofconventions/pages/default.aspx
IPIECA (2000) A guide to contingency planning for oil spills on water. http://www.amn.pt/DCPM/Documents/Contingency.pdf

IPIECA-IOGP (2015) The global distribution and assessment of major oil spill response resources. Oil spill response joint industry project. http://www.oilspillresponseproject.org/wp-content/uploads/2016/02/JIP-9-Response-resources.pdf
ITOPF (2014) TIP 02: fate of marine oil spills. http://www.itopf.com/knowledge-resources/documents-guides/document/tip-2-fate-of-marine-oil-spills/
ITOPF (2017a) Country profile: Vietnam—ITOPF. http://www.itopf.com/knowledge-resources/countries-regions/countries/vietnam/
ITOPF (2017b) Oil tanker spill statistics—ITOPF. http://www.itopf.com/knowledge-resources/data-statistics/statistics/
JCG (2017) http://www.kaiho.mlit.go.jp/e/english.pdf
Jeevan J, Chen SL, Pateman H (2016) Implications of one belt one road for Malaysian connectivity to international trade routes. In: Proceedings of the OBOR conference 2016, Melbourne, Victoria, Australia, 1–2 Dec 2016. ResearchGate, pp 1–14. https://www.researchgate.net/publication/326112034_Implications_of_One_Belt_One_Road_for_Malaysian_connectivity_to_international_trade_routes_Proceedings_of_the_OBOR_Conference_2016_1-2_December_2016_Melbourne_Victoria_Australia_pp_1-14_2016
KCG (2008) National oil and HNS contingency plan. http://merrac.nowpap.org/plan/connector/1/data/plan/basic/Glist/1/
Kim A-R, Lee S-W, Seo Y-J (2022) How to control and manage vessels' ballast water: the perspective of Korean shipping companies. Mar Policy 138:105007. https://doi.org/10.1016/j.marpol.2022.105007
Lakshmi E, Priya M, Sivanandan Achari V (2021) An overview on the treatment of ballast water in ships. Ocean Coast Manag 199:105296. https://doi.org/10.1016/j.ocecoaman.2020.105296
Lee BG (2001) Sea prince incident and changes of response scheme after the incident, 16
Liu J, Wang P, Liu G, Dai J, Xiao J, Liu H (2021) Study of the characteristics of ballast bed resistance for different temperature and humidity conditions. Constr Build Mater 266:121115. https://doi.org/10.1016/j.conbuildmat.2020.121115
MONRE (2016) Environmental status report of Vietnam in period 2011–2015. http://tcmt.monre.gov.vn/pages/article.aspx?item=B%C3%A1o-c%C3%A1ohi%E1%BB%87n-tr%E1%BA%A1ng-m%C3%B4itr%C6%B0%E1%BB%9Dng-qu%E1%BB%91c-gia-giai-%C4%91o%E1%BA%A1n-2011---2015-42030
Neill PE, Arim M (2019) Human health link to invasive species. In: Encyclopedia of environmental health, pp 570–578. https://doi.org/10.1016/B978-0-12-409548-9.11731-2
Nguyen HN, Nguyen KH (n.d.) The experiment of the OilSAS model: an assistant tool, for responding to oil spill incidents in Van Phong bay, Khanh Hoa province
Nguyen ATT, Jacobson C, Ross H (2017) Changes in the nature of the Cat Ba forest social-ecological systems, chap 11. In: Redefining diversity & dynamics of natural resources management in Asia, vol 2. Elsevier, pp 177–192. https://doi.org/10.1016/B978-0-12-805453-6.00011-5
Perera LP, Mo B (2016) Emission control based energy efficiency measures in ship operations. Appl Ocean Res 60:29–46. https://doi.org/10.1016/j.apor.2016.08.006
Phan VH (2017) Estimating maximum amount of oil discharge in Vietnamese water—basis to build oil spill response. J Mar Sci Technol 52:67–71
Phung CS (2005) Report on the subject of scientific: study and propose prevention and response measures for oil spills level 1 in Da Nang City, Ho Chi Minh City, pp 10–130
Rigby G, Hallegraeff G (2001) Ballast water exchange and marine plankton distribution trials on the M.V. Iron Whyalla. In: Australian Quarantine and Inspection Service (AQIS) ballast water series report
Rigby GR, Hallegraeff GM, Sutton CA (1999) Novel ballast water heating technique offers cost-effective treatment to reduce the risk of global transport of harmful marine organisms. Mar Ecol Prog Ser 191:289–293
Russo MA, Leitão J, Gama C, Ferreira J, Monteiro A (2018) Shipping emissions over Europe: a state-of-the-art and comparative analysis. Atmos Environ 177:187–194. https://doi.org/10.1016/j.atmosenv.2018.01.025

Salleh NA, Rosli FN, Akbar MA, Yusof A, Sahrani FK, Razak SA, Ahmad A, Usup G, Bunawan H (2021) Pathogenic hitchhiker diversity on international ships' ballast water at West Malaysia port. Mar Pollut Bull 172:112850. https://doi.org/10.1016/j.marpolbul.2021.112850

Shafiquzzaman M (2021) Effect of pre-aeration on the removal of arsenic and iron from natural groundwater in household based ceramic filters. J Environ Manage 291:112681. https://doi.org/10.1016/j.jenvman.2021.112681

Shang L, Hu Z, Deng Y, Liu Y, Zhai X, Chai Z, Liu X, Zhan Z, Dobbs FC, Tang YZ (2019) Metagenomic sequencing identifies highly diverse assemblages of dinoflagellate cysts in sediments from ships' ballast tanks. Microorganisms 7(8):250. https://doi.org/10.3390/microorganisms7080250

Shapoori M, Gholami M (2014) Effect of a ballast water treatment system on survivorship of natural populations of marine plankton in Persian Gulf, Iran. Mar Sci 4(2):44–48. http://article.sapub.org/. We investigated the viability of phytoplankton from ballast water of commercial and petroleum ships berthed at the ports of Kharg Island, Iran. Ballast water was carried in regional seas in order to assess the efficiency of this type of ballast water management at reducing the abundance and diversity of phytoplankton. Although there was an overall reduction in the average abundance and number of taxa after exchange this was not consistent between tanks and voyages. On some occasions there were changes in species composition after exchange. Mentioned research was a part of Regional and National plan and protection for environmental impact assessment and also control and prevention of navigation activities. The result showed that species were dominant in the samples of *Ceratium sp.*, *Rhizosolenia sp.*, *Coscinodiscus jonesianus*, *Eucampia sp.*, *Chaetocerus diversicurvatus*, *Thalassionema nitzschoides*. Factors such as the depth of the water during the exchange process, the season and the method of exchange influenced the efficacy of the exchange process. The variability in the results after exchange mean that this is unlikely to be a ballast water management method that would give consistent results and careful consideration would have to be given to the suitability of using this method in regional seas as a means of reducing the risk of introducing non-native species

Su DS (2015) Looking ahead: the port industry, 53

Thach ND, Van Hung P (2023) Optimal UV quantity for a Ballast Water Treatment System for compliance with Imo Standards. Pol Marit Res 30(4):31–42. https://doi.org/10.2478/pomr-2023-0056

Thach ND, Van Hung P (2024) Development of UV reactor controller in ballast water treatment system to minimize the biological threat on marine environment. J Sea Res 198:102465. https://doi.org/10.1016/j.seares.2023.102465

UNCTAD (2020) Statistics | UNCTAD. https://unctad.org/statistics

USCG (2008a) 33CFR155 Appendix B, determining and evaluating required response resources for vessel response plans. https://www.law.cornell.edu/cfr/text/33/appendix-b_to_part_155

USCG (2008b) 33CFR155, subpart D, tank vessel response plans for oil

Van Hung P (2022) Potential effects of the Kra Canal on Vietnam's maritime industry. Transp Res Interdiscip Persp 14:100622. https://doi.org/10.1016/j.trip.2022.100622

Van Hung P, Viet DQ, Duc NM, Le TD (2023) Ship routing optimization using bacterial foraging optimization algorithm for safety and efficient navigation. Int J Electr Comput Eng 13(2):2309–2315. https://doi.org/10.11591/ijece.v13i2.pp2309-2315

VASI (2016) Vietnam administration of sea and islands, database oil spills in 2015, 18

VAWR (2008) Vietnam Academy for Water Resources. http://www.vawr.org.vn/images/Image/IMAGE575.jpg

VINASARCOM (2007) National committee for search and rescue, general report on the state of oil spills drifted to coast of the central provinces and southern province of Vietnam in 2007, 4

Yang X, Huang G, Zhang P, An C, Yao Y, Li Y, Zhou S (2021) Life cycle-based water footprint analysis of ceramic filter for point-of-use water purification in remote areas. Sci Total Environ 786:147424. https://doi.org/10.1016/j.scitotenv.2021.147424

Zhen L, Li M, Hu Z, Lv W, Zhao X (2018) The effects of emission control area regulations on cruise shipping. Transp Res Part D Transp Environ 62:47–63. https://doi.org/10.1016/j.trd.2018.02.005

Chapter 12
"Where the Land Meets the Sea"—ICZM Implementation in Estonia

Siiri Külm, Anton Shkaruba, and Kalev Sepp

Abstract The chapter introduces the Integrated Coastal Zone Management and process of developing ICZM in Estonia. In this chapter will be described co-developed different approaches for land-sea interactions in two case areas in Estonia based on the defined characteristics and key interests, which are derived from local or regional challenges. In doing so we integrate spatial data on human activities and ecology. Paragraph on "Data integration" gives overview on different possibilities on data integration from nature and human activities to present integrated human-ecological networks. Especially produced GIS information about environmental research projects and of participatory planning were utilised. Outcomes of visualized thematic maps covering the case areas are presented. The section "Stakeholder involvement" describes how different interests were integrated in the preparation of the plans considering the sea-land interface (e.g. human and nature values, recreation areas, maintenance of blue-green networks, construction of housing). That included such activities as deployment of e-platforms and working group sessions with different stakeholders and end-users, as well as involvement of local residents, stakeholders and steering group members. The section "Results" gives an overview of ICZM plans developed for different coastal areas, and describes how these plans are used by spatial planners dealing with coastal accessibility and coastal tourism. The description of ICZM plans provide an ample set of integrated knowledge and links to experts to support local-level planning. Finally, the chapter offers conclusions about the implementation of socially, environmentally, and economically sustainable coastal and spatial plans, which are based on the diversification of coastal land use.

Keywords Integrated Coastal Zone Management · Land-sea interaction · Coastal planning · Stakeholder involvement

---

S. Külm (✉) · A. Shkaruba · K. Sepp
Estonian University of Life Sciences, Tartu, Estonia
e-mail: siiri.kylm@emu.ee

V. Maccarrone and M. Fadzil Akhir (eds.), *Coastal Sustainability*, Coastal Research Library 39, https://doi.org/10.1007/978-3-031-75749-5_12

12.1 Introduction

The marine and coastal areas in Europe are extremely diverse because of a wide range of natural, social, economic and cultural factors shaping them. As pressures from urbanization, climate change, and economic activities grow, the need for effective management of coastal areas becomes more obvious. Several measures have been developed in Europe with an aim to find opportunities for more sustainable management and protection of coastal areas. Integrated Coastal Zone Management (ICZM) and Maritime Spatial Planning (MSP) emerge as key frameworks in Europe to address the complex challenges associated with coastal and marine environments.

To promote sustainable development of coastal zones, the European Commission communicated "Integrated Coastal Zone Management: A Strategy for Europe" (COM/2000/547) in 2000 (European Commission 2000), as well as, the Recommendation concerning the implementation of Integrated Coastal Zone Management in Europe (2002/413/EC) (European Parliament and Council 2002). In its previous communication it defined ICZM as "*a continuous process with the general aim of implementing sustainable development in coastal zones through optimal management of human activities in this area to improve the state of the coastal environment and maintain its diversity*" (European Commission 1999). This journey has started in 2007, when the European Commission adopted the Blue Paper proposing an Integrated Maritime Policy for the EU (European Commission 2007), and a detailed Action Plan (SEC(2007) 1278). The European Council endorsed the IMP and the Action Plan on 14 December 2007 (Meiner 2010).

Until 2010, the European Union (EU) Member States reported their progress in implementing ICZM, and this resulted in a proposal for a Directive establishing a framework for maritime spatial planning and integrated coastal management, adopted by the European Commission (EC) in 2013 (European Commission 2013). Maritime Spatial Planning (MSP) gained prominence as a complementary framework to ICZM. The European Union's Maritime Spatial Planning Directive (2014/89/EU) established the legal basis for MSP, encouraging Member States to develop spatial plans for their marine areas. The directive emphasized the need for coherent planning, stakeholder involvement, and cross-border cooperation, reflecting the interconnected nature of marine spaces. MSP aims to optimize the use of marine space, preventing conflicts between different sectors and promoting sustainable development. Through the comprehensive planning, MSP addresses the growing demands for maritime resources, shipping lanes, renewable energy installations, and marine conservation areas. Europe has been at the forefront of MSP, with the European Union (EU) providing a policy framework that encourages Member States to develop and implement MSP strategies. Therefore in the context of spatial planning, ICZM is formally considered separate from MSP. The development of Marine Spatial Planning is obligatory for Member States, while the implementation of the Integrated Coastal Zone Management is optional for countries and local authorities.

The purpose of the ICZM plan is the integration of natural and cultural values with the sustainable and balanced economic development processes in the coastal zone, which brings together the environments, processes, and practices related to the coastal sea and land. The management plan makes it possible to give meaning to the common elements of coastal lifestyles, ecosystems and environmental changes, as well as to promote participatory planning methods.

Estonia is in the north-eastern part of Europe, by the Baltic Sea. The length of the Estonian coastline is 3794 km of which 1242 km are on the mainland and 2552 km are divided among the islands. The ICZM plans in Estonia were drafted under the SustainBaltic (2016–2018) programme, which was financed through Interreg's Central Baltic Programme 2014–2020 of the European Regional Development Fund (ERDF). In the course of the project, two Integrated Coastal Zone Management (ICZM) plans or land use zonings were completed in Estonia. The project aimed to improve the share of the managed coastal networks in the Central Baltic area by the cross-border preparation of the ICZM plans for a total of four case areas with their public assessment in Estonia and Finland. This was planned to be achieved by (1) integrating multidisciplinary human-ecological data on the whole project area; (2) selecting the themes of ICZMs and case study areas; (3) making four digital ICZM plans, which were first tested with the current regional land use of the project area; (4) public assessment of the plans using both e-platforms and arranging working groups with stakeholders and end-users moderated by the participants of the project.

This chapter provides an overview of the drafting process of ICZM plans within the framework of project SustainBaltic, along with an examination of the results derived from the two project reports (Kull et al. 2018; Kuusik et al. 2018) and authors' experience with the process. In order to develop two ICZM plans, two case areas were selected in different parts of the whole Estonian coast (Fig. 12.1). One case area is located in northern part of Estonia (Lääne-Viru County) and the completed ICZM plan was named Lääne-Viru Integrated Coastal Zone Management Plan. The second case area is located in the western part of Estonia (Läänemaa County) and completed ICZM plan was named 'Spatial development plan for the coastal area between Hara peninsula and Ristinina'. Two different approaches and methodologies were followed to develop these ICZM plans. In the northern case area the management plan focuses on the contact-nodes of the land and the sea by linking them spatially and by making proposals for developing cross-sectoral cooperation. In the western case area we have applied the methodology for land use zoning. It included the analysis of current land-uses, as well as an overview of land use zoning and objectives in terms of the problems, needs and stakeholder interests identified for the area. All this represented a base for laying down the land-use zoning for the area. Two drafted ICZM plans are indicative documents that municipalities can use for the preparation of master plans or other local development documents (development plans, projects), as well as for the development of national-level marine spatial plans.

Fig. 12.1 Two ICZM case areas are marked by a purple line: 1. Lääne-Viru case area; 2. Läänemaa case area

12.2 Drafting ICZM Plans—Data Integration, Methodology and Stakeholder Involvement

12.2.1 Selection of Case Areas

The selection of case areas was based on the analysis of regional GIS layers and following the criteria:

(a) Availability of and proximity to nature areas of high nature value;
(b) Diversity of land use specific to coastal zones;
(c) Representative trends in population patterns and dynamics (e.g. marginalisation and sparsely populated areas);
(d) Abundance of cultural heritage;
(e) Cross-border locations.

As a result of an analysis based on these criteria, regions with moderate human impact in Läänemaa and Lääne-Viru counties were selected, which would reflect more generally the socio-ecological tendencies and value-based challenges of coastal zones in Estonia.

After the case area selection, two little bit different approaches were followed to achieve finalizing ICZM plans and are described as follows.

12.2.2 *Lääne-Viru Case Area*

Based on the GIS analysis (see above), the border of Lääne-Viru case study area was initially set at 5 km inland and 5 km offshore. On a later stage the inland border was modified following up consultations with various stakeholders (e.g. by considering the use of the marine area by coastal fishermen), so the coastal zone that is up to a few kilometres wide and extends to a larger road running along the coast was set. The total length of the case area's coastline was 93 km, while the area of the ICZM case site (marine and coastal zone combined) was 457 km^2.

The management plan development process was based on desktop analysis of such development documents as plans and spatial development strategies (Fig. 12.2). Also very important role played the outcomes of interviews with stakeholder representatives from the governmental bodies, business and NGOs to understand better their views to the current situation and possible future expectations.

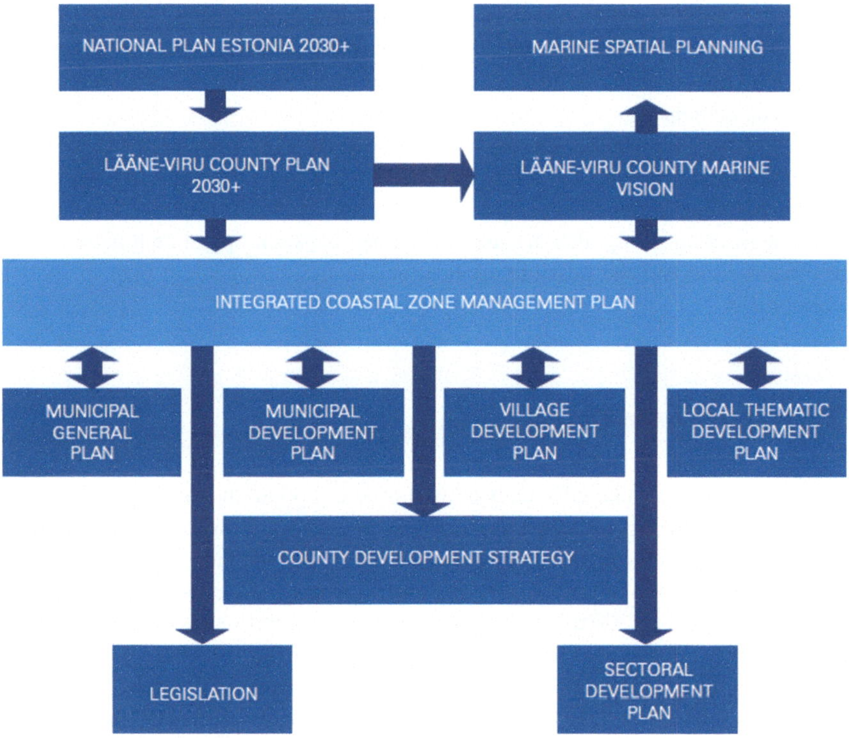

Fig. 12.2 The connections of ICZM plan and its possible applications in the spatial planning (Kuusik et al. 2018)

The development of the management plan (Kuusik et al. 2018) included a massive field work program, as well as several meetings, surveys and consultations with representatives of rural municipalities, industry specialists, and several other stakeholder groups, such as coastal fishermen and port operators.

After being completed, the management plan was available online for several months for people to give feedback. Yet only a few people had left their comments, although the plan was promoted through local newspapers. During public discussions held on site a number of new topics were addressed, as well as possible solutions. In addition, the public was invited to provide their feedback though the electronic public participatory geographic information system (PPGIS) questionnaire. It was promoted through direct mailing, Facebook, newspapers of rural municipalities and advert columns, and attracted more responses: total 130 people responded, of which about 70 answered every question (all the questions were non-mandatory); by the end of the long questionnaire, about 50 people answered each question. The PPGIS questionnaire was prepared in the Maptionnaire software that supports open and multiple-choice questions, assessment scales, and also has functionality of uploading map layers (for example, planned residential building areas or forests with increased public interest) where online users can chose or add locations. The information collected on the feedback stage was analysed and used for the revision of the management plan.

In addition to stakeholders, an important role in the development of IZCM plans was played by the members of the regional steering group. They provided feedback on such elements of the management plan as topics, structure, and proposed activities. The steering group also discussed the links between operational activities in the coastal zone and the possible applications of the management plan in spatial development.

12.2.3 Läänemaa Case Area

In the case of Läänemaa, GIS analysis was used to collate the layers reflecting on case study selection criteria (see above), and to identify an area with the highest concentration of the desired features and properties. For the purposes of case study analysis, the area 3 km inland and 300 m offshore was recognised as the coastal zone, and the zone between 300 m and 3 km offshore was recognised as the marine zone. The total length of the case area's coastline was 48 km and the area of the ICZM case site (marine and coastal zone combined) was 218 km^2.

Similarly to the previous case, we have used the land use zonation methodology to develop ICZM plan in Läänemaa case area. As a first step, we mapped stakeholders connected to the case area and the local problems requiring solutions. On the second step, the solution-related objectives identified for the area have been used as an input to regional development visions and to land use zoning principles. The final statements of objectives, vision, and principles were discussed and confirmed with the members of the project steering group and local governments' representatives.

On the third step, building on the inputs from previous steps, we have created a classification of land use zones and mapped them. The zones were distinguished on the map in as compact and comprehensive areas as possible and named after the primary usage of the zone. In addition to the main land use activity within a respective zone, auxiliary ones have been allowed as long as they did not conflict with the main activity.

The draft scheme of land use zoning was presented to and discussed with the representatives of local governments and other stakeholder groups at public hearing events and through other meeting formats. Any interested parties were welcome leave their feedback. We were collecting responses in writing, and through the web-based map application created under this project. The latter was an ArcGIS Web application with the following features: (1) viewing of designated zones; (2) adding comments; (3) adding points, polygons and lines accompanied with suggestions. The feedback was discussed and considered for the final version of the land use zoning as long as it was consistent with the principles of the land use zoning.

12.3 Drafting ICZM Plans—Results

The aim of the SustainBaltic project was to develop solutions for the integrated management of coastal zones, which would involve consistent and communicative planning of connections between the sea and land. The ICZM plans have considered the natural and cultural values in case study areas as well as their current state. The plans provide an assessment of how different areas interact, descriptions of current activities, and possible conflicts. ICZM Incorporating input from online submissions and face-to-face meetings, primarily from local communities and stakeholder representatives, the ICZM plans integrate visions and proposals. Consequently, they offer recommendations and visionary descriptions on achieving the objectives established initially, aiming for sustainable and balanced development. This section provides details on two ICZM plans based different methodologies.

12.3.1 Lääne-Viru Case Area

In the Lääne-Viru case area, the results of the ICZM plan (Kuusik et al. 2018) are based on future visions and outlined management activities. To achieve these goals, three main objectives of spatial development were identified, serving as the foundation for drafting the ICZM plan:

1. Emphasizing the significance of diverse maritime transport, ports, and maritime tourism.
2. Fostering the sustainability of historical coastal villages, characterized by active communities, cultural heritage, identity, and traditions. This includes enhancing

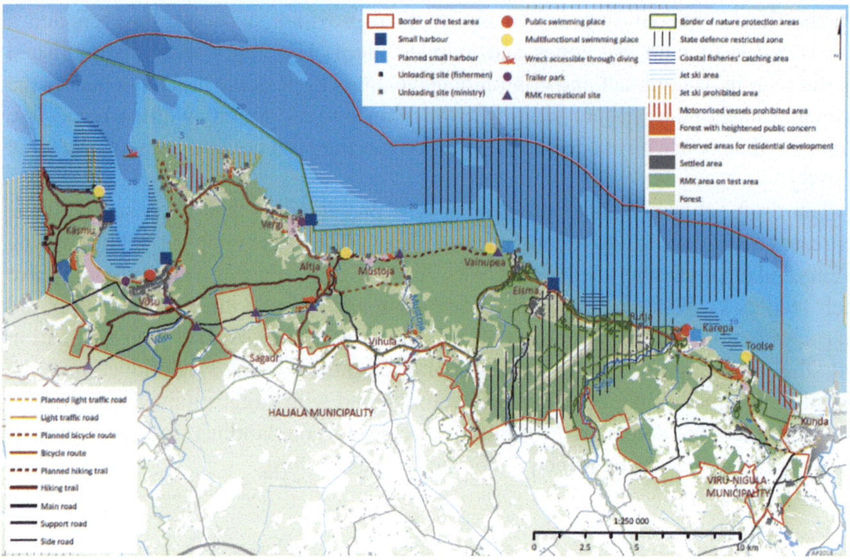

Fig. 12.3 The scheme of ICZM plan of Lääne-Viru case area (Kuusik et al. 2018)

the quality and diversity of infrastructure and both public and private services to attract and retain permanent residents.

3. Ensuring the sustainable utilization of natural resources in the maritime and coastal areas, alongside the implementation of innovative renewable energy solutions.

This ICZM plan (refer to Fig. 12.3) comprises detailed descriptions of various topics such as recreation, mobility, communities, and entrepreneurship, tailored to the regional specifics. Additionally, it outlines overarching development visions and provides recommendations for future actions.

The Lääne-Viru case area is characterised by varied coastal landscapes, including heritage-rich villages, forest areas, sandy beaches, nature protection areas, small harbours, etc. Lahemaa National Park, where the case area is partially located, is already a well-known tourist destination. Locally cherished values include the sea, sandy beaches, pine forests, and ensuring public access to the coastline. Residents prefer swimming in private beaches or those near their homes or vacation spots. Although current situation of coastal bathing areas is satisfactory, there is a need to focus on waste management and the availability of amenities. Parking issues arise, particularly during peak beach days and cultural events, suggesting the potential implementation of seasonal parking fees in popular coastal areas. The surge in tourist arrivals also imposes environmental pressures, necessitating proactive measures from local authorities. Encouraging greater involvement from tourism companies, such as through cooperation funds or targeted taxes, could further support the region's promotion and sustainability efforts.

The Lääne-Viru case exhibits largely unimpeded access to the sea and coastline, primarily due to a significant portion of the land being state-owned, with approximately 75% covered in forest. As a result, the right to roam along coastal paths generally remains intact. However, occasional issues arise, such as roads being obstructed by coastal landowners. Enhancing cycling and pedestrian tracks would not only improve cycling opportunities, but also ensure safety within the coastal zone. During the summer season, establishing better public transport connections between various villages and surrounding cities, local hubs, and Tallinn, the capital of Estonia, is crucial. This would facilitate smoother travel for residents and visitors, enhancing accessibility to amenities and services. Small-craft harbours are evolving into multifunctional hubs, offering amenities such as accommodation, catering, and entertainment facilities beyond their traditional navigation role. The development of such small-craft harbours is anticipated to attract more people to the sea, potentially necessitating port expansion or the introduction of new functions to extend the operating season. However, harbour development remains a contentious issue within the community, significantly impacting established village structures and community dynamics. Planning for ports involves more than just constructing berths; it requires careful consideration of complex spatial and social factors. As a result, the planning process for ports tends to be prolonged, primarily due to the necessity of conducting extensive public discussions to address various stakeholder concerns and ensure community input is adequately considered.

To preserve the heritage and sustainability of the community, it is vital to maintain and increase the number of permanent residents in the coastal zone. This can be achieved by encouraging small businesses tailored to the region's specificities and promoting teleworking opportunities. While tourism-related jobs offer opportunities, their seasonal nature and the need for service diversification pose challenges. Coastal fishing, integral to the cultural heritage, provides some income, yet faces declining catch volumes and the necessity for supplementary activities during lean seasons. Balancing the interests of professional and recreational fishermen, while promoting fish stock sustainability, is crucial. Moreover, the aging population presents opportunities for the silver economy, necessitating the development of tailored services and infrastructure. Initiatives supporting activities on or towards the sea, such as sailing and fishing camps, should be prioritized. It is imperative to maintain community gathering places like village shops and libraries. Residential development in coastal areas must be carefully planned, considering climate change impacts and varying perspectives on heritage preservation. Community engagement and integration of cultural practices into hobby and tourism activities are essential. Involving communities in voluntary maritime rescue and landscape stewardship enhances local engagement. Future planning should consider demographic shifts, such as aging populations and increased teleworking, as well as the potential for blue growth trends like renewable energy and sustainable tourism. This requires comprehensive planning that accounts for social infrastructure and environmental sustainability in coastal and marine areas.

12.3.2 Läänemaa Case Area

In this case area, land use zoning (Kull et al. 2018) primarily aimed to enhance the integration of coastal sea and land, creating opportunities their diverse use. The following zones (Fig. 12.4) were delineated in the coastal zone: (1) recreation zone, (2) recreation development zone, (3) port and industrial zone, (4) traditional settlement zone, (5) residential development zone, (6) agricultural zone, (7) nature zone. In the marine zone, the following zones were identified: (1) general marine use zone, (2) targeted marine management zones: targeted marine management zone for Natura 2000 habitats, targeted water transport zone, targeted marine management zone for water sports, targeted marine management zone for dumping, and targeted marine management zone for military use. These zones are delineated on the map as compact and comprehensive areas, with the designated use of each zone based on its primary purpose. While other activities are permitted within the zones, they must not conflict with the designated primary use (Kull et al. 2018). As this IZCM plan based on zonation is of recommendatory nature, it outlines the main indicative uses of specific areas without excluding any other smaller scale uses of the area.

While developing ICZM plans, the objective of this land use zoning was established with specific goals in mind reflecting on the character of the area, land-use trends and stakeholder interests. These included creating a high-quality living environment, promoting the economic use of coastal natural resources, striking a balance between the interests of the local community and holidaymakers, ensuring that human activities do not have significant or irreversible negative effects on the coast, and maintaining or increasing the value of the coast. To achieve these objectives, descriptions were provided, and for each designated zone, various recommendations were suggested. These recommendations included restrictions, utilisation of recreational value, landscape management strategies, and guidelines for the development and planning of buildings. These measures were aimed at preserving or enhancing the value of each zone.

The prevalence of nature areas is notably high due to the presence of unfavourable soils resulting from the tectonic uplift in the northern part of the region, historically maintaining these areas as forests. Given that the case area includes only coastal areas, it is pertinent to focus forest management efforts preserving the recreational value of these areas rather than intensive management. The designation of the 'Nature zone' aims to safeguard the natural values of the area and promote environmentally friendly usage practices. Conversely, the southern part of the region comprises more agricultural areas. To encourage environmentally sustainable agricultural practices, the 'Agricultural zone' was established. The objectives of this zone include preserving the characteristic landscapes and communities of coastal areas while fostering employment opportunities. Additionally, this zone encompasses protected areas and other naturally preserved areas.

Given that most households in this case area are situated in old farmhouses, the 'Traditional settlement zone' promotes the preservation of local milieu and cultural

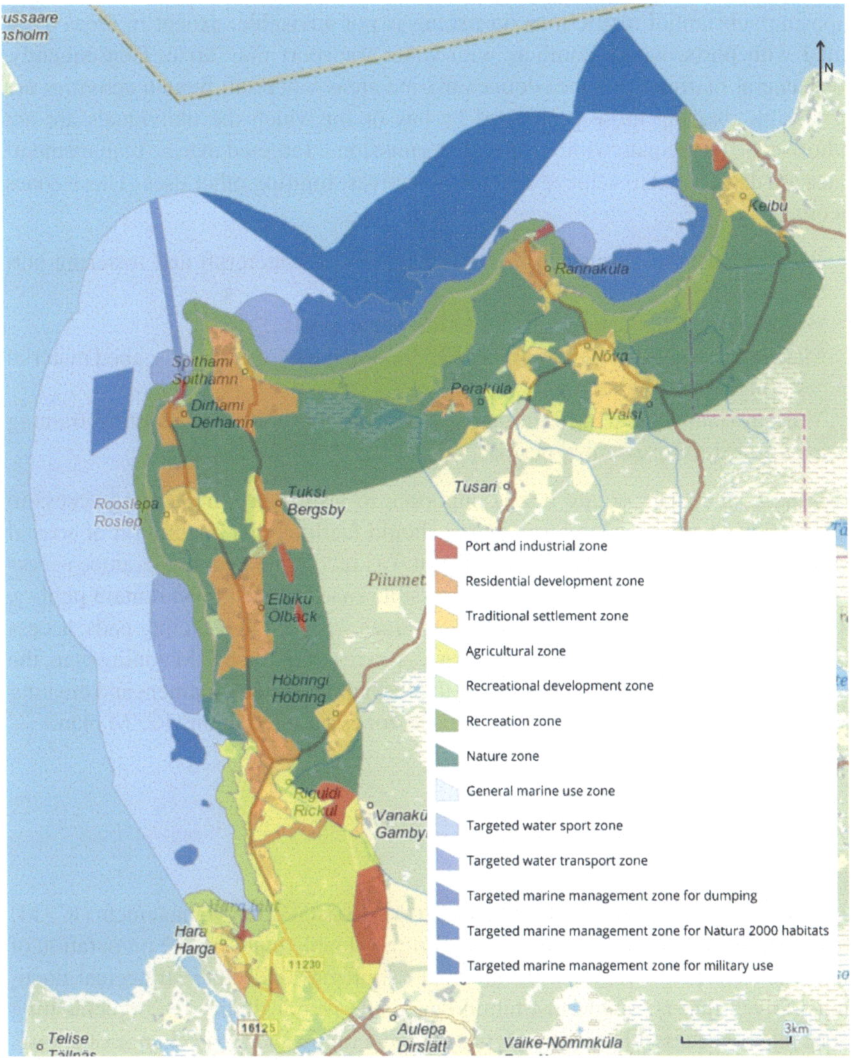

Fig. 12.4 The scheme of ICZM plan of Läänemaa case area (Kull et al. 2018)

heritage while providing recommendations on enhancing the quality of life in proximity to recreational and agricultural areas. In order to propose additional opportunities for residential construction for the local community and capitalize on the recreational potential of the region through the development of summer recreational areas, the 'Residential development zone' was designated.

When zoning the sea area, there was no significant need to restrict its usage extensively. The sea in the region is a vital area for coastal fishing; therefore, imposing substantial restrictions on fishing is not advisable, except in areas associated with ports, where conflicts with water transport may arise. Consequently, the 'General marine use zone' defines marine areas where all human activities are permissible, barring those prohibited by law or for which the individuals are not authorised to participate without special permission. 'Targeted marine management' zones are designated to achieve specific objectives, limiting other uses. These zones encompass:

1. Water Transport: Ensuring the safe navigation of watercraft and fostering port development.
2. Water Sport: Establishing safe areas for water sports.
3. Marine Management for Dumping: Facilitating the movement of seabed material resulting from sea dredging.
4. Marine Management for Military Use: Reserving areas for shooting training exercises during specified periods.

In the Läänemaa case are it was important to also focus on ensuring access to coastal and sea areas, as land use zoning should facilitate the integration of coastal sea and land. In line with this ICZM plan, it was recommended to guarantee access to the coast and sea through publicly accessible roads (which must remain public), forest trails and hiking paths in recreational areas, walking and cycling pads, access to landing sites, and maritime rescue access. As a result of the ICZM zoning plan, the local authority has recommendations for the further region development and creating diverse opportunities for utilising designated zones outlined in this ICZM plan.

12.4 Conclusions

It is important to acknowledge that the Integrated Coastal Zone Management (ICZM) plan serves as an indicative document, offering valuable input for the preparation of comprehensive plans, regional development strategies and other strategical documents. Given that no single institution bears sole responsibility for implementing ICZM plans, successful execution relies on cooperation and the exchange of knowledge among various stakeholders. Project SustainBaltic facilitated the testing of two distinct approaches in completing ICZM plans across two case areas located in different parts of Estonia. The principles and methodologies applied in land use zoning are also transferrable and applicable to other coastal regions.

The process of drafting and completing ICZM plans underscored the significance of stakeholder engagement through collaboration. Public meetings conducted in case areas proved instrumental in understanding the interest, specific trends, and challenges encountered by local stakeholders, such as coastal fishermen and small-craft harbour managers. Utilising two web-based participatory applications demonstrated the potential for stakeholders to express their views, opinions and suggestions in a prompt and coherent manner. This diversified use of participatory methods facilitated the analysis of environmental issues arising during the drafting of ICZM plans.

Consultations conducted towards the end of 2019 with Estonian Marine Spatial Planning (MSP) coordinators revealed that the finalised Estonian ICZM case plans were integrated into the national MSP process to address land-sea interactions and underscore the regional specificity of regional development strategies. At the municipal level, Estonian ICZM plans are currently utilised by spatial planners as thematic background material to inform work related to coastal accessibility and coastal tourism. These plans offer a comprehensive repository of integrated knowledge and links to experts, thereby supporting local-level planning endeavours (Hietala et al. 2021).

References

European Commission (1999) Towards a European integrated coastal zone management (ICZM) strategy: general principles and policy options. Office for Official Publications of the European Communities, Luxembourg

European Commission (2000) Communication from the Commission to the Council and the European Parliament on integrated coastal zone management: a strategy for Europe (COM/2000/547), adopted 27 September, 2000

European Commission (2007) Communication from the Commission to the European Parliament, the Council, the European economic and social committee and the Committee of the Regions—an integrated maritime policy for the European Union. Brussels COM, 575 final

European Commission (2013) Impact assessment accompanying the document proposal for a Directive of the European Parliament and of the Council establishing a framework for maritime spatial planning and integrated coastal management. Brussels

European Parliament and Council (2002) Recommendation of the European Parliament and of the Council of 30 May 2002 concerning the implementation of integrated coastal zone management in Europe (2002/413/EC). http://eur-lex.europa.eu/legal-content/

Hietala R, Ijäs A, Pikner T et al (2021) Data integration and participatory process in developing integrated coastal zone management (ICZM) in the northern Baltic Sea. J Coast Conserv 25:47. https://doi.org/10.1007/s11852-021-00833-4

Kull A, Palginõmm V, Semm M, Tomson P (2018) Spatial development plan for the coastal area between Hara peninsula and Ristinina. Tartu

Kuusik M, Pikner T, Printsmann A, Raet R (2018) Lääne-Viru integrated coastal zone management plan. Tallinn

Meiner A (2010) Integrated maritime policy for the European Union—consolidating coastal and marine information to support maritime spatial planning. J Coast Conserv 14:1–11. https://doi.org/10.1007/s11852-009-0077-4

Chapter 13
Coastal Protection in the Mediterranean Sea: An Overview of International and European Coastal Defense Regulations

Ludovica De Benedetti and Marianna Marzano

Abstract Coastal defense is a critical issue worldwide, as coastal erosion caused by both natural and anthropogenic causes poses significant risks to coastal communities and infrastructure. The term coastal defence refers to any activity aimed at protect a coastal area from erosion and/or flooding; examples consist of engineering schemes such as sea walls and embankments. International and European regulations play a fundamental role in guiding coastal defense practices in the Mediterranean Sea, and ensuring sustainable, effective measures, compatible with both human development needs and environmental protection objectives. This document provides an overview of the main international and European regulations relating to coastal defense with a specific focus on the Mediterranean Sea, and aims to highlight the principles, objectives and implementation mechanisms that underpin these frameworks. The document also intends to highlight the importance of collaboration and cooperation at an international and regional level between States in order to effectively respond to the challenges and seize the opportunities associated with the implementation of coastal defense measures.

Keywords UNCLOS · Barcelona Convention · EU regulations and directives · Marine environment strategy · Maritime spatial planning

---

L. De Benedetti (✉)
National Research Council of Italy—Institute of International Legal Studies, Via dei Taurini 19, 00185 Rome, Italy
e-mail: ludovica.debenedetti@cnr.it

M. Marzano
Pisa University, Lungarno Pacinotti 43, 56126 Pisa, Italy

V. Maccarrone and M. Fadzil Akhir (eds.), *Coastal Sustainability*, Coastal Research Library 39, https://doi.org/10.1007/978-3-031-75749-5_13

13.1 Introduction

Nowadays, coastal areas increasingly suffer from coastal pollution, coastal erosion and flooding, and numerous hazards threaten the coastal environment, such as coastal storms and flooding, posing significant risks to coastal physical, economic and social systems. Coastal areas are becoming increasingly urbanized, which increases the concentration of people and assets at risk. This is making coastal communities more vulnerable to coastal risks. For these reasons, a coastal defense strategy and a legal framework are essential to ensure the conservation and protection of coastal environments. This chapter will provide an overview of the main international and European provisions relevant for coastal protection. Coastal protection concerns all regulations and strategies aimed at defending coastal areas from different stressors, resulting in new challenges to minimize coastal vulnerability through mitigation strategies. Both at an international and European level there are no specific rules regarding coastal defense. Over time, however, international conventions have been approved both on the environment and on climate change and standards, regulations, directives and guidelines have been approved at an international, regional, European and national level in specific sectors which require the implementation of measures for the coastal protection, think of those concerning environmental impact assessments, the protection of biodiversity, pollution, etc. It should be noted that international agreements and regulations provide the general framework for coastal protection, while European and Italian strategies and regulations provide more specific guidance and requirements. In the first paragraph the international legal framework will be analyzed; the analysis will start from the international framework on climate change, then the regulation of the law of the sea, environmental conventions and finally a focus will be on the Mediterranean region. In the second paragraph the main European regulations and strategies will be analyzed. Finally, the third paragraph will focus on some of the measures and techniques used for coastal management in the context of the Mediterranean Sea.

13.2 Climate Change Framework

For coastal defense it is crucial to consider, in a climate change perspective, the impact of the exponential global average sea levels. Climate change is increasing the frequency and intensity of coastal hazards such as flooding, erosion, and storm surges. As sea levels rise, coastal communities become more vulnerable to flooding and erosion. Climate change is driving sea levels rise through two phenomena. First, the rapid melting of ice at the poles: this extra water flows into the oceans, directly contributing to sea level rise. Moreover, the increased concentrations of greenhouse gases in the atmosphere. As the ocean absorbs heat due to greenhouse gases, it takes up more space, causing the overall sea level to rise. The combination of these two factors is having a significant impact on coastal communities around the world

(Griggs and Reguero 2021). According to the World Meteorological Organization (WMO), global average sea levels have risen faster since 1900 than in any preceding century over the past 3000 years.[1] The sea level could rise by 65 cm by the end of the century, a figure in line with estimations by the UN Intergovernmental Panel on Climate Change. Several international agreements, conventions and strategies focus on climate change including provisions on sea level rise. Combating climate change is a cardinal goal of the well-known an action agenda of 17 Sustainable Development Goals (SDGs) for people, planet and prosperity signed in September 2015 by the governments of the 193 member countries of the UN. In particular, Goal 13 concerns actions to "take urgent action to combat climate change and its impacts", also making reference to sea level rise.[2] Goal 14.2 requires to "sustainably manage and protect marine and coastal ecosystems to avoid significant adverse impacts, including by strengthening their resilience, and take action for their restoration in order to achieve healthy and productive oceans". The United Nations Framework Convention on Climate Change[3] (UNFCCC) in the preamble recognizes the heavy effects of climate change on the coastal areas. In particular, UNFCCC recalls the provisions of General Assembly resolution 44/206 of 22 December 1989 on the possible adverse effects of sea-level rise on islands and coastal areas, particularly low-lying coastal areas and the pertinent provisions of General Assembly resolution 44/172 of 19 December 1989 on the implementation of the Plan of Action to Combat Desertification[4] (Sands 1992). Between Commitments of the Convention, quoted in Article 4, Parties are required to the promotion of sustainable management of all ecosystems, including coastal ecosystems, the conservation and enhancement of coastal areas and the cooperation for plans for coastal zone management are fundamental aspects. Indeed, All Parties shall "cooperate in preparing for adaptation to the impacts of climate change; develop and elaborate appropriate and integrated plans for coastal zone management, water resources and agriculture, and for the

[1] Sea levels continued to rise in 2022, reaching a new record since satellite measurements in 1993. Moreover, the rate of global mean sea-level rise has doubled in the past decade—from 2.27 mm per year in 1993–2002 to 4.62 mm per year in 2013–2022. Global mean sea level, 1993–2022 (millimetres).

[2] Goal 13 includes strengthen resilience and adaptive capacity to climate-related hazards in all countries; integrate climate change measures into national policies, strategies and planning; improve education and awareness-raising on climate change tools; strategies to support developed countries and small island developing States, including the goal of mobilizing jointly $100 billion annually by 2020 from all sources to address the needs of developing countries.

[3] The 198 countries that have ratified the Convention are called Parties to the Convention. The UNFCCC is a "Rio Convention", one of two opened for signature at the "Rio Earth Summit" in 1992. Its sister Rio Conventions are the UN Convention on Biological Diversity and the Convention to Combat Desertification. The three are intrinsically linked. It is in this context that the Joint Liaison Group was set up to boost cooperation among the three Conventions, with the ultimate aim of developing synergies in their activities on issues of mutual concern.

[4] Moreover it "Recognizing further that low-lying and other small island countries, countries with low-lying coastal, arid and semiarid areas or areas liable to floods, drought and desertification, and developing countries with fragile mountainous ecosystems are particularly vulnerable to the adverse effects of climate change".

protection and rehabilitation of areas, particularly in Africa, affected by drought and desertification, as well as floods". Dealing with legal sources on climate change it is essential to mention the Paris Agreement. The Paris agreement is the first universal, legally binding global climate agreement, adopted at the 21st Conference of the Parties to the UN Framework Convention on Climate Change (COP21) in Paris on 12 December 2015 aimed at reducing climate-related risks by reducing greenhouse gas emissions to net zero and limiting global-mean temperature increase. As seen above, the global temperature goal is crucial also for overall sea level to rise, limiting the rapid melting of ice at the poles and the increased concentrations of greenhouse gases in the atmosphere. It has to be considered that sea level rise is not the only way climate change affects coastlines. Coastal zones are affected by climate changes processes and dynamics through interannual and long-term changes in winds, storm surges or wave action. Overall, it is important to increase our knowledge on climate change processes to know how, and at what rate, coasts are changing, in order to support knowledge-based decision-making to direct our energy and creative thinking towards adaptation strategies and sustainable management which can prepare the coastlines and cities of the world. Guidance was created to ensure coastal hazards and climate change are being factored into planning in coastal areas.[5]

13.2.1 Law of the Sea Framework

Among the international legal regime on coastal protection it is fundamental to consider the UNCLOS Convention,[6] which establishes a legal framework for the use and management of marine resources and areas, including coastal zones.[7] Under the UNCLOS Convention, there is a general obligation to protect the marine environment and to ensure that activities conducted under the jurisdiction or control of a party do not cause harm beyond its marine waters, to avoid transferring harm or risk from one area to another, and to avoid turning one type of pollution into another. UNCLOS also requires states to enact and, wherever possible, to enforce their own marine conservation laws. Important principles are the protection of the marine environment, including the ecosystem-based approach, the precautionary approach and sustainable

[5] For example, the Coastal hazards and climate change guidance is a tool to help decision makers consider the potential effects of climate change, now and in the future, based on the latest scientific information, including a 10-step process to assess the risks and determine what actions to take: https://environment.govt.nz/assets/publications/Coastal-hazards-and-climate-change-guidance-2024-ME-1805.pdf.

[6] The United Nations Convention on the Law of the Sea (UNCLOS), opened for signature in 1982 in Montego Bay, currently has been ratified by 168 states plus the European Union.

[7] In fact, the Convention defines a comprehensive regime of rights and responsibilities of states in the use of the seas and oceans, setting out guidelines governing negotiations, the environment and the management of natural resources.

development.[8] In classical international law, the sea was traditionally divided into an area under the sovereignty of the coastal state (territorial sea) and an area in which the principle of freedom of use for all states applies (high seas). The Montego Bay Convention on the Law of the Sea provides for a more structured legal regime. The areas identified by UNCLOS are as follows: Inland waters: the sea space within the baseline.[9] In this area, the laws of the coastal state regulating resource use and vessel passage apply in a binding manner; Territorial sea: it includes the sea space from the baseline to 12 nautical miles. The laws of the coastal state still apply in this area, but within the territorial waters there is the right of every vessel to so-called innocent passage.[10] Moreover, the Convention identifies other marine areas such: Archipelagos: the internal waters of states comprising archipelagos are identified by drawing a baseline joining the outermost points of the outermost islands, where these points are reasonably close together; Contiguous zone: it extends from the territorial sea no further than 24 nautical miles from the baseline. In this area, the coastal state can both punish violations committed within its territory or territorial sea and prevent violations of its customs, tax, health and immigration laws or regulations. This makes the contiguous zone a hot pursuit area; Exclusive economic zone: also known by the acronym EEZ, it is the area of sea extending 200 nautical miles from the baseline in which the coastal state can exercise the right of exclusive exploitation of natural resources. This principle was created to curb indiscriminate exploitation of fisheries. In addition, considering new technologies that allow oil to be drilled in very deep waters, this principle has also recently been used for exclusive exploitation of mines. If there are no more than 400 nautical miles between the coasts of two different states, it is necessary for the two coastal states to conclude a bilateral agreement. Other marine areas are the Continental Shelf: the continental shelf is regarded as the natural extension of a state's territory, which can then exploit its mineral or otherwise non-living resources exclusively. The continental shelf may exceed 200 NM but

[8] Article 193 of UNCLOS is similar to Principle 21 of the Stockholm Declaration recognizing the sovereign right of states "to exploit their natural resources pursuant to their environmental policies and in accordance with their duty to protect and preserve the marine environment." Article 194 requires states, individually and jointly, "to take all measures consistent with this Convention that are necessary to prevent, reduce and control pollution of the marine environment from any source, using... the best practicable means at their disposal and in accordance with their capabilities." Article 194(3) enumerates four specific sources of pollution: "the release of toxic, harmful and noxious substances... pollution from vessels... pollution from installations and devices used in exploration or exploitation of the natural resources of the seabed and subsoil... pollution from other installations and devices operating in the marine environment..." Finally, Article 194(5) mandates measures necessary "to protect and preserve rare and fragile ecosystems as well as the habitat of depleted, threatened or endangered species and other forms of marine life."

[9] The baseline, so called because it is the starting point for defining inland and international waters, is defined as a broken line joining notable points on the coastline, generally keeping to shallow waters, but where the coastline is particularly rugged or in cases where islands are very close to the coast, the baseline may cut across and encompass large tracts of sea.

[10] Harmless passage is defined as the crossing of marine areas in a continuous and expeditious manner that does not affect the peace, good order and security of the coastal state. Fishing, dumping of garbage, armed activities and espionage are not considered harmless actions; submarines and submarines must also sail in surfacing while showing the flag;

not exceed 350, or it may be calculated by measuring 100 NM from the 2500 m isobath; High sea: open sea, specifically that which does not fall under any national jurisdiction; Enclosed or semi-enclosed sea: a gulf, basin, or sea surrounded by two or more states and communicating with another sea or ocean by a narrow passage, or consisting wholly or mainly of the territorial seas and exclusive economic zones of two or more coastal states. UNCLOS Convention provides and lists the rights of coastal States,[11] defined as sovereign, (exploration, exploitation, conservation and management of natural resources, biological and otherwise) on the resources contained on the seabed and in the marine subsoil, as well as in the water column above, to be extended also to the activities necessary to carry them out (Treves 2015). Under the jurisdictional profile, the Convention recognizes the jurisdiction over the installation and use of artificial islands, plants and structures, the conducting of scientific research and the protection and preservation of the environment to the State that holds the EEZ. As for the continental shelf (Part VI of UNCLOS), the coastal state exercises sovereign rights with respect to the exploration and exploitation of its natural resources, i.e. mineral resources (e.g. hydrocarbons) and sedentary biological resources (living organisms that remain motionless on the shelf or move about while remaining in contact with the seabed. A very delicate problem, given the vastness of the marine areas engaged by the powers of coastal states on the continental shelf, is that of the 'delimitation' of the shelf between coastal states that face each other or are adjacent, as the Mediterranean States. It is necessary to consider that the rise in sea level, generating an acceleration of coastline erosion and flooding of extensive low-lying coastal areas, will have a significant impact on the delimitation of the maritime zones (Rizal 2016). Indeed maritime zones are determined by direct reference to shorelines or coastlines, which are expressed in the law of the sea as 'baselines. The fact that low-tide elevations will be permanently submerged by the waters will lead to a significant reduction in the extent of the territorial sea and the internal waters of a coastal state. There may in the future be some economic and geopolitical consequences such as the loss of part of the sovereignty of a coastal state resulting from a reduction in the extent of its internal or archipelagic waters or territorial sea; the modification or disappearance of an EEZ generating a loss of exploitation of the coastal state's natural resources (fish, minerals, oil, etc.); or even the modification of the rules governing passage through different maritime zones (for instance when a zone part of the territorial sea where the right of innocent passage applies becomes part of the EEZ where freedom of navigation and overflight applies). These modifications could lead to political tensions, in particular as regards the freedom of navigation and access to maritime resources (Caron 2015).

[11] Art. 56 of the UNCLOS Convention.

13.2.2 Other Relevant Environmental Conventions

In the international regime on coastal protection it is significant to include the Convention on Biodiversity (CBD). The Convention on Biological Diversity of 1992 aims to halt biodiversity loss, ensuring the conservation and sustainable use of marine biodiversity, and to create a global network of marine protected areas (MPAs). Coastal ecosystems such as estuaries, wetlands and mangrove forests are highly valuable for coastal communities since they contain significant diversity (De Fontaubert et al. 1996). Human activities everywhere are wearying marine and coastal living resources and degrading marine and coastal ecosystems in harmful, and sometimes irreversible, ways. In the oceans, as on land, the scope of this degradation has no precedent in human history. It is generally agreed that such biodiversity loss—the decimation of stocks of living resources, widespread appearance of ecosystem imbalances and the debilitation of ecological processes—may well undermine the adaptive potential of those systems and their ability to meet future human needs. In view of their common concern regarding marine and coastal biodiversity, and recognizing the significant differences between marine and terrestrial biodiversity, the Parties to the Convention on Biological Diversity' agreed on a program of action for implementing the Convention with respect to marine and coastal biodiversity at their second Conference of the Parties (COP), held in Jakarta in 1995. This program, the Jakarta Mandate on Marine and Coastal Biological Diversity demonstrates that the Biodiversity Convention is crucial legal instrument for promoting the conservation of marine and coastal biodiversity and the sustainable use of living marine and coastal resources. Although it includes some qualifications and additions, the Jakarta Mandate draws heavily on the recommendations of the first meeting of the Convention's Subsidiary Body on Scientific, Technical and Technological Advice (SBSTTA), held in Paris in September 1995. For coastal protection it is important to mention the Convention on Wetlands, an intergovernmental treaty also called Ramsar Convention because was adopted in the Iranian city of Ramsar in 1971 and came into force in 1975.[12] The Convention on Wetlands provides the framework for the conservation and wise use of wetlands and their resources. The Convention's mission is "the conservation and wise use of all wetlands through local and national actions and international cooperation, as a contribution towards achieving sustainable development throughout the world". International cooperation, policy making, capacity building and technology transfer are crucial aspects to reach this goal (Matthews 1993). Wetlands are being degraded and lost due to the increasing demands of the growing human population and over-exploitation. To counter this, it is crucial to ensure the wise use of wetlands, which can be seen as the conservation and sustainable use of wetlands and all the services they provide, for the benefit of people and nature.[13] Under the three pillars of the

[12] Almost 90% of UN member states, from all the world's geographic regions, have acceded to become "Contracting Parties".

[13] The Convention defines the wise use of wetlands as "the maintenance of their ecological character, achieved through the implementation of ecosystem approaches, within the context of sustainable development".

Convention, the Contracting Parties commit to: work towards the wise use of all their wetlands through national plans, policies and legislation, management actions and public education; designate suitable wetlands for the list of Wetlands of International Importance (the "Ramsar List") and ensure their effective management; cooperate internationally on transboundary wetlands, shared wetland systems.

Also relevant for the coastal protection are global conventions protecting marine environment from pollution.[14] The International Convention for the Prevention of Pollution from Ships—known as Marpol—adopted by the International Maritime Organization (IMO) in 1973, aims at reducing marine pollution from ships as a result of the routine operation of vessels and minimizing pollution as a result of marine accidents. The Convention replaced the OILPOL 1954 Convention as the main global treaty covering prevention of marine pollution by ships.[15] The Protocol of 1978 was adopted in response to a spate of tanker accidents in 1976–1977. As the 1973 MARPOL Convention had not yet entered into force, the 1978 MARPOL Protocol absorbed the parent Convention and the combined instrument entered into force on 2 October 1983. MARPOL has robust and evolving Annexes,[16] strict enforcement and continuous involvement of the International Maritime Organization and its Marine Environment Protection Committee in its operation. These factors contribute to the consideration of it as a successful environmental Convention (Fitzmaurice 2023).

In accordance with MARPOL, it must be considered the International Convention on Oil Pollution Preparedness, Response and Cooperation (OPRC) which was created in response to a call at a conference in Paris in 1989 by leading industrial nations to develop further measures to prevent pollution from ships. The International Maritime Organization (IMO) answered that call by adopting the OPRC Convention in November 1990. The Convention entered into force in May 1995.[17] The OPRC

[14] Other marine environment global Conventions: Convention on the Prevention of Marine Pollution by Dumping of Wastes and Other Matter (London Convention), London, 1972; International Convention for the Prevention of Pollution from Ships, 1973, as modified by the Protocol of 1978 relating thereto (MARPOL 73/78), London, 1973 and 1978; International Convention for the Prevention of Pollution of the Sea by Oil, London, 1954, 1962 and 1969; International Convention on Civil Liability for Oil Pollution Damage (CLC), Brussels, 1969, 1976, 1984 and 1992; International Convention on the Establishment of an International Fund for Compensation for Oil Pollution Damage (FUND) 1971 and 1992, Brussels, 1971/1992; International Convention Relating to Intervention on the High Seas in Cases of Oil Pollution Casualties Intervention Convention, Brussels, 1969.

[15] The Protocol of 1978 was adopted in response to a series of tanker accidents in 1976–1977. Since the Convention had not yet entered into force, the 1978 Protocol absorbed the parent Convention. The combined treaty entered into force in October 1983. It was subsequently amended by another Protocol adopted in 1997, and kept updated through other relevant amendments.

[16] The Convention currently includes six technical Annexes. Special Areas with strict controls on operational discharges are included in most Annexes.

[17] The Protocol on Preparedness, Response and Cooperation to Pollution Incidents by Hazardous and Noxious Substances (OPRC-HNS Protocol) aims to establish national systems for preparedness and response, and to provide a global framework for international cooperation in combating major incidents or threats of marine pollution. It ensures that ships carrying hazardous and noxious substances are covered by preparedness and response regimes similar to those already in existence for oil incidents. For the purposes of the HNS Protocol, a Hazardous and Noxious Substance is

Convention is designed to improve the ability of nations to cope with an oil spill. Aside from ensuring that each country establishes a rapid response emergency system to deal with spills, it also establishes a mechanism for international cooperation to deal with these incidents.

13.2.3 Focus on Mediterranean Region

Recent studies have found that the Mediterranean Sea is also suffering the effects of global climate change: its conditions, even, are changing even faster than those of the oceans because, being a semi-enclosed sea, it has little water exchange. The most important legal instrument for the protection of the marine environment and the Mediterranean coast is the Barcelona Convention, adopted in Barcelona on 16 February 1976 and amended on 10 June 1995. The Barcelona Convention constitutes the principal regional legally binding Multilateral Environmental Agreement (MEA) in the Mediterranean. The aim is to protect the Mediterranean marine and coastal environment while boosting regional and national plans to achieve sustainable development. Several Protocols, strategies and action plans exist to implement the Barcelona system. The Convention together with its 7 Protocols, makes up a legislative framework of the Action Plan for the Mediterranean (MAP).[18] The MAP also has another important legal document, "Regional Climate Change Adaptation Framework for the Mediterranean Marine and Coastal Areas", approved in February 2016. With its approval the Mediterranean countries agreed on the strategic objectives, directions and priorities, for administrators and stakeholders, on climate change for marine and coastal areas. The Regional CC Adaptation Framework focuses on the use of Integrated Coastal Zone Management as a climate adaptation measure. The main objective of the Framework is to set a regional strategic approach to increase the resilience of the Mediterranean marine and coastal natural and socioeconomic systems to the impacts of climate change, assisting policy makers and stakeholders at all levels across the Mediterranean in the development and implementation of coherent and effective policies and measures by identifying strategic objectives, strategic directions and priorities.[19]

defined as any substance other than oil which, if introduced into the marine environment is likely to create hazards to human health, to harm living resources and marine life, to damage amenities or to interfere with other legitimate uses of the sea.

[18] In line with Article 17 of the Barcelona Convention, UNEP provides secretariat services to the Contracting Parties through its MAP Coordinating Unit, established in Athens in 1982 on the basis of a Host Country Agreement between Greece and UNEP. The overall mission of the Coordinating Unit, is to promote and facilitate the implementation of the Barcelona Convention, its Protocols and Strategies, and of the Decisions and Recommendations of the Contracting Parties. It ensures the functioning of the MAP system and facilitates the work of the Contracting Parties to meet their commitments under the Barcelona Convention and its Protocols.

[19] Furthermore, there are agreements that deal with specific aspects relating to excavation activities and therefore only marginally involve the topic in question in its entirety. This is especially the case with the 1996 Protocol to the London Convention.

In a more focused way, the aforementioned Barcelona Convention is accompanied by the on Integrated Coastal Zone Management in the Mediterranean (signed in Madrid on 21 January 2008), ratified by the EU on 13 September 2010 with Council Decision 2010/631/EU and entered in force on 24 March 2011. The ICZM Protocol provides the legal framework for the integrated management of the Mediterranean coastal zone. According to what is indicated in the ICZM Protocol, 'integrated coastal zone management' means a

> dynamic process for the sustainable management and use of coastal zones, taking into account at the same time the fragility of coastal ecosystems and landscapes, the diversity of activities and uses, their interactions, the maritime orientation of certain activities and uses and their impact on both the marine and land parts. (Alterman and Pellach 2016)

The aim was to establish a common framework for the integrated management of the Mediterranean coastal zone and shall take the necessary measures to strengthen regional cooperation for this purpose.[20] Under the Protocol, Parties are called to take the necessary measures to strengthen regional cooperation in order to meet the objectives of integrated coastal zone management. Such measures include those aimed at protecting the characteristics of certain specific coastal ecosystems (e.g. wetlands and estuaries, marine habitats, coastal forests and woods and dunes), those aimed at ensuring the sustainable use of the coastal zone, and those aimed at ensuring that the coastal and maritime economy is adapted to the fragile nature of coastal zones.

To implement the ICZM Protocol, the Contracting Parties adopted in 2012 the Action Plan for the implementation of the IZCM Protocol (2012–2019). Protocol on Integrated Coastal Zone Management in the Mediterranean (ICZM Protocol) requires countries to preparate National Integrated Coastal Zone Management (ICZM) Strategies. In 1995 UNEP published the Guidelines for integrated management of coastal and marine areas to provide general guidelines for national policy-makers, managers and professional involved in the development and implementation of integrated coastal and marine areas management (ICAM) programs.[21] The Guidelines for integrated management of coastal and marine areas started from the assumption that policies have been based on the sectoral approach and therefore failed to take into account the overall impact of coastal development on resources. On the contrary, it is now generally accepted that coastal systems are far too complex to be managed through independent sectoral policies, but resolved within a comprehensive environmental, social and economic management framework.[22]

[20] Projects like "Mare Nostrum Project" help to bridge the policy-implementation gap in the legal aspects of ICZM along the Mediterranean. Key to the project is the comparison of different jurisdictions and their approach to integrated coastal zone management.

[21] The document was prepared by the Priority Actions Programme Regional Activity Centre (PAP/RAC) of the Mediterranean Action Plan—UNEP under the guidance of the Oceans and Coastal Areas Programme Activity Centre (OCA/PAC) of UNEP. The Priority Actions Programme Regional Activity Centre (PAP/RAC) assists Contracting Parties in meeting their obligations under the ICZM Protocol.

[22] Sustainable development requires that the quantity and quality of coastal resources are safeguarded in order that they not only satisfy the present needs but provide a sustained yield of economic

13.3 Coastal Defence in EU Context

The previous paragraph analyzed international regulations and highlighted the absence of specific provisions regarding coastal erosion. However, it noted the existence of more general conventions, treaties, and strategies related to the protection of the marine environment and the division of competences that can also be applied to this sector. Similarly, at the EU level, general conventions, regulations, and directives address environmental protection, biodiversity conservation, and environmental impact assessment (Pranzini et al. 2015).

The coastal zones of the Mediterranean are continually subjected to altering natural processes, erosion and marine submersions. The impacts of these processes and of these events vary from one coastline to the next and depend on the geological structure of the coastline and its exposure to the impact of waves (Benassi et al. 2015).

Coastal erosion presents substantial challenges for coastal communities, ecosystems and infrastructure, particularly in the Mediterranean region. In response to these pressing concerns, the European Union (EU) has developed a regulatory framework, which, although it does not provide, as already underlined, regulations directly dedicated to the problem, deals with the main causes underlying coastal erosion and promoting a more sustainable development of coastal areas to safeguard their ecosystems and resilience. This paragraph delves into the development of European regulations relating to defense against coastal erosion, analyzing the fundamental directives, policies and initiatives that have influenced the continent's strategies for coastal management with a particular focus on the case of the Mediterranean Sea.

13.3.1 European Regulations

The EU's involvement in coastal defense traces back to its early environmental initiatives, notably Directive 85/337/EC (amended by Directive 97/11/EC and subsequent Directives 2011/92/EU and 2014/52/EU)[23] which concern the assessment of environmental impacts from specific public and private projects. Annex II of this

---

and environmental services for future generations. "Guidelines for Integrated Management of Coastal and Marine Areas, UNEP".

[23] The Directive was approved after the issuing of the first action programs of the European Communities on the environment in 1973, 1977 and 1983. The general outlines of the programs, approved by the Council of the European Communities and by the representatives of the governments of the Member States, underlined, in fact, the importance of avoiding pollution and other nuisances at source, rather than fighting their effects later and therefore immediately taking into account the environmental repercussions of decision-making and planning activities. Furthermore, the directive had the aim of harmonizing the legislation in force in the Member States regarding the environmental impact assessment of public and private projects to guarantee equal conditions in the context of competition in the common market.

Directive encompasses activities subject to Environmental Impact Assessment (EIA), including "reclamation of land from the sea" and

> coastal works aimed at combating erosion and altering maritime structures, such as dykes, jetties, and other sea defense installations, excluding maintenance and reconstruction. (MATTM-Regioni 2018)

The foundation of the EU policy relating to coastal zones can be captured more organically in the Fifth Environment Action Program, and subsequently in the Sixth Environment Action Program (from 2002 to 2012), adopted as a response to the European Council's call for a global strategy on integrated coastal zone management.

The Sixth Environment Action Program, in fact, was based on a global approach, by theme, rather than on specific pollutants or types of economic activity as had been done until then, and, for this reason, provided for the adoption of seven thematic strategies covering air pollution, the marine environment, sustainable use of resources, waste prevention and recycling, sustainable use of pesticides, soil protection and the urban environment.[24]

From what has been said so far, it is clear that an important change in the EU's action has been to adopt organic strategies on the environment and biodiversity, to indicate to Member States the objectives to be pursued and the deadlines within which to pursue them. Within these strategies there are various objectives that influence coastal protection and which we will analyze shortly. On the other hand, the EU has also continued to adopt detailed rules and directives focused on individual areas (e.g. protection from flood risk) which establish binding rules or objectives for Member States and which, although not focused on coastal defense, establish rules that also have an impact in this sector. The analysis will start from the general strategies and then focus on the detailed legislation.

As regards specifically the protection of the marine and coastal environment, in 2007 the Commission launched "An integrated maritime policy for the European Union" to respond to the growing concerns linked to the awareness of the fact that all maritime activities were interconnected, while corresponding maritime policies and decision-making processes were still rather fragmented. The EU's Integrated Maritime Policy (IMP) which represents a holistic approach to all EU policies related to the sea and which aims to have a greater development of sea-related activities, but with a lower impact on environment, provided 5 cross-cutting policies: blue growth, marine data and knowledge, maritime spatial planning, integrated maritime surveillance, and sea basin strategies (Grbec 2023a, b; Sacco et al. 2017).[25]

[24] The program clarified that it was necessary to go beyond a strictly legislative approach and adopt a strategic approach in order to face the challenges posed by environmental problems, influencing not only legislative decisions, but also the decisions of businesses, consumers, political actors and citizens. The communication focused on four priority areas for action: climate change; biodiversity; environment and health; and sustainable management of resources and waste.

[25] In 2009 and 2012, the Commission presented two reports on the progress of the objectives set by the "Integrated Maritime Policy for the European Union" which describe the main results achieved and the corresponding maritime sectoral policies. Regulation (EU) no. 508/2014 was also approved: it repealed Regulation (EU) no. 1255/2011, which established an intermediate program to support the further development of an integrated maritime policy.

Blue growth is a long-term strategy adopted by the Commission in 2012 to harness the potential of the blue economy and support the development of sustainable marine and maritime economic activities (Touri 2021). It focuses mainly on the development of renewable energy, and on the development of the blue economy. As regards coastal defense, there are several relevant measures envisaged by the strategy, in particular the objectives relating to sustainable coastal tourism. In particular, in the strategy the Commission plans to promote ecotourism and encourage connections with other sustainability actions; promote the implementation of the Protocol to the Barcelona Convention on Integrated Coastal Zone Management and the Council Recommendation on the same topic and promote maritime spatial planning and green infrastructure; promote strategies regarding waste prevention and management and waste dumping at sea, to support sustainable coastal and maritime tourism.

Within the scope of the EU's Integrated Maritime Policy, the Directive 2014/89/EU on Maritime Spatial Planning was adopted in 2014.[26] This directive entrusted Member States with the task of developing plans for sustainable maritime management, which include, among other things, the identification of areas where the implementation of coastal defense measures is necessary. The Directive also provides for the involvement of stakeholders (Gómez-Ballesteros et al. 2021; Keijser et al 2018) in the efficient management of maritime activities in a transparent manner to avoid potential conflicts and create synergies between the different activities.

Also relevant is the Marine Strategy Framework Directive (MSFD) adopted in 2008 by the EU to maintain clean, healthy, productive and resilient marine ecosystems, while ensuring a more sustainable use of marine resources. The directive, in fact, requires Member States to develop national strategies for the marine environment in order to obtain or maintain, where it exists, a "good environmental status" (this status should have been achieved by 2020[27]). Marine strategies must include periodic assessments of the state of the marine environment, set objectives to be achieved and establish monitoring programs. In this context, it is also important to maintain a good state of the subsoil and coasts to guarantee the good marine ecological status of coastal waters and the conservation of biodiversity.

The new EU Biodiversity Strategy for 2030,[28] adopted in 2020 as a key part of the European Green Deal, is a long-term action plan that sets out specific commitments

[26] The MSP Directive aims to reverse environmental degradation of the seas and facilitate the sustainable use of marine resources, including mature uses such as fishing and shipping, as well as emerging uses such as renewable energy and mariculture by establishing a framework for maritime spatial planning to promote sustainable development that can ensure the use of maritime space for diverse maritime uses and manage spatial uses and conflicts in marine areas, and at the same time protect biodiversity. In 2022, the Commission produced a first report on progress in the implementation of the directive (EU Commission Report 2022).

[27] In 2020, the Commission adopted a report on the first implementation cycle of the MSFD. The report underlined that, although the EU's marine environmental protection system is one of the most sophisticated in the world, further improvements are needed especially in fishing control, plastic waste, nutrient inputs excess, underwater noise and other types of pollution.

[28] The EU Biodiversity Strategy 2030 is a global, long-term plan that aims to reverse the degradation of ecosystems. For these reasons it is also the proposal for the EU contribution to the next international negotiations on the post-2020 global biodiversity framework.

and actions that states must implement by 2030 to protect nature and recover of degraded ecosystems (Spiliopoulou et al. 2023; Hermoso et al. 2022). The degraded ecosystems to be protected include coastal ecosystems which also often fall within protected areas in which specific measures are envisaged for the reintegration and protection of habitats. To tackle climate change, biodiversity loss and pandemics, the strategy supports the creation of protected areas covering at least 30% of Europe's land and sea and strictly protected areas covering at least 10%. For the implementation of this strategy, the mobilization of funds from various sources is envisaged, including European, national and private funds. As part of this plan, the Commission proposed the EU's first ever Nature Restoration Law which includes an overarching restoration objective for the long-term recovery of nature in the EU's land and sea areas, with binding restoration targets for specific habitats and species. The regulation combines a general restoration objective for the long-term recovery of nature in the EU's terrestrial and marine areas with binding restoration objectives for specific habitats and species which should cover at least 20% of the EU's terrestrial and marine areas by 2030 and, ultimately, all ecosystems that need restoration by 2050. Specific objectives include the restoration of marine habitats including coastal ecosystems and sediment beds which bring significant benefits, also for the mitigation of changes climate.[29]

An indispensable tool, complementary to strategies such as the Marine Environment Strategy and Maritime Spatial Planning, is Integrated Coastal Zone Management, introduced in Europe through the "Recommendation on the implementation of Integrated Coastal Zone Management in Europe (2002/413/EC)" (Ernoul and Wardell-Johnson 2015; Billé and Rochette 2015). The EU has adopted two key documents that serve as primary references for community action on coastal management: the Commission Communication on Integrated Coastal Zone Management: a strategy for Europe (27 September 2000) and the Recommendation of the European Parliament and of the Council (30 May 2002) on the implementation of Integrated Coastal Zone Management in Europe.

The European Commission Communication COM/2000/547 illustrates the community work program to promote the integrated management of coastal zones through the use of community tools and programs, proposing to member states a collaborative approach for the planning and management of coastal zones. Recommendation 2002/413/EC contains an invitation to Member States to adopt a strategic approach to the management of their coasts based on a series of actions for the

[29] The Article 4 of the EU Nature Restauration Law on "Restoration of terrestrial, coastal and freshwater ecosystems" provides that: "Member States shall put in place restoration measures for the terrestrial, coastal and freshwater habitats of the species listed in Annexes II, IV and V to Directive 92/43/EEC and of the terrestrial, coastal and freshwater habitats of wild birds falling within the scope of Directive 2009/147/EC that are, in addition to the restoration measures referred to in paragraphs 1 and 4 of this Article, necessary to improve the quality and quantity of those habitats, including by re-establishing them, and to enhance connectivity, until sufficient quality and quantity of those habitats is achieved".

protection of coastal ecosystems, the development of employment and economic-social development that respects values environmental actions carried out by local communities.

As seen in the second paragraph, as regards the specific case of the Mediterranean, in 2008 a protocol of the Barcelona Convention on Integrated Coastal Zone Management in the Mediterranean was signed. The protocol was subsequently ratified by the EU in 2010 and provides the legal framework for the integrated management of the Mediterranean coastal zone.

There is also a complex system of EU key legislation and policies which concern specific aspects of environmental, maritime protection or the management of human activities linked to the coastal environment which are directly relevant with respect to phenomena linked to coastal protection.

Important are, first of all, the directives that address the prevention of the risk of erosion in coastal areas, including Directive 2007/60/EC on the assessment and management of flood risk.[30] Directive 2007/60/EC for coastal areas provides for the adoption of specific risk management plans depending on the level: low, medium or high. The directive contains provisions that concern the issue of coastal erosion from a "precautionary" point of view, with the aim of anticipating possible damage to the coastal strip through a scientific risk analysis (Nones and Pescaroli 2016; Nones 2015).

The EU Coastal and Marine Policy outlines objectives for sustainable management of coastal and marine environments, emphasizing an integrated approach to coastal defense considering environmental, social, and economic factors.

Additionally, directives such as the EU Environmental Impact Assessment Directive and the EU Habitats Directive ensure the environmental sustainability of coastal defense projects, minimizing their impact on habitats and species. The Habitats Directive (92/43/EEC) requires member states to protect habitats and species of community interest, including coastal ones, by establishing protected areas and ensuring that any coastal defense intervention does not compromise the integrity of these habitats. In parallel, the EIA Directive (2011/92/EU) obliges member states to assess the environmental impact of coastal defense projects. This assessment process ensures that environmental effects are considered and mitigated, involving the public and relevant authorities in the final decision. This requires careful planning to balance coastal protection needs with biodiversity conservation. Overall, these directives require European states to take an integrated and sustainable approach to coastal management and protection.[31]

[30] The Floods Directive requires Member States to assess whether all watercourses and coastlines are at risk of flooding, to map the extent of flooding and the property and humans at risk in these areas and to take appropriate and coordinated measures to reduce the risks that flooding and inundation pose to human health, the environment, cultural heritage and economic activity. The Directive also strengthens the public's right to access all information collected and to participate in the planning process.

[31] Under the EU Environmental Impact Assessment (EIA) Directive (2011/92/EU as amended by 2014/52/EU), major construction or development projects in the EU must receive an assessment of the impact they may have on environment before the project can begin. For some projects such as

Finally, noteworthy is the European Regions Charter for the promotion of a common framework for strategic actions in the Mediterranean coastal areas (Bologna Charter 2012), underscoring coastal protection, integrated management of coastal and marine systems, and climate change adaptation. The Joint Action Plan on Mediterranean coasts (JAP), a key deliverable of the EU-funded COASTGAP project, operationalizes the Bologna Charter, outlining joint actions for coastal administrations across Europe.

13.3.2 Case Study: Coastal Defense in the Mediterranean

As we have mentioned in the previous paragraphs, the Mediterranean region presents a unique set of challenges and opportunities for coastal defense against erosion. Indeed, the Mediterranean coast features very diverse coastal ecosystems, cultural heritage sites, densely populated coastal cities, and high concentration of economic activity along the coast and therefore requires tailored approaches for erosion management. At the same time, the Mediterranean appears to be one of the regions of the world most vulnerable to climate change (Cramer et al. 2018; Ferragina and Quagliarotti 2008; Brochier and Ramieri 2001). The coasts of the Mediterranean basin are subject to the continuous impacts of climate change and human activities. Beach retreat, coastal erosion and marine flooding especially affect low-lying coastal plains, river deltas and lagoons. These areas are especially vulnerable to sea level rise (SLR), which is projected to increase by up to 1 m by 2100 AD,[32] due to global warming, according to the Intergovernmental Panel on Climate Change (IPCC) projections (Anzidei et al. 2024; Loizidou et al. 2024).

The Mediterranean coasts are a very successful tourist destination and this has led both to an increase in the population concentrated along the coastal regions, and to a strong concentration of the tourist population in the summer periods with an increase in anthropic pressures resulting from the greater construction of infrastructure to support residential, tourist, commercial and transport activities with a consequent modification of the coastline (Motta Zanin et al. 2024; EEA 2020; Malek et al. 2018; Salvati 2014; Perry 2003).

These human-induced pressures contribute to coastal erosion processes, which currently affect almost 30% of the coastal areas of the Mediterranean basin (EEA

those involving nuclear power plants, plants for the disposal of hazardous waste, dams of a certain capacity, the EIA is always required, for other projects, however, including urban or industrial development projects and, as far as we are interested, works necessary against flooding, it is up to individual EU Member States to decide whether an EIA will be carried out on a case-by-case basis or by defining specific criteria (such as location, size or type of project). The EIA assesses the significant direct and indirect impact of a project based on a wide range of environmental factors, including population and human health, biodiversity, water, climate, landscape, material assets, cultural heritage.

[32] "There could be a 48% increase in the land area globally susceptible to flooding by 2100, posing a threat to 52% of the world's population and 46% of its assets" (Motta Zanin et al. 2024).

2020). Furthermore, climate changes have led to an increase in the incidence of storms, rising sea levels and, consequently, coastal flooding which lead to further erosion and modification of the coastline (Ali et al. 2022; Sarkar et al. 2022; UNEP/MAP and Plan Bleu 2020; Satta 2017).

To address these issues it is important that Mediterranean countries develop strategies and collaborate on common projects to develop Nature-Based Solutions and integrated coastal management. In particular, it is therefore important to understand what is meant by Nature-Based Solutions and what initiatives are underway in this sense and for integrated coastal management. Finally, the areas in which cooperation is important and the active projects in this sense will be analysed.

13.3.2.1 Nature-Based Solutions

The European Commission defines nature-based solutions as "Solutions that are inspired and supported by nature, which are cost-effective, simultaneously provide environmental, social and economic benefits and help build resilience. Such solutions bring more, and more diverse, nature and natural features and processes into cities, landscapes and seascapes, through locally adapted, resource-efficient and systemic interventions" (Gómez Martín et al. 2020).

Also in the context of coastal defense against erosion the EU's biodiversity strategy promotes nature-based solutions (NbS) like wetland restoration and dune stabilization to address coastal erosion and enhance resilience in the Mediterranean. Traditionally, coastal protection relied on "hard" engineering structures such as dikes and groynes. However, these methods are insufficient against the growing risks from climate change (Pranzini et al. 2015), leading to a demand for adaptable, cost-effective, and sustainable alternatives (Eggermont et al. 2015).

NbS include enhancing artificial structures with ecological features and improving natural coastal habitats, such as beaches, dunes, and wetlands, to mitigate climate risks (UNEP/MAP and Plan Bleu 2020).

Despite their potential, the adoption of NbS varies, with the Mediterranean region lagging (Su 2021). Success depends on environmental conditions and socio-ecological processes, with different implementation methodologies affecting outcomes (Cohen-Shacham et al. 2019). Initiatives like LIFE Blue Natura[33] already used NbS to protect Mediterranean coasts from erosion.

Nature-based solutions provide a holistic approach to mitigate coastal erosion and marine flooding, therefore their greater use in the Mediterranean is necessary with cooperation between coastal countries and sharing of good practices to improve resilience to change climate change and at the same time offer significant socio-economic and environmental benefits.

[33] The LIFE program was a financial instrument of the European Union dedicated exclusively to the environment. Its overall objective for the period 2004–2020 was to contribute to sustainable development and the achievement of the objectives and targets of the European Strategy 2020 and the relevant Union strategies and plans on the environment and climate.

13.3.2.2 Integrated Coastal Management

As we have already anticipated in the previous paragraphs, integrated coastal management (ICM) in the Mediterranean Sea is a crucial strategy to mitigate coastal erosion and preserve the region's unique and diverse ecosystems. This approach includes a comprehensive framework that integrates environmental, socioeconomic and governance aspects to manage coastal zones sustainably. Considering the population density, economic activities and climate change impacts of the Mediterranean, the ICM prioritizes the protection of coastal landscapes through a combination of hard and soft engineering solutions, habitat restoration and community engagement (Papatheochari and Coccossis 2019).

By harmonizing policies between nations and promoting collaboration between stakeholders, the ICM aims to improve the resilience of coastal areas, safeguard livelihoods and preserve the natural heritage of the Mediterranean basin. Effective implementation of the ICM involves continuous monitoring, adaptive management practices and the use of scientific research to inform decision-making, ensuring that coastal protection measures are sustainable and effective in the long term.

Several initiatives exemplify the application of ICM principles in the Mediterranean region. Two positive examples are the MEDSEA Foundation and the MEDCOAST Foundation.

The MEDSEA[34] Foundation, in collaboration with local stakeholders, promotes integrated approaches to coastal management, including erosion monitoring, beach nourishment and the development of sustainable tourism starting from the specific case of the Sardinian coasts. Their efforts focus on using scientific data to track erosion patterns and implement measures that restore natural habitats while protecting coastal infrastructure.

The MEDCOAST Foundation[35] supports various projects that aim to balance development and conservation in coastal areas. It provides training, conducts research and facilitates policy dialogue between Mediterranean countries to promote sustainable coastal management practices.

Also crucial are the local and regional efforts of municipalities and regional governments in Mediterranean countries that are already adopting the principles of integrated coastal zone management to address their specific coastal issues. For example, efforts in the Venice Lagoon include extensive monitoring and restoration projects aimed at mitigating the effects of sea level rise and preserving the lagoon's unique ecosystem.

Through these and other initiatives, Mediterranean countries are demonstrating a commitment to integrated coastal management, recognizing its importance in addressing coastal erosion and ensuring the long-term sustainability of their coastal

[34] The MEDSEA foundation was born from the will of a group of researchers who decided to commit themselves to the protection and sustainable management of the marine and coastal resources of Sardinia and the Mediterranean Sea.

[35] MEDCOAST aims to contribute to coastal and marine conservation in the Mediterranean and the Black Sea, through improved coastal management practices.

areas. By combining scientific research, community engagement and cross-border collaboration, ICM efforts in the Mediterranean are paving the way for more resilient and adaptive coastal management practices, but it is important to share more data and good practices to ensure more harmonized and inclusive actions the coastal areas of the Mediterranean.

13.3.2.3 Transnational Cooperation

Countries bordering the Mediterranean increasingly recognize the need for collaborative efforts to safeguard their coasts, which are vital to tourism, biodiversity and local economies. This growing awareness has pushed Mediterranean countries to emphasize international cooperation, recognizing that the challenge of coastal erosion transcends national borders. Such cooperation involves sharing scientific research, best practices and technological innovations to develop sustainable coastal management strategies. By pooling resources and expertise, Mediterranean nations can implement comprehensive measures that not only protect their coasts but also improve resilience against future environmental challenges.

Regional initiatives play a key role in facilitating this dialogue and promoting joint projects. The Mediterranean Action Plan (MAP), initiated by the United Nations Environment Program (UNEP), serves as a crucial framework for collective environmental management efforts (Grbec 2023a, b). Through the MAP, Mediterranean countries engage in coordinated actions to mitigate the impacts of coastal erosion and other environmental threats, promoting sustainable development in the region. Another significant initiative is the Union for the Mediterranean (UfM), which aims to strengthen regional cooperation and integration. The UfM supports various projects relating to coastal management, adaptation to climate change and disaster risk reduction. By promoting synergies between Member States, the UfM strengthens the region's capacity to respond to erosion and other coastal challenges. Transnational initiatives, such as the COASTGAP project under the Bologna Charter, exemplify the effectiveness of collaborative efforts in addressing the common challenges of coastal erosion. COASTGAP focuses on sharing knowledge, best practices and resources between Mediterranean regions, promoting a collective approach to improve coastal resilience. The project highlights the importance of Integrated Coastal Zone Management (ICZM) and the development of innovative solutions to protect vulnerable areas from erosion and sea level rise.

By working together through these initiatives, Mediterranean countries can leverage their collective strengths to safeguard their coasts. This cooperation not only helps mitigate the immediate impacts of coastal erosion, but also creates the foundation for long-term sustainability and resilience. As the Mediterranean region continues to face the evolving challenges posed by climate change and human activities, international cooperation remains a key strategy to preserve the health and vitality of its coastal areas (Randazzo et al. 2013).

13.4 Conclusion: Challenges and Future Directions

Despite the progress achieved through International and EU regulations and policies, coastal defense remains a complex and evolving challenge. Climate change, demographic pressures, and competing land uses continue to exert significant pressures on coastal ecosystems.

From the analysis of international and European legislation has emerged a wide range of provisions concerning coastal protection, although not grouped systematically and not always in a direct manner. It is key to consider climate change as a significant and urgent threat to coastal systems, affecting coastlines through sea level rise process and other climate dynamics such as storm surges, interannual and long-term changes in winds, wave action. Therefore, a climate change perspective is fundamental at any level and in any strategy which aims to protect coastal environment. Looking ahead, addressing these challenges will require a multifaceted approach that leverages technological innovation, strengthens international cooperation, and fosters adaptive governance structures. As seen above, an appropriate integrated plan for coastal zone management is crucial in several legal sources dealing with coastal defense, both at international and at European level. Coasts are very complex systems which require a management based on a holistic approach which consider environmental, socio-economic, and urban planning aspects. Cooperation is also a key aspect for an effective coastal defense. States need to cooperate at international and regional level now all the more to set environmental strategies and national plans harmoniously, applying the principles of international and European legislations which call for cooperation as a fundamental effort to achieve common goals. Common efforts will be decisive for the achievement of future results and to avoid geo-political imbalances which could emerge due to climate change processes, increasing urbanization and many risks from which coastal ecosystems already suffer.

Moreover, the effectiveness of regulatory measures depends on robust enforcement mechanisms, adequate funding, and stakeholder engagement at all levels.

Initiatives such as the European Green Deal and the upcoming EU Biodiversity Strategy for 2030 provide opportunities to mainstream coastal defense objectives into broader sustainability agendas, signaling a continued commitment to safeguarding coastal heritage for future generations.

By embracing integrated approaches and proactive strategies, it will be possible to strike a balance between protecting vulnerable coastal areas and fostering sustainable growth.

References

Ali W et al (eds) (2022) Climate change 2022: impacts, adaptation and vulnerability. In: Contribution of working group II to the sixth assessment report of the Intergovernmental Panel on Climate Change. Cambridge University Press, Cambridge, UK and New York, NY, USA, pp 2233–2272

Alterman R, Pellach C (2016) Legal-Institutional Instruments for Integrated Coastal Zone Management (ICZM) in the Mediterranean

Anzidei M, Alberti T, Vecchio A, (2024) Sea level rise and extreme events along the Mediterranean coasts: the case of Venice and the awareness of local population, stakeholders and policy makers. Rendiconti Lincei

Barcelona Convention, adopted on 16 February 1976 in Barcelona and entered into force in 1978

Benassai G, Di Paola G, Aucelli PPC (2015) Coastal risk assessment of a micro-tidal littoral plain in response to sea level rise. Ocean Coast Manag 104:22–35

Billé R, Rochette J (2015) The Mediterranean ICZM protocol: Paper treaty or wind of change? Ocean Coastal Manage 105:84–91

Brochier F, Ramieri E (2001) Climate change impacts on the Mediterranean coastal zones

Caron DD (2015) Climate change, sea level rise and the coming uncertainty in oceanic boundaries: a proposal to avoid conflict. Maritime boundary disputes, settlement processes, and the law of the sea, pp 1–17

Cohen-Shacham et al (2019) Core principles for successfully implementing and upscaling Nature-based Solution. Environ Sci Pol 98:20–29

Convention on Biological Diversity, opened for signature on 5 June 1992. entered into force on 29 December 1993. https://www.cbd.int/doc/legal/cbd-en.pdf

Convention on Wetlands, adopted in the Iranian city of Ramsar in 1971 and came into force in 1975

Council Directive 92/43/EEC of 21 May 1992 on the conservation of natural habitats and of wild fauna and flora. OJ L 206, 22.7.1992, pp 7–50

Cramer W et al (2018) Climate change and interconnected risks to sustainable development in the Mediterranean. Nat Clim Change 8:972–980

De Fontaubert A, Downes D, Agardy T (1996) Biodiversity in the seas, implementing the convention on biological diversity in marine and coastal habitats. IUCN Environ Policy Law 32:50–57

EEA (2020) Towards a cleaner Mediterranean: a decade of progress. Monitoring Horizon 2020 regional initiative, Joint EEA-UNEP/MAP Report

EEC Directive 85/337/EEC of 27 June 1985 on the assessment of the effects of certain public and private projects on the environment, OJ L 175, 5.7.1985, pp 40–48

EC Communication to the European Parliament, the Council, the European Economic and Social Committee and the Committee of the Regions (2012) Blue Growth opportunities for marine and maritime sustainable growth. Brussels, 13.9.2012 COM (2012) 494

EC Communication to the Council, the European Parliament, the Economic and Social Committee and the Committee of the Regions (2001) on the Sixth Environment Action Programme of the European Community, "Environment 2010: our future, our choice"

EC Communication to the European Parliament, the Council, the European Economic and Social Committee and the Committee of the Regions (2007) An Integrated Maritime Policy for the European Union

EC Communication to the Council and the European Parliament on integrated coastal zone management: a strategy for Europe, COM/2000/547

EC Directive 2008/56/EC of the European Parliament and of the Council of 17 June 2008 establishing a framework for community action in the field of marine environmental policy (Marine Strategy Framework Directive), OJ L 164, 25.6.2008, pp 19–40

EC Regulation (EC) No 413/2002 of 5 March 2002 establishing the standard import values for determining the entry price of certain fruit and vegetables. OJ L 63, 06/03/2002

EC Report to the European Parliament and the Council (2022) outlining the progress made in implementing Directive 2014/89/EU establishing a framework for maritime spatial planning. COM/2022/185 final

Eggermont H et al (2015) Nature-based solutions: new influence for environmental management and research in Europe. GAIA-Ecol Perspect Sci Soc

Ernoul L, Wardell-Johnson A (2015) Environmental discourses: Understanding the implications on ICZM protocol implementation in two Mediterranean deltas. Ocean Coast Manag 103:97–108

EU Directive 2014/89/EU of the European Parliament and of the Council of 23 July 2014 establishing a framework for maritime spatial planning, OJ L 257, 28.8.2014, pp 135–145
EU Directive 2011/92/EU of the European Parliament and of the Council of 13 December 2011 on the assessment of the effects of certain public and private projects on the environment (codification) Text with EEA relevance. OJ L 26, 28.1.2012, p. 1–21
EU Directive 2008/56/EC of the European Parliament and of the Council of 17 June 2008 establishing a framework for community action in the field of marine environmental policy (Marine Strategy Framework Directive). OJ L 164, 25.6.2008, pp 19–40
EU Directive 2007/60/EC of the European Parliament and of the Council of 23 October 2007 on the assessment and management of flood risks. OJ L 288, 6.11.2007, pp 27–34
EU Recommendation of the European Parliament and of the Council of 30 May 2002 concerning the implementation of Integrated Coastal Zone Management in Europe. OJ L 148, 6.6.2002, pp 24–27
EU Regulation n. 508/2014 of the European Parliament and of the Council of 15 May 2014 on the European Maritime and Fisheries Fund and repealing Council Regulations (EC) n. 2328/2003, (EC) No 861/2006, (EC) n. 1198/2006 and (EC) n. 791/2007 and Regulation (EU) n. 1255/2011 of the European Parliament and of the Council. OJ L 149, 20.5.2014, pp 1–66
Ferragina E, Quagliarotti D (2008) Climatic change in the Mediterranean Basin: territorial impact and search for a common strategy. New Med 7(4):4–12
Fitzmaurice M (2023) The International Convention for the Prevention of Pollution from Ships (MARPOL): 91–108
Gómez-Ballesteros M et al (2021) Transboundary cooperation and mechanisms for Maritime Spatial Planning implementation. SIMNORAT project. Marine Policy 127, article n. 104434
Gómez E et al (2020) An operationalized classification of nature based solutions for water-related hazards: from theory to practice. Ecol Econ 167, Article 106460
Gorjanc S et al (2022) Evaluating the progress in achieving good environmental status in the Mediterranean: a methodology to assess the effectiveness of Marine Strategy Framework Directive's Programmes of Measures. Marine Policy 136, Article n. 104889
Grbec M (2023) Marine area-based conservation under European Union Law. Legal aspects of marine protected areas in the Mediterranean Sea: an Adriatic and Ionian perspective, pp 91–117
Grbec M (2023) The sustainable development of the wider Mediterranean: multi-stakeholder settings, ecosystem approach, and maritime spatial planning. Legal aspects of marine protected areas in the Mediterranean Sea: an Adriatic and Ionian perspective, pp 205–216
Griggs G, Reguero B G (2021) Coastal adaptation to climate change and sea-level rise in water. Water, 7–17
Hermoso V et al (2022) The EU Biodiversity Strategy for 2030: opportunities and challenges on the path towards biodiversity recovery. Environ Sci Policy 127:263–271
International Convention on Oil Pollution Preparedness, Response and Cooperation (OPRC), adopted on 30 November 1990 and entered into force on 13 May 1995
International Convention for the Prevention of Pollution from Ships - MARPOL, adopted on 2 November 1973 (the 1978 MARPOL Protocol absorbed the parent Convention)
Keijser X et al (2018) Stakeholder engagement in maritime spatial planning: the efficacy of a serious game approach. Water (Switzerland) 10(62), article n. 724
LIFE Blue Natura EU Program Web site. https://life-bluenatura.eu/en/the-project/. Accessed 09 July 2024
Loizidou XI et al (2024) Adapting to sea level rise: participatory, solution-oriented policy tools in vulnerable Mediterranean areas. Environ Syst Decis
Loizidou XI, Loizides MI, Orthodoxou DL (2017) Marine strategy framework directive: innovative and participatory decision-making method for the identification of common measures in the Mediterranean. Mar Policy 84:82–89
MATTM-Regioni (2018) Linee Guida per la Difesa della Costa dai fenomeni di Erosione e dagli effetti dei Cambiamenti climatici. Document developed by the National Table on Coastal Erosion MATTM-Regions with the technical coordination of ISPRA, 305 pp

Malek Ž et al (2018) Global change effects on land management in the Mediterranean region. Global Environ Change 50:238–254
Matthews G (1993) The Ramsar Convention on Wetlands: its history and development. Ramsar Convention Bureau, pp 1–87
MEDCOAST web site. http://www.medcoast.net. Accessed 28 June 2024
MEDSEA Foundation web site. www.medseafoundation.org/index.php/it. Accessed 03 July 2024
Motta Zanin G, Muwafu SP, Costa MM (2024) Nature-based solutions for coastal risk management in the Mediterranean basin: a literature review. J Environ Manage 356
Murillas-Maza A et al (2020) Programmes of measures of the marine strategy framework directive: are they contributing to achieving good environmental status in the Mediterranean? Mar Pollut Bull 161, article n. 111715
Nones M, Pescaroli G (2016) Implications of cascading effects for the EU Floods Directive. Int J River Basin Manage 14(2):195–204
Nones M (2015) Implementation of the floods directive in selected EU member states. Water Environ J 29(3):412–418
Papatheochari T, Coccossis H (2019) Development of a waterfront regeneration tool to support local decision making in the context of integrated coastal zone management. Ocean Coast Manag 169:284–295
Perry A (2003) Impacts of climate change on tourism in the Mediterranean: adaptive responses. In: Giupponi C, Shechter M (eds) Climate change in the Mediterranean: socio-economic perspectives of impacts, vulnerability and adaptation. Edward Elgar Publishing Ltd
PNAS Proceedings of the American Academy of Sciences (2018) Climate-change-driven accelerated sea-level rise detected in the altimeter era. Report. https://www.pnas.org/doi/epdf/10.1073/pnas.1717312115. Accessed 12 April 2024
Pranzini E, Wetzel L, Williams AT (2015) Coastal erosion and protection in Europe. J Coast Conserv Plan Manage 19(4):445–459
Randazzo G, Raventos JS, Stefania L (2013) Coastal erosion and protection policies in Europe: from EU Programme (Eurosion and Interreg Projects) to local management. In: Finkl C (eds) Coastal hazards. Coastal Research Library, vol 1000. Springer, pp 443–487
Rizal AK (2016) Coastal erosion and its impact on some aspects of maritime boundaries, ISPSW strategy series: focus on defense and international security coastal erosion and its impact on some aspects of maritime boundaries, vol 410, pp 1–6
Sacco P et al (2017) Toward a better European Maritime governance: The EUCISE2020 project. In: Proceedings of the international astronautical congress, IAC, vol 3, pp 1784–1792
Salvati L (2014) Looking at the future of the Mediterranean urban regions: demographic trends and socioeconomic implications. Rom J Reg Sci 8(2):74–83
Sands P (1992) The United Nations framework convention on climate change. Rev Eur Compar Int Environ Law 270:271–275
Sarkar E N et al (2022) A literature review of climate-related coastal risks in the Mediterranean, a climate change hotspot. Sustainability 14(23), Article 15994
Satta A (2017) Assessment of coastal risks to climate change related impacts at the regional scale: the case of the Mediterranean region. Int J Disaster Risk Reduct, 284–296
Schiffman H (2021) International law and the protection of the marine environment. Int Law Inst, 1–22
Spalding M et al (2014) Mangroves for coastal defence. Guidelines for coastal managers & policy makers. Wetlands Int Nat Conserv, 1–4
Spiliopoulou K et al (2023) Protected areas and the ranges of threatened species: towards the EU Biodiversity Strategy 2030. Biol Conserv 284, Article n. 110166
Su YC et al (2021) Enhancing uptake of nature-based solutions for informing coastal sustainable development policy and planning: a Malaysia case study. Front Ecol Evol 9
Touri M (2021) Communication for sustainable development and blue growth: towards new theoretical and empirical directions. In: The Palgrave handbook of international communication and sustainable development, pp 77–100

Treves T (2015) Coastal states' rights in the maritime areas under UNCLOS. Brazilian J Int Law 12:40–49
UNEP/MAP and Plan Bleu (2020) United Nations Environment Programme/Mediterranean action plan and plan Bleu, State of the Environment and Development in the Mediterranean. Nairobi
United Nations Framework Convention on Climate Change (UNFCCC), Signed in 1992, entered into force on 21 March 1994
United Nations Convention on the Law of the Sea, adopted in 1982

Chapter 14
The Strategic Nexus of Cultural and Scientific Diplomacy in International Relations and in Contemporary Models of Governance Towards Sustainability in Southeast Asia

Francesca Tolve

Abstract This work examines the role of scientific diplomacy in promoting coastal sustainability in Southeast Asia. It also briefly explores the relationship between science diplomacy and geopolitics, also emphasizing the crucial contribution of scientific diplomacy to achieving the Sustainable Development Goals (SDGs), with a particular focus on coastal sustainability in the region. By integrating cultural and scientific diplomacy into modern governance structures, this work aims to provide a more effective approach to addressing environmental challenges on a global scale.

Keywords Science diplomacy · International cooperation · Sustainability · Cultural diplomacy · Governance

14.1 Introduction

When delving into the concept of "sustainability," we are confronted with a complex web of interconnected elements. Biodiversity, human cultural diversity, geo-economy, and the myriad ecosystems that comprise our world are all integral components of this intricate system, constantly evolving and adapting in response to human activities and natural phenomena (Ibish 2010).

In the coastal regions of Southeast Asia the challenges to sustainability are particularly pronounced.

F. Tolve (✉)
Office for International Relational of National Research Council of Italy, Rome, Italy
e-mail: francesca.tolve@cnr.it

V. Maccarrone and M. Fadzil Akhir (eds.), *Coastal Sustainability*, Coastal Research Library 39, https://doi.org/10.1007/978-3-031-75749-5_14

Climate change, significantly impacting in the region (Global Climate Risk Index 2018; ASEAN State of Climate Change Report 2021) exacerbates these challenges, posing significant threats to the foundational biodiversity and to the delicate ecosystems that define these coastal landscapes (Strategic Plan for Biodiversity 2011–2020).

In the face of these challenges, it goes without saying that urgent action is imperative.

Climate change mitigation and adaptation, migration patterns, the transition to renewable energy, and the overarching goal of global sustainable development through the Agenda 2030, demand significant and constant attention on national, regional and global scales.

Effectively meeting these challenges requires, therefore, the development of innovative governance models that fully integrate culture to sustainability and scientific knowledge, grand political and diplomatic strategies, and international cooperation.

The present exploration seeks to navigate these complexities with a balanced perspective, acknowledging the urgency of the issues at hand with new adaptable responses and strategies, through cultural and science diplomacy permeating contemporary models of governance and international relations towards sustainability.

14.2 Navigating Complexities of Governance Through Cultural and Scientific Diplomacy

Understanding the complexities of governance within diverse and dynamic global systems requires an appreciation of its multifaceted nature. Governance, broadly defined as the process of decision-making and policy implementation within systems or organizations, involves a wide range of structures, processes, and mechanisms through which authority and control are applied to manage resources, address challenges, and achieve objectives efficiently and effectively (Stolker 2004; Vymetal 2007). Over time, governance models have evolved significantly across various contexts, influenced by cultural, political, and economic factors.

The need for governance to adapt and evolve is critical, given the contemporary shifts in societal, political, and economic landscapes on both national and global scales. Addressing these complexities requires a nuanced understanding of systems characterized by non-linear interactions and dynamic behaviour, rather than simplistic, linear cause-and-effect relationships. This complexity necessitates inclusive, accountable, and effective decision-making processes to drive societal progress.

In the twenty-first century, progressing towards sustainability demands the systematic incorporation of science diplomacy, alongside cultural diplomacy, into governance models. Cultural diplomacy acts as a vital conduit for fostering mutual understanding, dialogue, and cooperation among diverse societies, enriching governance processes with diverse perspectives and insights. Concurrently, science diplomacy promotes evidence-based decision-making and international collaboration, providing solutions to complex global challenges.

Strategically amplifying cultural and scientific diplomacy within decision-making frameworks signifies a pivotal shift towards more nuanced, holistic, and adaptive approaches to governance. This recalibration is indispensable for effectively navigating the intricate and multifaceted challenges characteristic of contemporary governance, and for fostering sustainability and development agendas within a globally interconnected milieu.

14.3 Culture to Sustainability: An Insight into the Concept and in Southeast Asia

Before delving into the role and potential of cultural and scientific diplomacy for sustainability, we need to briefly better explore the concept of culture to sustainability, with the understanding that culture, in our case, refers to being educated or instilled with sustainability principles (Wiktor-Mach 2018).

As emphasised in the Introduction, sustainability encompasses a complex system of interconnected elements, interplaying between biodiversity, human cultural diversity, geo-economy, and ecosystems, all of which continuously evolve in response to human activities and natural phenomena.

When speaking of *culture to sustainability*, we are highlighting the importance of embedding sustainability principles, values, and practices into the cultural fabric of society. In this scenario, *culture* refers to the collective mindset, attitudes, behaviours, and norms that shape human interactions with the environment and influence decision-making processes.

Informing and educating individuals and communities about sustainability principles is crucial for fostering a culture that prioritizes environmental stewardship, social equity, and economic prosperity. This entails promoting awareness, understanding, and appreciation of the interconnectedness between human activities and the health of ecosystems, biodiversity, and planet as a whole.

Cultural transformation towards sustainability involves instilling values of conservation, resilience, and responsibility towards future generations.

This transformation requires rethinking societal norms and practices to align with sustainable development goals, such as reducing carbon emissions, minimizing waste, promoting renewable energy and preserving eco-systems and water below land.

Incorporating sustainability education into both formal and informal learning environments—schools, universities, workplaces, and communities—is essential for nurturing a culture of sustainability. This also involves empowering individuals to take action towards positive environmental and social change.

Furthermore, cultural shifts towards sustainability also involve fostering collaboration, innovation, and collective action at all levels of society. This requires building partnerships between governments, businesses, civil society organizations, and grassroots movements to address complex sustainability challenges effectively.

Ultimately, *culture to sustainability* signifies a transformation in mindset and behaviour, where sustainability becomes deeply ingrained in the values, beliefs, and practices of individuals and communities.

In 2016, the European Commission, in collaboration with the High Representative of the Union for Foreign Affairs and Security Policy, conceptualized a new vision for culture to mainstream sustainability with cultural policies and international relations through a joint Communication titled "Towards an EU Strategy for International Cultural Relations" (European Commission 2016).

The European Union, perhaps for the first time, officially framed culture not only as the foundation of human development but also as a strategic component within its external policies, proposing a cross-cutting approach where culture impacts global sustainability, sustainable growth, mutual understanding, conflict resolution, and security.

Within Southeast Asia, the Mekong-Lancang Cooperation (MLC) serves as a pertinent example of how cultural diplomacy can effectively contribute to sustainability.

Encompassing six riparian countries—Cambodia, China, Laos, Myanmar, Thailand, and Vietnam—the MLC endeavours to foster cultural diplomacy for climate resilience, where indigenous communities residing in the Mekong River Basin harbour profound knowledge of traditional water management systems.

The Mekong-Lancang Cooperation actively supports the documentation and preservation of these systems, which incorporate sustainable agricultural practices, flood management techniques, and community-based adaptation strategies.

Furthermore, indigenous agricultural practices in the Mekong-Lancang region demonstrate resilience to climate change impacts. For instance, agroforestry systems seamlessly integrate trees, crops, and livestock, thereby enhancing soil fertility, conserving water, and mitigating climate-related risks. Additionally, the MLC acknowledges the cultural significance of natural landscapes and cultural heritage sites in the region (MLC Secretariat 2020) where conservation efforts targeting sacred sites, traditional villages, and historic monuments contribute not only to preserving cultural identity but also to fostering social cohesion in the region.

The MLC framework serves as a compelling demonstration of how regional cooperation mechanisms can effectively integrate cultural diplomacy with climate resilience efforts in Southeast Asia.

By leveraging traditional cultural practices and indigenous knowledge, the MLC facilitates cross-cultural understanding, strengthens social cohesion, and promotes sustainable development in the Mekong-Lancang region.

14.4 The Evolution and Strategic Role of Science Diplomacy in Addressing Global Sustainability Challenges

Having briefly delved into the concept of sustainability and having emphasised the need to incorporate cultural diplomacy within the innovative models of governance, attention is now directed to the pivotal role of science diplomacy by, first of all, delving into the concept's historical foundations, theoretical underpinnings, and practical applications, to highlight how science diplomacy may offer the ground for effective responses to sustainability.

The concept and practice of Science Diplomacy, well established in America through the American Association for the Advancement of Science (AAAS) and disseminated in Europe with the Report "New Frontiers in Science Diplomacy - Navigating the Changing Balance of Powers" (Royal Society and AAAS 2010), establishes, indeed, a crucial, dynamic, diplomatic nexus between science, political action, and international cooperation.

This nexus, which is evidently not new, has progressively gained considerable attention in scientific literature (Copeland 2011; Gluckman 2016; Ruffini 2017; Gluckman et al. 2017; Kaltofen and Acuto 2018b; Berkman 2019; Flink and Ruffini 2019; Arnaldi 2023) and significance at both the European Union and global levels (European Commission 2012; European Commission 2021) in shaping relations among Countries, as well as in constituting a more effective, science—based systematic approach to respond to global transitions and to the complex, multifaced actual emerging global challenges.

The etymological foundation of Science Diplomacy, where "scientia" derives from the Latin scire (knowledge), and "diplomacy" from the French "diplomatie" (constructive approach to international relations), intertwines with its theoretical foundation which assumes that science should always consistently represent an open and non-ideological space for the sharing and free exchange of ideas and knowledge among individuals (Bonn Declaration on Freedom of Scientific Research 2020) regardless of political, cultural, or religious affiliations.

Thus science, whose conceptual nucleus is given by "relations", becomes the aggregating factor, the element of cohesion and the space for peaceful and constructive dialogue among Countries which may be historically divided by conflicts of economic, ethnic/religious, or political nature.

For these reasons, Science Diplomacy may be seen as a common space oriented towards understanding between parties, expanding, and enriching the very concept of cooperation which is linked to the process of growth and development for individuals and for society that, the cooperative and communicative act, through the effort of mediation and convergence, is always capable to realize (Cerruti 2018).

In the realm of international relations, Science Diplomacy can be interpreted as a response to the shift from a linear bi-polar post-war model prevalent throughout the latter part of the twentieth century, to the current condition of high complexity in the intertwining relationships between States, Nations, non-State players, and different ethnic communities (Mazzei 2016).

This dynamic articulates a new multipolar global order/disorder where ties between Countries are based not only on geographical proximity and spatial contiguity but, most importantly, on transversal global interests continuously redefining and modifying boundaries and (geo) political relations.

Moreover, the nature and complexity of current major challenges such as climate change and adaptation processes, migrations, the transition to clean renewable energy, and the promotion of long-term inclusive and global sustainable development through the implementation of the 2030 Agenda, clearly underlines the necessity the develop new governance models that integrate science and diplomacy.

The promotion of scientific cooperation becomes, therefore, an essential and systematic tool of foreign policy, building "bridges" between countries.

In essence, Science Diplomacy represents an adaptive and responsive strategic approach to complex circumstances and challenges, whose success is directly proportional to the ability to implement systemic cooperation between political actors and the research community, also fostering strategic partnerships and peaceful, long-lasting relations between countries, driven by both (geo) political considerations and approaches, and global social challenges (Tolve 2020).

Defined in this manner, Science diplomacy may play a pivotal role in responding to the multifaceted challenges posed by sustainability in Southeast Asia, where the development of sustainable solutions requires science-based policy responses (science in diplomacy), scientific cooperation to improve relations between Countries (science for diplomacy) and political action to facilitate and support international scientific cooperation (diplomacy for science).

These are the three complementary dimensions, according to the taxonomy proposed by the Royal Society and AAAS (Royal Society and AAAS 2010) to define science diplomacy.

It seems crucial also to clarify the distinction between science diplomacy and international scientific cooperation, as the former does not focus just on development and sharing of knowledge, but frames knowledge and scientific advancements within the development of context-specific solutions and as part of a broader strategy of national or international policy objectives (Turekian et al. 2015; Kaltofen and Acuto 2018b), where science diplomacy constitutes a clear vector of renewal and enrichment in the conduct of international relations (Ruffini 2019).

In 2012, the European Commission advocated a science diplomacy approach as part of the new strategic vision for international scientific cooperation (European Commission 2012). This approach aimed to integrate research and innovation into the broader framework of the European Union's external action, recognizing science diplomacy as a crucial tool for soft power and nuanced efficacy in enhancing relations with non-EU countries and regions.

The evolving global landscape, particularly shaped by the Covid-19 pandemic, further highlighted the interconnectedness of nations and the impracticality of scientific acquisitions and applications contained within national borders. This awareness, along with the demonstrated inequities in countries' vulnerability, prompted a refinement of the Community's approach to international cooperation in 2021 (European Commission 2021).

The European Commission, acknowledging the necessity of collaboration for the common good, adopted a more global approach to international cooperation in research and innovation on the topical issues which are shaping the actual changing world.

The new strategic approach prioritizes addressing challenges such as climate change, biodiversity crisis, pollution and resource depletion, underscoring the importance of global joint action and international partnerships.

This included enhancing regional cooperation with the Association of Southeast Asian Nations (ASEAN) under the ASEAN-EU dialogue on Science and Technology while also emphasizing evidence-informed policymaking and sharing of scientific insights for policy development.

Furthermore, as part of its global approach to research and innovation, the European Commission has gradually engaged into the definition of a proper European framework for science diplomacy, identifying as important step of this process 12 concrete suggestions to make science diplomacy more strategic and effective, in particular through the use of scientific evidence, anticipatory vision and foresight (Towards a European approach to science diplomacy 2023).

Among the overarching goals of this framework, there is leveraging the impact of research and innovation to address the systemic global challenges of climate change, biodiversity loss, and environmental pollution, also calling for alignment with the Sustainable Development Goals (SDGs, https://sdgs.un.org/goals).

Science assumes a foundational position, called to provide essential support to diplomatic and political endeavours as a critical element in the evolving landscape of global growth and sustainable development (United Nations General Assembly 2023).

14.5 Coastal Sustainability in the Context of SDGs in Southeast Asia and the Role of Scientific Diplomacy

The scope of the new global sustainable development goals (SDGs) extends beyond the Millennium Development Goals (MDGs) in geographical terms, as the objectives set concern all countries and not only developing nations. This reflects the acquired understanding that sustainable development must be pursued globally.

The Sustainable Development Goals (SDGs) play a pivotal role in guiding efforts towards more sustainable marine and coastal resources in Southeast Asia.

In this context, the nuanced interplay of challenges and opportunities that emerges warranties examination.

Goal 14, focusing on life below water, underscores the urgent need to conserve and sustainably utilize oceans, seas, and marine resources.

In Southeast Asia, characterized by remarkable marine biodiversity and the dependence of coastal communities on marine resources, the protection of marine ecosystems holds, therefore, not only ecological but also high social and economic significance.

Similarly, Goal 13, centred on climate action, assumes critical importance. The region faces escalating challenges from rising sea levels, intensifying storms, and ocean acidification, posing significant threats to coastal populations. Consequently, adopting mitigation and adaptation measures is essential to bolster the resilience of Southeast Asia's coastal ecosystems and the communities reliant upon them.

Additionally, Goal 6, focused on clean water and sanitation, is of paramount importance. Coastal waters are increasingly compromised by pollution stemming from agricultural runoff, industrial discharges, and inadequate waste management practices, posing risks to both marine biodiversity and human health. Ensuring access to clean water and sanitation facilities is fundamental for safeguarding coastal ecosystems and the well-being of local communities.

Lastly, Goal 11, promoting sustainable cities and communities, cannot be understated.

Rapid urbanization and coastal development exert significant pressures on coastal ecosystems, necessitating sustainable urban planning, infrastructure development, and coastal management strategies to mitigate adverse impacts.

In the context of coastal sustainability in Southeast Asia, Sustainable Development Goal 17—partnership for the goals—also assumes paramount significance as a catalyst for fostering partnerships and mobilizing resources essential for achieving sustainable development objectives. SDG 17 offers, indeed, a framework for collaborative action, encouraging multi-stakeholder partnerships, knowledge sharing, and capacity building initiatives to address the complex challenges of Southeast Asia's intricate network of coastal ecosystems. Furthermore, the dynamic geopolitical landscape of Southeast Asian region underscores the importance of SDG 17 in facilitating regional cooperation and dialogue, particularly in addressing issues such as transboundary pollution, maritime disputes, and resource management.

The implementation of these SDGs, however, poses significant challenges.

Insights drawn from the Asian experience on SDG implementation progress (Rahman et al. 2020) reveal a persistent need for high-quality data to assess several SDG indicators, with data available for only approximately one-third of the targets (50 out of 169 targets).

Moreover, achieving SDGs demands a comprehensive investment strategy that allocates resources more thoughtfully across economic, social, and environmental domains. This necessitates not only scientific research and innovation but also diplomatic endeavours and political commitment.

Furthermore, science plays a pivotal role in furnishing evidence-based solutions and guiding decision-making processes to address the multifaceted challenges of coastal sustainability. Thus, the robust endorsement of science within diplomatic and political spheres, along with its facilitation of cohesive regional cooperation, becomes indispensable for the successful implementation of SDGs.

Consequently, the strategic integration of science and its engagements emerges as a critical constituent within the dynamic framework of global progress and sustainable development, a notion reaffirmed by the United Nations General Assembly (2023).

In addition to the points already discussed, practical examples of ongoing projects and initiatives provide further insight into how the Sustainable Development Goals (SDGs) are being operationalized in Southeast Asia for coastal sustainability, and how science diplomacy plays a pivotal role in facilitating international collaborations and knowledge exchange, enhancing efforts in this context.

For instance, in the Philippines, community-based coastal management initiatives have established marine protected areas (MPAs) with the support of international scientists (Balgos 2005). This collaboration safeguards critical marine habitats and promotes sustainable fishing practices that benefit local livelihoods, exemplifying the impact of scientific partnership.

In Vietnam, science diplomacy fosters international collaborations that encourage the adoption of innovative aquaculture practices, such as integrated multitrophic aquaculture systems, balancing economic gains with ecological sustainability (United Nations Viet Nam 2023).

Likewise, in Indonesia, (Robertua 2019) science diplomacy supports green infrastructure development projects like mangrove restoration, providing coastal protection and contributing to climate action efforts in line with SDG 13.

Regional research collaborations, facilitated by science diplomacy, allow Southeast Asian countries to jointly monitor and assess marine and coastal ecosystem health, advancing conservation strategies through data sharing and harmonization (Science Diplomacy Review 2021).

Education and awareness campaigns, supported by scientific expertise, engage local communities and promote participation in sustainable coastal practices. This further underscores the role of science diplomacy in fostering cooperation and shared knowledge across the region.

Through collaborative approaches and innovative solutions, science diplomacy plays an instrumental role in advancing progress toward coastal sustainability and achieving SDG goals while addressing the region's unique challenges in which coastal sustainability is not just an environmental issue, but it is rather a matter of social equity, economic prosperity, and global responsibility (WWF Coastal Community Initiative. Impact Report 2020–2023).

14.6 The Complex Interplay Between Science Diplomacy, Geopolitics, and Coastal Sustainability in Southeast Asia

The complex interplay between science diplomacy, geopolitics, and coastal sustainability in Southeast Asia is a crucial aspect of international efforts aimed at addressing the sustainability of coastal regions. By acknowledging the potential obstacles stemming from conflicting geopolitical interests, this paragraph delves into the delicate balance necessary when forming scientific partnerships in a region characterized by numerous national concerns.

Southeast Asia is a diverse array of nations, each with its unique set of priorities and concerns. The region encompasses a variety of shared and distinct national apprehensions, notably exemplified by territorial disputes. The South China Sea serves as a focal point of maritime and territorial contentions involving several Southeast Asian nations, including China, Vietnam, the Philippines, Malaysia, and Brunei. The overlapping claims to islands and waters in this region have long been a pivotal geopolitical issue, affecting the stability of the region (Turker 2023).

Economic disparities are also evident among Southeast Asian nations, facing challenges related to economic growth, poverty, and income inequality. The region is further challenged by environmental degradation, manifested in issues such as deforestation, air and water pollution, and biodiversity loss. The accelerated pace of urbanization and industrialization exacerbates environmental deterioration, impacting both ecosystems and human health.

Moreover, Indonesia, alongside Malaysia and Singapore to some extent, confronts challenges associated with transboundary haze pollution. Seasonal forest fires, often linked to land clearance methods, lead to air quality concerns that not only impact health but also contribute to regional tensions. Southeast Asia is also vulnerable to a range of natural disasters, from typhoons to earthquakes and tsunamis, presenting variations in preparedness, response capabilities, and recovery efforts among nations, thereby posing challenges for regional collaboration during emergencies.

Additionally, public health challenges such as dengue, malaria, and emerging infectious diseases are prevalent across the region. The COVID-19 pandemic has underscored the significance of regional cooperation in addressing health crises. With a backdrop of cultural and ethnic diversity, managing cultural differences while fostering national unity remains a crucial consideration for countries like Indonesia and Malaysia, where substantial ethnic and religious diversity is inherent.

Migration and refugee issues, such as the Rohingya crisis in Myanmar, pose additional challenges for Southeast Asia, with implications extending to neighbouring countries like Bangladesh and Malaysia. It is crucial, nevertheless, to recognize that the concerns presented here are not exhaustive, and each Southeast Asian country possesses unique geopolitical, economic, social, and environmental circumstances and that, given the dynamic nature of the geopolitical landscape, issues may evolve over time.

The aim is, however, to emphasize the importance of successful instances where science diplomacy has effectively mitigated geopolitical tensions, demonstrate how strategic scientific collaboration can transcend political barriers, mitigate geopolitical tensions, and facilitate international cooperation by aligning scientific objectives and establishing a common ground for addressing shared environmental concerns.

Exploring the intersection of diplomacy, science, and geopolitics in Southeast Asia, can be exemplified through three relevant case studies: the Mekong River Commission (MRC), the ASEAN-European Union (EU) Science, Technology, and Innovation (STI) Cooperation Facility, and the Coral Triangle Initiative (CTI).

Each case exemplifies a distinct approach to leveraging diplomatic channels for scientific collaboration, highlighting the role of science in promoting regional cooperation and addressing shared environmental challenges towards sustainability.

Through these examples, we examine the diverse ways in which science and diplomacy intersect to promote peace, stability, and sustainable development in Southeast Asia.

Science in Diplomacy: The Mekong River Commission (MRC) serves as a paradigmatic example of science in diplomacy in Southeast Asia. Comprising Cambodia, Laos, Thailand and Vietnam, the MRC aims to promote the sustainable management of the Mekong River's resources. Scientists from member countries engage in collaborative research and data analysis to inform decision-making on critical issues such as hydropower development, irrigation, and flood control. By integrating scientific knowledge into diplomatic processes, the MRC fosters cooperation and mitigates conflicts over the shared river resources, thereby contributing to regional stability and development.

Diplomacy for Science: The ASEAN-European Union (EU) Science, Technology, and Innovation (STI) Cooperation Facility exemplifies the use of diplomacy to facilitate scientific collaboration between Southeast Asian countries and the European Union. Through diplomatic negotiations, agreements are forged to fund joint research projects, exchange scientific expertise, and establish innovation partnerships. This initiative enhances scientific capabilities in Southeast Asia by leveraging the EU's technological advancements and research networks, thereby promoting knowledge exchange and capacity building in the region.

Science in and for Diplomacy: The Coral Triangle Initiative (CTI) on Coral Reefs, Fisheries, and Food Security represents a unique approach to using science in and for diplomacy in Southeast Asia. Involving six countries in the region, the CTI aims to conserve marine biodiversity and promote sustainable fisheries management. Scientific research and monitoring are integral to the initiative's success, providing evidence-based solutions for policymakers to address environmental challenges *(science in diplomacy)*, while fostering regional cooperation and diplomatic relations. Through joint scientific efforts, the CTI strengthens ties among member countries and contributes to peace and stability in the region.

These case studies illustrate the diverse ways in which science and diplomacy intersect in Southeast Asia. Whether through integrating scientific knowledge into diplomatic processes, fostering scientific collaboration through diplomatic channels, or utilizing science to promote regional cooperation, these examples underscore the critical role of science in advancing diplomatic objectives and addressing shared challenges in the region.

As Southeast Asia continues to navigate complex geopolitical dynamics and environmental pressures, the synergy between science and diplomacy remains essential for promoting sustainable development, peace, and stability.

14.7 Conclusions

Science diplomacy remains a fluid concept subject to varied interpretations and applications. As a multipurpose cooperative methodology, it offers multiple potential uses and developments, though it faces numerous challenges. These challenges range from its intersection with contemporary global geopolitical complexities and its role in transformative scientific processes and innovative policymaking, to its influence on integration, peace, and sustainable development. These challenges underscore the inseparability of sustainability from peace and stability.

The strategic integration of cultural and scientific diplomacy in international relations and contemporary governance models in Southeast Asia highlights a transformative path toward sustainable development. The synergistic use of cultural diplomacy fosters mutual understanding and respect, bridging diverse ethnic and national identities within the region. This cultural cohesion is essential for creating a peaceful and collaborative environment, crucial for addressing shared environmental and socio-economic challenges.

In addition, scientific diplomacy leverages the power of collaborative research and technological innovation. A crucial aspect of this diplomacy is also its role in promoting equity in research, which is particularly significant in the Global South and thus in Southeast Asia. By ensuring fair participation and representation in scientific endeavours, scientific diplomacy must be also designed to support inclusive development and mitigates disparities in research capabilities and resources.

The convergence of cultural and scientific diplomacy thus acts as a catalyst for more inclusive and participatory governance models. It empowers local communities by integrating indigenous knowledge with scientific insights, ensuring that sustainability efforts are contextually relevant and ethically sound. Furthermore, this dual diplomatic strategy promotes stronger multilateral relationships and regional cooperation. Through initiatives like the ASEAN Centre for Biodiversity and the Mekong River Commission, countries in Southeast Asia can address transboundary environmental concerns more effectively.

In conclusion, the fusion of cultural and scientific diplomacy presents a robust framework for sustainable development in Southeast Asia. By embracing this holistic strategy, the region can navigate the complexities of globalization while safeguarding its rich cultural heritage and natural ecosystems. This integrated approach not only enhances regional stability and prosperity but also contributes to global sustainability objectives. Moreover, prioritizing equity in research through scientific diplomacy ensures more balanced and just progress, benefiting all segments of the diverse population. This model holds potential as a blueprint for other regions, emphasizing the importance of combining cultural and scientific insights in governance and international relations for sustainable futures.

References

Amaratunga M, Abeysinghe T (2021) Role of traditional knowledge in climate change adaptation: a case study of indigenous communities in the Mekong Delta, Vietnam. In: The role of traditional knowledge in climate change adaptation. Springer, Cham, pp 217–230

ASEAN State of Climate Change Report (2021) Current status and outlook of the ASEAN region Toward the ASEAN climate vision 2050 Jakarta, ASEAN Secretariat. https://asean.org/wp-content/uploads/2021/10/ASCCR-e-publication-Correction_8-June.pdf

Arnaldi S (ed) (2023) Science diplomacy. Foundations and practice, EUT Edizioni Università di Trieste. https://eut.units.it/it/catalogo/science-diplomacy-foundations-and-practice/5639

Asian Development Bank (2024) Asian development outlook. https://www.adb.org/sites/default/files/publication/957856/asian-development-outlook-april-2024.pdf

Balgos MC (2005) Integrated coastal management and marine protected areas in the Philippines: concurrent developments. Ocean Coastal Manage 48(11–12):972–995. https://doi.org/10.1016/j.ocecoaman.2005.03.003

Berkman PA (2019) Evolution of science diplomacy and its local—global applications. Eur Foreign Affairs Rev 24(Special Issue):63–80. https://par.nsf.gov/servlets/purl/10122500

Bonn Declaration on Freedom of Scientific Research (2020) https://www.bmbf.de/SharedDocs/Downloads/files/_drp-efr-bonner_erklaerung_en_with-signatures_maerz_2021.pdf?__blob=publicationFile&v=2#:~:text=Freedom%20of%20scientific%20research%20stands,a%20pillar%20of%20any%20democracy

Cerruti M (2018) Il tempo della Complessità, Raffaello Cortina Editore, Milano ISBN 9788832850000

Chaturvedi S, Balakrishnan B, Srinivas KC (eds) (2021) Science diplomacy review, vol 2, no 3. https://www.ris.org.in/sites/default/files/2021-09/SDR%20November%202020.pdf

Chayanov A, Sukhbaatar T (2019) Traditional knowledge for climate resilience: the case of indigenous communities in the Mekong River Basin. In: Resilient communities and renewable energy. Springer, Cham, pp 103–117

Copeland D (2011) Science diplomacy: what's it all about? CEPI—CIPS policy brief, vol 13, pp 1–4. https://www.cips-cepi.ca/wp-content/uploads/2011/11/Copeland-Policy-Brief-Nov-11-5.pdf

Dao NH (2020) Integrating indigenous knowledge into climate change adaptation: lessons from the Mekong River Delta in Vietnam. J Sustain Dev 13(5):139–151

Eckstein D, Künzel V, Schäfer L (2018) Briefing paper, global climate risk index. Who suffers most from extreme weather events? Weather-related loss events in 2016 and 1997 to 2016. Germanwatch. https://www.germanwatch.org/sites/default/files/publication/20432.pdf

European Commission (2012) Enhancing and focusing EU international cooperation in research and innovation: a strategic approach. COM (2012) 497 final. https://eur-lex.europa.eu/legal-content/EN/TXT/PDF/?uri=CELEX:52012DC0497&from=SK

European Commission (2016) Towards an EU strategy for international cultural relations, Join (2016) 29 final. https://eur-lex.europa.eu/legal-content/EN/TXT/PDF/?uri=CELEX:52016JC0029

European Commission (2021) The global approach to research and innovation. Europe's strategy for international cooperation in a changing world. COM (2021) 252 final. https://eur-lex.europa.eu/legal-content/EN/TXT/PDF/?uri=CELEX:52021DC0252

Flink T, Ruffin N (2019) The current state of the art of science diplomacy. In: Simon D, Kuhlmann S, Stamm J, Canzler W (eds) Handbook on science and public policy. Edward Elgar, Cheltenham, pp 104–121. https://www.academia.edu/38918284/The_Current_State_of_the_Art_of_Science_Diplomacy

Gluckman PD (2016) Science advice to governments: an emerging dimension of science diplomacy. Sci Diplomacy 5(2). https://www.sciencediplomacy.org/sites/default/files/science_advice_to_governments_0.pdf

Gluckman PD, Turekian VC, Grimes RW, Kishi T (2017) Science Diplomacy: a pragmatic perspective from the inside. Sci Diplomacy 6(4):1–12. https://www.sciencediplomacy.org/sites/default/files/pragmatic_perspective_science_advice_dec2017_1.pdf

Ibisch PL, Vega E A, Herrmann TM (eds) (2010) Interdependence of biodiversity and development under global change. Technical Series No. 54. Secretariat of the Convention on Biological Diversity, Montreal (second corrected edition)

Kaltofen C, Acuto M (2018a) Rebalancing the encounter between science diplomacy and international relations theory. Global policy 9(3):15–22. https://doi.org/10.1111/1758-5899.12620

Kaltofen C, Acuto M (2018b) Science diplomacy: introduction to a boundary problem. Glob Policy 9(3):8–14. https://doi.org/10.1111/1758-5899.12621

Mekong-Lancang Cooperation Secretariat (2020) Mekong-Lancang Cooperation: towards sustainable development and shared prosperity. https://mlcchina.cn/attach/0a5f246a77e9449cb1dc3af06dbed4a0.pdf

Mekong River Commission (2021) Basin development strategy for the Mekong River Basin 2021–2030 & MRC Strategic Plan 2021–2025. https://firebasestorage.googleapis.com/v0/b/mrc-mekhong.appspot.com/o/BDS-2021-2030-and-MRC-SP-2021-2025.pdf?alt=media&token=28011e1e-87a8-492f-84be-bb8c762f9f0b

Pandey R (2019) Indigenous knowledge for sustainable development: insights from Mekong River Basin. In: Indigenous knowledge systems and sustainable development. Springer, Singapore, pp 123–137

Rahman M, Islam Khan T, Sadique Z (2020) SDG implementation progress: what does the Asian experience reveal? In: Southern voice. Southern perspectives, Global debates, no 67. https://southernvoice.org/wp-content/uploads/2020/07/SDG-Implementation-Progress-Asia-Rhaman-Khan-and-Sadique-2020.pdf

Report about the 1st European science diplomacy conference: towards a European approach to science diplomacy, 18–19 Dec 2023, Madrid, Spain. https://eu-science-diplomacy.service-facility.eu/docs/Report_about_the_1st_European_Science_Diplomacy_Conference_Towards_a_European_Approach_to_Science_Diplomacy.pdf

Robertua V (2019) The reconstruction of Indonesia's mangrove diplomacy. Global Strategis 13(2):171–183. https://doi.org/10.20473/jgs.13.2.2019.171-183

Ruffini PB (2017) Science and diplomacy: a new dimension of international relations. Springer International Publishing, Cham

Ruffini PB (2019) Diplomatie scientifique. De quelques notions de base et questions-clès. Philosophia Scientiae 3:67–80

Ruffini PB (2020) Conceptualizing science diplomacy in the practitioner-driven literature: a critical review. In: Humanities and social science communication, Nature. https://www.nature.com/articles/s41599-020-00609-5

Strategic Plan for Biodiversity 2011–2020 and the Aichi Targets. https://www.cbd.int/doc/strategic-plan/2011-2020/Aichi-Targets-EN.pdf

Stolker G (2004) Designing institutions for governance in complex environments: normative rational choice and cultural institutional theories explored and contrasted. Economic and Social Research Council Fellowship. Paper n 1

The Royal Society, The American Association for the Advancement of Science (2010) New frontiers in science diplomacy. The Royal Society, London

Tolve F (2020) Lo sviluppo delle relazioni internazionali nella prospettiva della Science Diplomacy. In: Democrazia Culturale e Ricerca della Pace. Un confronto tra diritto, economia, etica, pedagogia e politica. Aracne Editrice, pp 215–225

Turker H (2023) Maritime chessboard: the geopolitical dynamics of the South China Sea. In: Geopolitical monitor. https://www.geopoliticalmonitor.com/maritime-chessboard-the-geopolitical-dynamics-of-the-south-china-sea/

Turekian VC, Macindoe S, Copeland D, Davis LS, Patman RG, Pozza M (2015) The emergence of science diplomacy. In: Davis LS, Patman RG (eds) Science diplomacy: new day or false dawn? World Scientific Publishing Company, Singapore, pp 3–24

United Nations, Sustainable development Goals. https://sdgs.un.org/goals

United Nations General Assembly (2023) Political declaration of the high-level political forum on sustainable development convened under the auspices of the General Assembly. https://documents.un.org/doc/undoc/ltd/n23/261/94/pdf/n2326194.pdf?token=kilhVK3to4Lz17OCs9&fe=true

United Nations Viet Nam (2023) 2022 UN Country annual results report. https://vietnam.un.org/en/264785-2023-un-country-annual-results-report-viet-nam

Vymětal P (2007) Governance: defining the concept, University of Economics, Prague Faculty of International Relations, Working Papers 1/2007. https://www.researchgate.net/profile/Petr-Vymet-al/publication/40345960_Governance_defining_the_concept/links/56273db308aeabddac93ec6d/Governance-defining-the-concept.pdf

Wiktor-Mach D (2018) What role for culture in the age of sustainable development? UNESCO's advocacy in the 2030 agenda negotiations. Int J Cultural Policy 26(3):312–327. https://doi.org/10.1080/10286632.2018.1534841

WWF Coastal Community Initiative. Impact Report 2020–2023. https://wwfint.awsassets.panda.org/downloads/wwf-coastal-communities-initiative-impact-report-2020-2023.pdf

Correction to: The Effects of Climate Change on Ocean Upwelling and Productivity in Regional Seas

Muhammad Naim Satar, Winfred Marshal, and Mohd Fadzil Akhir

Correction to:
Chapter 2 in: V. Maccarrone and M. Fadzil Akhir (eds.),
***Coastal Sustainability*, Coastal Research Library 39,**
https://doi.org/10.1007/978-3-031-75749-5_2

The original version of the book was updated with the belated correction in Chapter 2: reference for Bakun (1990) citation and inserted in the reference list, which has now been updated. The book and the chapter have been updated.

The updated version of this chapter can be found at
https://doi.org/10.1007/978-3-031-75749-5_2

V. Maccarrone and M. Fadzil Akhir (eds.), *Coastal Sustainability*, Coastal Research Library 39, https://doi.org/10.1007/978-3-031-75749-5_15

The manufacturer's authorised representative in the EU is Springer Nature Customer Service Centre GmbH, Europaplatz 3, 69115 Heidelberg, Germany. If you have any concerns regarding our products, please contact ProductSafety@springernature.com

Printed and bound by CPI Group (UK) Ltd, Croydon, CR0 4YY
01/10/2026
02225932-0004